General Bi y

Third Edition

Holly J. Morris
David T. Moat
Lehigh Carbon Community College

Cover images supplied by the Holly Morris.

www.kendallhunt.com
Send all inquiries to:
4050 Westmark Drive
Dubuque, IA 52004-1840

ISBN 978-1-7924-0202-9

Published in the United States of America

TABLE OF CONTENTS

NOTE: Due to the customization method chosen by the Authors of this content, brackets are used to identify the various material borrowed from other Kendall Hunt authors. The number to the right of the bracket corresponds to the sources listed within the Credits section.

Lab Safety Rules

Photo courtesy of Holly J. Morris

The biology lab has many potential hazards for students working there. You can protect yourself from these hazards by following the appropriate procedures, being aware of the dangers of the chemicals you use, and taking commonsense safety precautions. This section is designed to introduce you to the basic lab safety rules and procedures, how to avoid accidents, and what you should do if an accident occurs.

PROTECTING YOURSELF

1. Food or drink is not allowed in the lab.
2. Wear long pants, closed-toed shoes, and tie back long hair.

   Wear clothing that you don't mind becoming stained, wet, or torn. It is a good idea to wear old worn clothing.

   Long pants must be worn in the lab. Clothing should be close-fitting especially in the sleeves, and it should cover as much of your body, arms, and legs as possible.

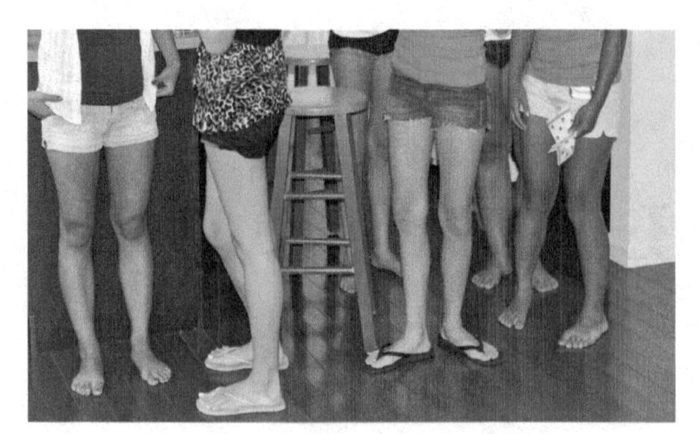

You will not be allowed to enter the lab if you are dressed like this. (Photo courtesy of Holly J. Morris)

Shoes must be close-toed. Open-toed shoes or sandals, or high heel shoes are not permitted in the lab. Be aware that cloth shoes such as tennis shoes are a potential hazard because they can absorb spilled liquids and hold the liquids next to your skin.

Not appropriate! (Photo courtesy of Holly J. Morris)

Tie back long hair. Long hair can dangle into beakers or other pieces of apparatus becoming contaminated.

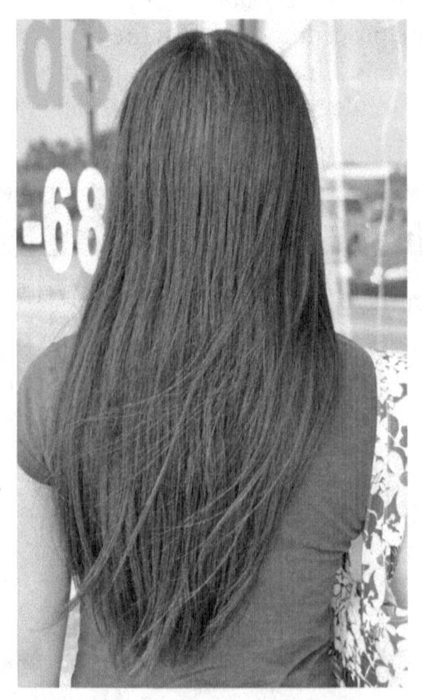

Needs to be tied back! (Photo courtesy of Holly J. Morris)

Do not wear caps or hats in the lab.

Remove all jewelry, ties, and scarves.

Consider wearing a lab coat and/or apron.

3. Wear eye protection in the lab.

   Your eye protection should protect both the front and sides of your eyes, and fit over your prescription glasses if you wear them. Do not wear contact lenses in the lab. Contact lenses present special hazards to your eyes because they can concentrate chemicals and vapors next to your eyes, trap foreign particles next to your eyes, and interfere with the effectiveness of the eyewash fountain.

4. Place your personal belongings away from your work area.

   Book bags, binders, purses, etc., should be placed in the cubby hole below the lab bench or under your lab bench. Only items that you need to conduct the actual experiment should be on the lab bench.

Working in the lab

1. Keep your work area clean while working in the lab.

Not like this! (Photo courtesy of Holly J. Morris)

2. Work on only the experiment being performed.

   Do not perform any unauthorized or modified experiments: Follow the directions provided. Do not work alone in the lab.

3. Know where the safety equipment is located and how to use it.

   Safety equipment includes first-aid kit, fire extinguisher, safety shower, eyewash station, fire blanket, and gas shut-off valve.

   Be sure to know where the appropriate lab exits are located and what to do in the event you need to evacuate the lab.

4. Use the fume hood when necessary.

   Experiments using toxic and/or irritating substances **MUST** be performed under the fume hood. When using the fume hood, turn the fan on, lower the glass cover, and place only your hands and arms in the working area.

5. Dispose of waste and excess reagents appropriately.

   You are responsible for disposing of your waste and excess materials in an appropriate fashion. You will be told where and how to dispose of such materials for each lab.

   Do not put paper or other materials in the broken glassware container.

6. Use only equipment that is in good condition.

   Examples of damaged equipment include:

   a. Glassware with chipped or broken rims.

   b. Glassware with linear or star-shaped cracks.

   c. Glassware with severely scratched bases.

   d. Glassware with sharp edges.

   e. Pipettes, or funnels with chipped tips.

   f. Clamps, rings, or other supports that don't work.

   g. Any other equipment that doesn't work properly.

7. Wash all of your glassware before you start working.

8. Avoid touching hot objects.

   Burns are one of the most common injuries in the lab. Be careful when handling hot objects. Use the appropriate clamps or pads. Allow objects to cool because hot glassware stays hot for a long time.

9. Handling Chemicals

   Handle chemicals with caution by:

   a. Being sure that you use the specified chemicals.

   b. Reading the labels carefully. Different chemicals can have similar names.

   c. Using only the amount needed.

   d. Taking only the amount of a chemical that is required and putting the chemical in a separate labelled container. If you are not sure how much of a chemical you will need, be conservative because you can always return for more.

   e. Never returning excess chemicals to their original containers.

   f. Leaving chemicals in their proper places.

   g. Cleaning up spills immediately and appropriately.

10. NEVER PIPETTE BY MOUTH

11. If an Accident Occurs

    a. If you hurt yourself in any way (or think you may have), no matter how slightly, you **MUST** tell the lab instructor immediately.

    b. If you burn yourself, flush the area of the burn with cold water for at least 15 minutes.

    c. If you get something in your eyes, immediately flush them at the eyewash station. This should never happen since you are wearing eye protection but if it does, alert someone close to you and proceed to the eyewash fountain. If you get a chemical on your face but not in your eyes, **DO NOT REMOVE YOUR EYE PROTECTION**. Proceed immediately to the eyewash station and flush your face completely. Use paper towels to dry off. Then remove the eye protection.

    d. If a small amount of chemical comes into contact with your skin, immediately wash your hands with soap and water. Remove any jewelry that might prevent total cleansing. If you are covered with a large amount of chemicals, use the safety shower.

    e. If your clothing becomes contaminated, remove it immediately and wash the affected areas of your body. Discard all contaminated clothing.

    f. Use the fire blanket to extinguish any fires involving lab personnel or students.

At the end of lab

1. WASH AND DRY ALL GLASSWEAR AND RETURN IT TO ITS PROPER PLACE.
2. Return all other materials to their proper place.
3. Wash and dry the lab bench.
4. Wash your hands before leaving the lab.

Maintaining a Lab Notebook

Photo courtesy of Holly J. Morris

A properly maintained laboratory notebook is one of the most important pieces of equipment that a scientist has. Laboratory notebooks have been examined in court cases, determined who is credited with a discovery, or who will be awarded a patent, and have been the starting point for an invention. **It is important that laboratory notebooks are complete and neat. It should be possible for you, or someone else, to go back to an experiment at a later date and understand what procedure was followed and what results were obtained.** Although you are unlikely to be developing patentable products in this course, and we are hopeful that you are not hauled into court over one of our lab experiments, it is important for you to establish good laboratory habits now.

As a result of variable, and sometimes questionable, lab practices, including documentation, the Food and Drug Administration established the Good Laboratory Practice (GLP) regulations, which became effective in June 1979. These regulations establish a protocol ensuring that all work is done according to agreed procedures and that the data are properly recorded. The protocol may sound cumbersome or inefficient in this digital world, but it will give you a taste of the real world.

General Information

Different laboratories or companies may have slightly different procedures or acceptable practices, but they should all fall within GLP. For example, some labs use both left and right pages, some labs allow scientists to write personal notes or do calculations on the left-hand page, other labs insist that the left-hand page be left blank.

We will follow a modified form of good laboratory practice that is appropriate to an undergraduate course.

1. Laboratory records are to be kept in a bound notebook (preferably sewn). They are not kept in a spiral notebook or a loose-leaf binder.
2. The pages are consecutively numbered. No pages are ever removed. Do not leave blank pages to be filled in later. Entries are made only on the right-hand page. The left-hand page is always left blank.
3. **Entries (with the exception of computer generated data printouts or graphs) must be handwritten in black or blue pen.** This is because notebooks are prone to getting liquids on them. Neither a pencil nor a Sharpie should ever be used. A pencil can be erased, and a Sharpie bleeds though the paper. If, in the future, you were to try to establish your procedure or discovery as the first of its kind, and your data were in pencil, you would be laughed out of the courtroom/patent office. **It is not appropriate to type notes or procedures into a word processor and paste the page into your notebook**. However, it is appropriate to glue computer generated data printouts in the **Results** section or computer generated graphs containing plotted data in the **Graphs** section. You may need to leave space as you continue onto your summary. Do not paperclip, staple, or tape data/graphs onto a page.
4. If you make a mistake, draw a single line through the error, so that the words can still be read.
5. Everything related to the laboratory work must be recorded in the notebook in an organized and neat manner. **If it cannot be easily read, it is not adequately recorded.** It is critical that the material is intelligible and understandable to the notebook author and any trained biologist who reads the records, attempts to reproduce these results, or endeavors to finish an incomplete analysis. This concept is often known as "traceable" in the industrial world.

Lab notebooks will be collected and graded multiple times during the semester, and dates will not necessarily be announced ahead of time. Failure to bring your notebook to class will result in a grade of zero for that day.

Getting Started

1. Write your name and the starting date on the outside of the notebook. On the inside cover, write your name, lab section and time, lab instructor's name, and some sort of contact information, such as your email address or phone number. This is in case you misplace your notebook and someone is attempting to return it to you.
2. Number the pages in the top right and the top left corners. Yes, you number the left-hand pages, even though you are leaving them blank.
3. Reserve the first four pages for a table of contents, which will be filled in as you go. For each entry in the Table of Contents, write the name of the experiment (or lab exercise), and the beginning page number.
4. The experiments (or lab exercises) in the notebook should be listed in chronological order.
5. There may be times when you begin a new experiment (or lab exercise) before the last one is finished. The procedure is to continue in chronological order, starting with the next page for the new experiment. When you resume the prior experiment, you continue on the next available page, and write in the Table of Contents, [Title of Experiment] continued, Page x.

For Each Experiment / Lab Exercise

Before class:

1. Record the **Date** and **Time** on the top right of the first page of the experiment/lab exercise.
2. Write the **Title** for the experiment/lab exercise, e.g., "Introduction to the Scientific Method."
3. Write a **Purpose** for the experiment/lab exercise. The **Purpose** includes the activities being performed and the significance of those activities. For example, the purpose of the "Scientific Method" exercises might be: "To use the scientific method for several exercises to illustrate its use because the scientific method is the standard protocol that scientists follow when conducting research."
4. The next section lists **Materials. Materials is a bulleted one column list of the items being used in the exercise**. In some cases, the materials may be listed at the beginning of the exercise. In others, you may have to search through the exercise to find everything that you need. For example, for the "Using Phenolphthalein" exercise in Chapter 5 "Acids, Bases, and pH", the **Materials** list would be:
   - Distilled water
   - 500 ml beaker
   - Phenolphthalein solution in dropper bottle
   - Sodium carbonate
   - Clean straw
   - Stirring rod
5. The next section lists the **Methods** (procedures). **Methods is a numbered list or bulleted points rather than paragraph-type narrative.** You need to record how the experiment is to be performed in your own words. This section may be the most important part of your notebook. Draw a vertical line 1/3 of the way from the right-hand side of the page. Write the procedure on the left 2/3 of the page. You will use the right-hand 1/3 during the experiment in case you need to make changes in your procedure as it was written. **You may not print from a computer or photocopy from some other source.**

During class:

1. Record all data gathered during the course of the experiment in the **Results** section. Included in this section are data tables, images viewed with a microscope, and other data obtained during the experiment/lab exercise.
2. Document your **Observations**. **Observations** generally fall into two categories:
   a. Things that were noted during the exercise that are not immediately explainable but can be interpreted later.
   b. Ways to improve the procedure.
3. Provide a sample of each type of calculation performed in the experiment/lab exercise in the **Calculations** section. If there are none, indicate **None** (or **NA**).

After class:

1. Draw or glue any computer generated graphs in the **Graphs** section. If there are none, indicate **None** (or **NA**).
2. Provide a **Summary** of your work that includes a summary of your results and discusses the significance/implications of your results.
3. At the end of the write-up, draw a horizontal line just below the **Summary**, draw an "X" through the unused portion of the page and write your initials.

CHAPTER 1

Scientific Method

Photo courtesy of Holly J. Morris

OBJECTIVES

After completing these exercises, you will be able to:

- Understand the scientific process of investigation.
- Apply this structured approach to examine a naturally occurring event.

INTRODUCTION

What is science? Science is a systematic way of looking at and understanding the natural world through observation and experimentation. The process that is used is called the scientific method, which is simply a way to look at and organize information in a consistent and methodical manner. Observable physical evidence, that is, data, are key to developing an understanding of natural phenomena.

Suppose you go out to your garage one morning and your car would not start. You ask yourself, "Why isn't the car starting?" As you think about possible reasons, you may start with, "Perhaps the battery is dead." You have made an observation, asked a question, and formed a hypothesis. Now you need to test your hypothesis so you try turning on the headlights. They work. You conclude that the battery is not the problem.

You performed an experiment (turning on the headlights), and collected data (observing that the headlights work). When you analyze your data, you realize that it does not support your hypothesis, so you reject that hypothesis and form a new one. You continue this process until you have determined why your car won't start. You have followed the scientific method.

The scientific method is the process by which scientists work. It is the process that scientists use to attempt to construct an accurate representation of the world.

The steps of the Scientific Method are:

1. Make an observation and ask a question.
2. Formulate a hypothesis to explain the event.
3. Test the hypothesis by performing an appropriate experiment.
4. Analyze the data from your experiment.
5. Either accept or reject your hypothesis. If you accept your hypothesis, move onto the next step. If you reject your hypothesis, formulate a new hypothesis.
6. Report your results (both positive and negative).

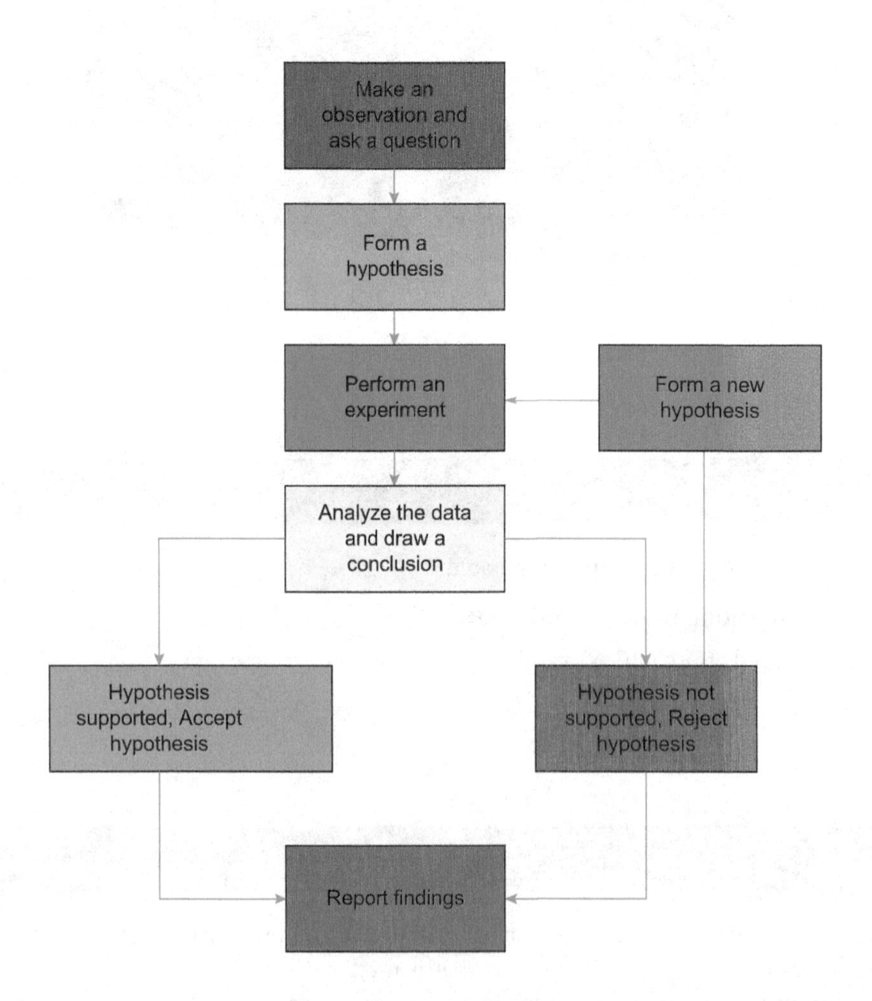

One important point to add—when you accept your hypothesis, that does not mean that you can claim your hypothesis is true. It just means that you have not found it to be false. For example, you see gray squirrels and so you claim that all squirrels are gray. Since you cannot see all squirrels, you cannot prove they are all gray. If, however, you see one black squirrel, you can reject your hypothesis that all squirrels are gray. Bottom line, you cannot prove a hypothesis, you can only disprove it.

DESIGNING AN EXPERIMENT

Once you have formed a hypothesis, your next step is to test it. Scientists test hypotheses by designing experiments to look for cause and effect relationships. Looking for relationships in a methodical and quantifiable way is critical. When designing an experiment, it is important to change, or vary, only one factor at a time. This is referred to as a variable, but there are other types of variables. The results you get from different treatments are a variable. So is using different starting materials. All of these variables have specific names, and it is important for you to understand how they fit into the whole process.

An *independent variable* is the one factor that is changed in order to test your hypothesis. Suppose your hypothesis is that if you use fertilizer on your flowers, the flowers will grow faster. In this case, the independent variable is fertilizer.

When you add fertilizer, you measure how tall your flowers have grown over specific time intervals. The height of the plants, which depends on whether or not fertilizer was used, is the *dependent variable.*

There are many types of flowers that you could grow in your garden, so for this experiment it is important that you control that variability to one type of flower. This is called a *controlled variable.* Other controlled variables could be the amount of water, sunlight, or weed control that the plants receive. In order for the results to be meaningful, all of these variables should be controlled so that all of the plants receive the same conditions, or as similar as possible.

If you want to examine whether there is an optimal amount of fertilizer to use when you grow your flowers, you can treat different groups of flowers with different amounts, or levels, of fertilizer. This manipulation of the independent variable is called *treatment level.* For example, you may mix one tablespoon of fertilizer in a gallon of water as treatment level one, three tablespoons in a gallon of water as treatment level two, and six tablespoons in a gallon of water as treatment level three, then compare growth rates.

One component of your experiment that is vital is the *control treatment*, often referred to simply as "the control." This is the group for which all controlled variables are the same as the treatment groups, for example, the same flowers, water, sun, and so forth, but it does not have an independent variable. In this example, it receives water without fertilizer.

Another important factor when designing an experiment is to reduce, as much as possible, experimental bias. One way to do this is to use *randomization* when assigning objects or individuals to treatment groups. In this example, you buy four flats of flowers. If you used all of the flowers in flat 1 for treatment 1, all of the flowers in flat 2 for treatment 2, and so forth, you might have biased results. It's possible that flat 1 had better sunlight, or more even watering at the nursery. The plants within each flat should be randomly assigned to a treatment group. This way if the flowers in one of the flats have an advantage over the other flats, they are distributed randomly through the treatments and the control.

There is one final factor when designing and performing experiments that needs consideration. If the total number of objects or individuals in each treatment group is small, variations between treatment groups are less reliable. Repeating the experiment multiple times will improve the reliability of the results. This is called *replication.*

In the following exercises, you will be introduced to the use of the Scientific Method.

EXERCISE 1 Practicing the Use of the Scientific Method

Cardiovascular fitness can be determined by measuring a person's pulse rate before and after a given period of cardiac exercise. A person who is more fit should have their pulse rate return to "normal" more quickly than a person who is less fit.

Purpose

To utilize the scientific method to perform an investigation of cardiovascular fitness.

Materials

- Steps or platforms (at least 8 inches high)
- Clock or watch with a second hand

Methods

- Members of the lab are separated into groups of 5 or 6.
- Each group determines the question regarding cardiovascular fitness that they will investigate.
  - Possible questions:
    - Is cardiovascular fitness greater in athletes than nonathletes?
    - Is cardiovascular fitness greater in nonsmokers than smokers?
    - Is cardiovascular fitness associated with body type (such as tall vs. short)?
- Each group develops a hypothesis that they will test.
- Each group member measures and records their "resting" pulse rate in the **Individual Results** table before the exercise.
  - Pulse rate should be taken when the group member is sitting quietly.
- Each group member performs the cardiovascular exercise described below:
  - Step up and down on a step or platform for 4 minutes at a rate of 30 steps per minute.
  - Step up and down needs to be maintained at a constant rate.
- Immediately after the exercise is complete, each group member measures and records their pulse rate in the **Individual Results** table.
  - Pulse rate should be taken when the group member is sitting quietly.
- Each group member continues to measure their pulse rate at 30 second intervals until their pulse rate returns to their "resting" pulse rate.
  - Pulse rate should be taken when the group member is sitting quietly.
- Determine and record number of minutes for each group members' pulse rate to return to their "resting" pulse rate (i.e., their recovery time) in the **Individual Results** table.
- Record the results in the **Treatment Level Averages** table for each treatment level (use average recovery rate for treatment levels having more than 1 member).

EXERCISE 2 Indirect Observations

Scientists make observations while performing experiments. Observations may be direct or indirect. A direct observation occurs when the scientist observes what is happening himself. An indirect observation occurs when the scientist determines what is happening by utilizing their other senses.

Purpose

To practice using nonvisual senses to determine what is occurring in an experiment.

EXERCISE 3 Graphing

Scientists use graphs to analyze and present their data. Graphs provide a way to visualize data in order to determine relationships between the variables in an experiment. The independent variable, the variable that the scientist deliberately manipulates in an experiment, is plotted on the *x*-axis. The dependent variable, the variable that corresponds to changes in the independent variable, is plotted on the y-axis. Together they form an ordered set of coordinates (x, y).

Other elements of an effective graph include:

- A title that describes what is being graphed.
- Axis labels that describe what is contained on the x-axis and the y-axis.
- Axis values.
- "Best fit" line when using a line graph.

Purpose

To develop the necessary skills to prepare effective graphical displays.

Name:

Exercise 1 Results

1. For the cardiovascular fitness experiment, list:
   - Hypothesis:

     ____

   - Independent variable:

     ____

   - Dependent variable:

     ____

   - Controlled variables:

     ____

   - Control treatment:

     ____

   - Treatment levels:

     ____

   - Replication:

     ____

   - Prediction (predict the results of your experiment based on your hypothesis):

     ____

2. For the cardiovascular fitness experiment, list in numerical order each step in your procedure:

____

____

____

____

____

____

3. For the cardiovascular fitness experiment, record the results of your experiment:

Individual Results						
	Student 1	Student 2	Student 3	Student 4	Student 5	Student 6
Treatment level						
Before step test (pulse rate)						
After step test (pulse rate)						
Recovery time (minutes)						

Treatment Level Averages						
	Treatment 1	Treatment 2	Treatment 3	Treatment 4	Treatment 5	Treatment 6
Treatment average						

4. For the cardiovascular fitness experiment, prepare a graph of your results:

5. For the cardiovascular fitness experiment, prepare a summary of your experiment that

(1) Contains your hypothesis, analyzes your results, states whether your results support or falsify your hypothesis, explains why your results support/do not support your hypothesis, and,

(2) Lists weaknesses in your experiment and ways to improve these weaknesses.

___

___

___

6. The following are brief descriptions of experiments and the hypotheses proposed for them. For each:

   (a) Identify the independent variable.

   (b) Identify the dependent variable.

   (c) Identify an appropriate control treatment.

   (d) Draw a graph that would be constructed if the hypothesis is supported by the data.

   I. Antacids are used to relieve the effects of heartburn, the major symptom of gastroesophageal reflux disease. **Hypothesis:** Increasing the amount of antacid taken decreases the time for heartburn relief to occur.

   Independent variable:

   Dependent variable:

   Control treatment:

   Graph:

II. Various fertilizers are tested with corn to determine which fertilizer increases yield per acre the most. **Hypothesis:** R5 fertilizer increases the yield per acre of corn the most.

Independent variable:

Dependent variable:

Control treatment:

Graph:

III. Marigold seeds are planted in pots and placed in environments with varying levels of light. The number of marigold seeds germinated is counted after 7 days. **Hypothesis:** Low light levels inhibit marigold seed germination.

Independent variable:

Dependent variable:

Control treatment:

Graph:

Exercise 2 Results

1. For each Ob-Scertainer, draw your hypothesized internal structure in the "Hypothesis" circle. If your initial structure is incorrect, after retesting, draw your revised structure in the "Retest" circle. Draw the actual internal structure in the "Actual Model" circle.

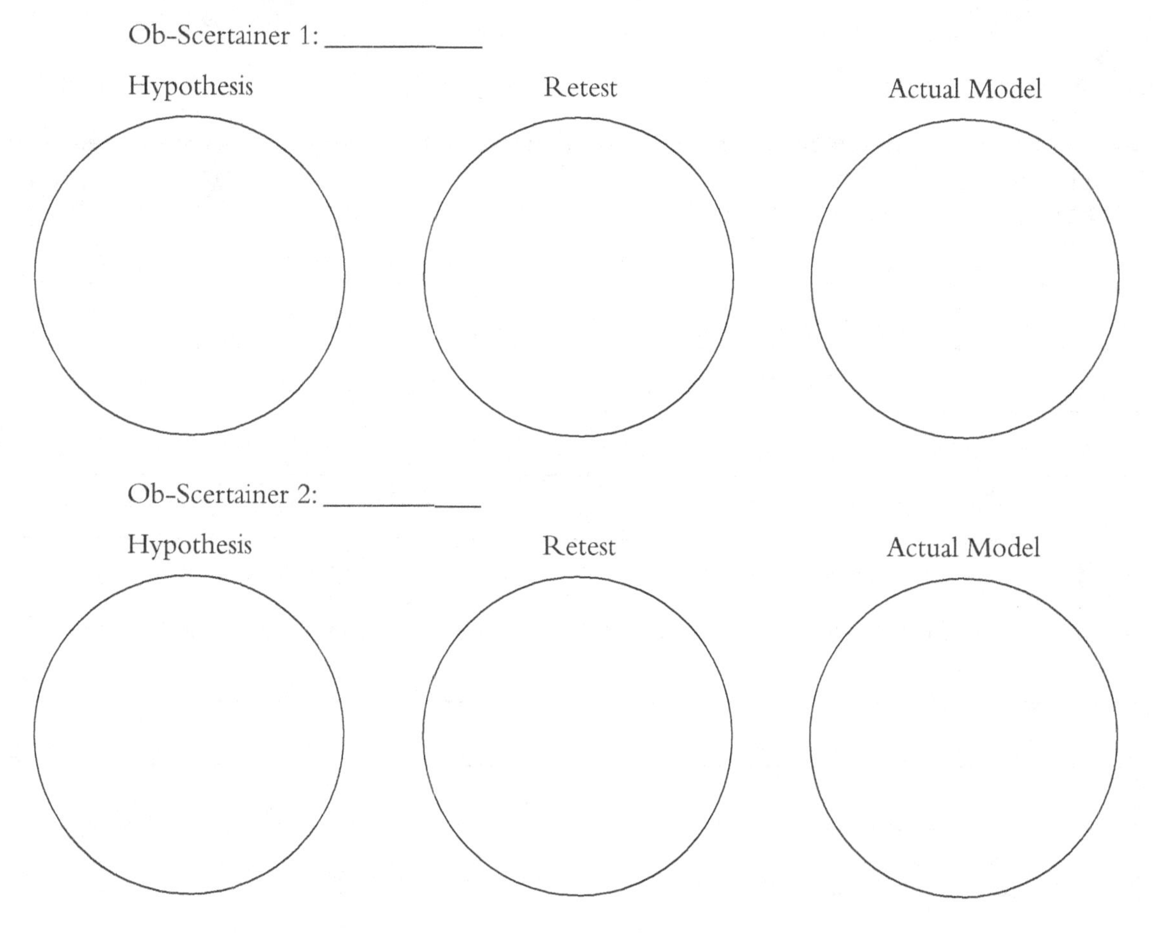

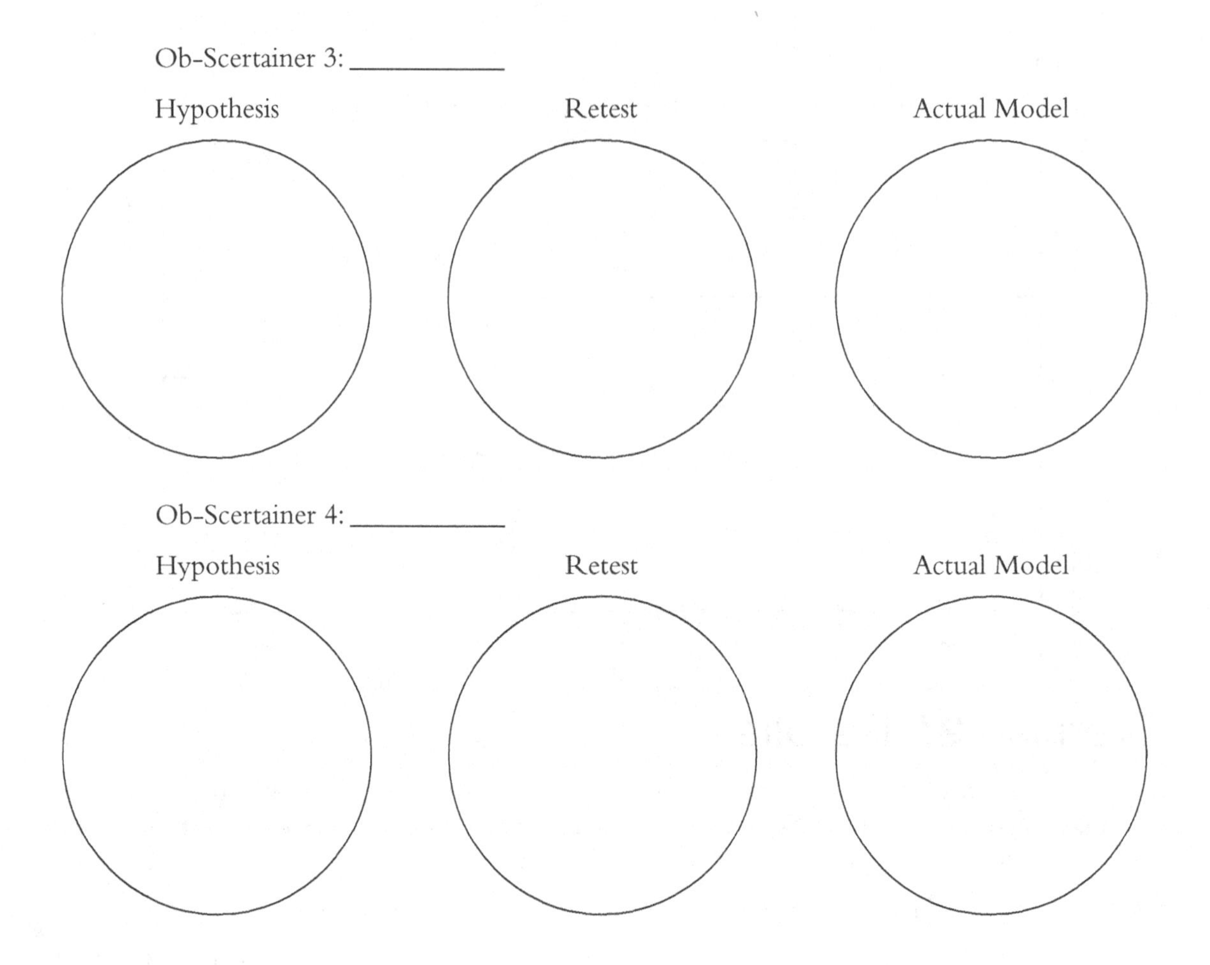

2. Choose one of your Ob-Scertainers and write a summary of the actual Ob-Scertainer configuration and your hypothesis for it. List those things that you were able to determine and those you were unable to determine.

____

____

____

____

____

3. Are there any reasons why you were successful for certain characteristics but not for others?

____

____

____

____

____

Exercise 3 Results

1. Prepare a bar graph of the following data regarding the "importance" of the various types of trees encountered while performing a terrestrial ecology review:

Tree	Importance
White Pine	875
Red Cedar	150
Hemlock	750
White Ash	450
White Oak	1550
Red Maple	1475
Buckeye	125

2. Prepare a bar or line graph of the following data regarding drug release time for aspirin and ibuprofen:

Release Time Since Administration (minutes)	Cumulative Release % (aspirin)	Cumulative Release % (ibuprofen)
5	60	40
10	70	58
15	75	66
20	80	74
25	83	82
30	85	90

Microscopy

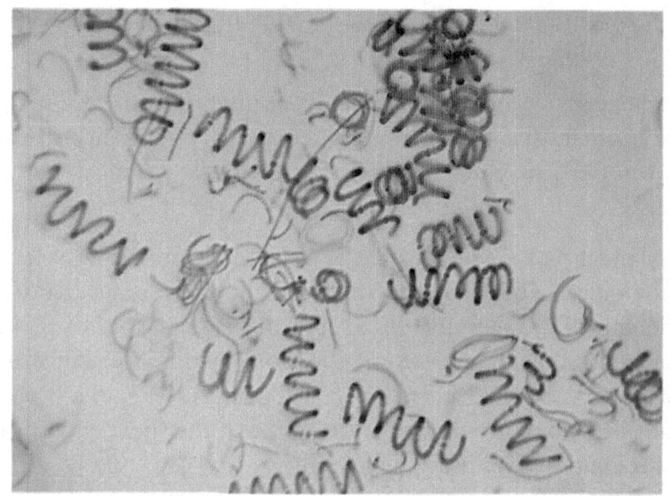

Photo courtesy of Holly J. Morris and David T. Moat

OBJECTIVES

After completing these exercises, you will:

- Be able to identify the parts of a stereoscopic microscope and a compound light microscope and list their functions.
- Describe the appropriate procedures to care for and use a stereoscopic microscope and a compound light microscope.
- Practice the skills necessary to become proficient in using a stereoscopic microscope and a compound light microscope.
- Prepare wet mounts.

INTRODUCTION

One of the most important techniques used in biology is microscopy. It is, therefore, important to become proficient in both the use and care of the two most commonly used microscopes, the compound light microscope and the stereoscopic microscope.

All microscopes enlarge the appearance of objects too small to be seen by the naked eye, but just magnifying an object does not necessarily mean that details will be visible. This depends on the resolving power of the microscope. These two concepts, magnification and resolution, need further discussion.

Magnification is how much bigger an object appears under the microscope than that object is in real life. Polished lenses bend the light in such a way as to enlarge the appearance of an object. Think about how a magnifying glass, which is a single polished lens, makes objects look bigger. Microscopes use multiple lenses, thus compounding the effect of magnification.

Light microscopes typically magnify objects between 40× and 1000×. The overall magnification of an object depends on the multiplicative magnification of the lenses, that is, the magnifying power of the eyepiece times the magnifying power of the objective. Eyepieces (oculars) typically have a magnification of 10×. When combined with an objective magnification of 4×, the overall magnification will be 40×. The maximum magnification of an objective on a typical compound light microscope is 100×, therefore the overall magnification when using that objective is 1000×.

Merely magnifying the appearance of an object is not enough if you cannot distinguish between two points that are near each other, that is, you cannot see detail. The ability to distinguish between two points is called resolution. Both magnification and resolution are critical factors to observing objects under the microscope.

Resolution, that is, how much detail you can see, is dependent on the wavelength of the illumination source and the size of the object. The physics formula for resolution states that if the size of the object is less than half the wavelength of the illumination source, the object will not be visible. White light has a wavelength between 400 and 700 nanometers (nm), therefore anything smaller than about 200 nm will not be visible. Examples of objects this size are most typical bacteria. Very often you will find a blue filter between the microscope's light source and stage. Blue light is at the short end of the wavelength spectrum and, therefore, improves resolution over using white light.

Proper illumination is necessary for seeing detail at high magnification. The higher the magnification, the more illumination that is needed for resolution. Focusing light from the light source to one small segment of the stage is the job of the condenser. As the light waves travel up and hit the condenser, they are bent by the lens of the condenser and focused toward the hole in the center of the stage. The position of the condenser determines how effectively the light is focused. The maximum focus of light occurs just above the surface of the condenser, so the condenser should be in the highest-most position. If you lower the condenser, light rays over-shoot the focal point on the stage, and resolution decreases.

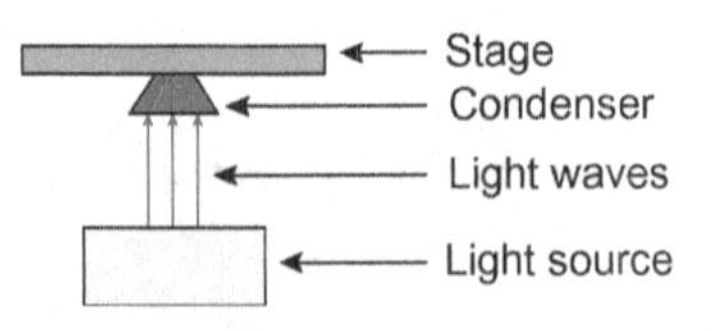

As a general guideline for viewing objects under the microscope, lower magnification needs less light, higher magnification needs more light. Ways to reduce or increase light are to change the amount of light emitted from the light bulb, and opening or closing down the iris diaphragm. Both methods change the amount of light entering the condenser.

Light travels at different speeds through different media. Think of a spoon in a glass of water.

© funkyfrogstock/Shutterstock.com

As light waves travel from air to water, they slow down slightly, making it look as though the spoon is broken. The same thing happens as light travels through glass (the microscope slide), then air, then glass again (the objective lens). The distortion of the light is imperceptible at lower magnifications, but at magnifications around 1000× and higher, the distortion of the light waves makes it difficult to resolve details on small objects. A small drop of oil, that has the same refractive index as glass, is used to displace the air between the lens and the slide, reducing distortion and improving resolution.

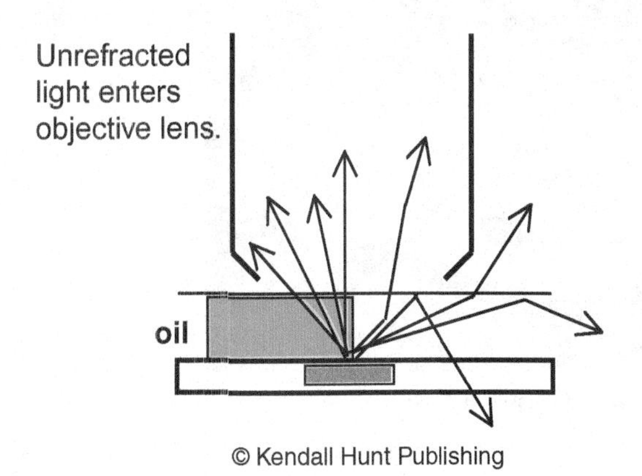

© Kendall Hunt Publishing

You will not be using the 100× objective, called the oil immersion lens, in this exercise, but you will use it in the future, so it is useful to understand why you need to use oil.

Three of the most commonly used microscopes are:

- Stereoscopic microscope
- Compound light microscope
- Electron microscope

We will be using stereoscopic microscopes and compound light microscopes.

STEREOSCOPIC MICROSCOPE

A stereoscopic microscope, also known as a stereoscope or a dissecting microscope, magnifies small three-dimensional objects such as flowers, fossils, or insects at a relatively low power. It is not intended for use with prepared slides. The stereoscope uses two optical pathways, with two oculars and two objectives, giving each eye a slightly different view. This is what produces the three-dimensional appearance of the samples being viewed. The ocular magnification is 10× and the objective magnification ranges from 0.9× to 4×, giving a total magnification ranging from 9× to 40×.

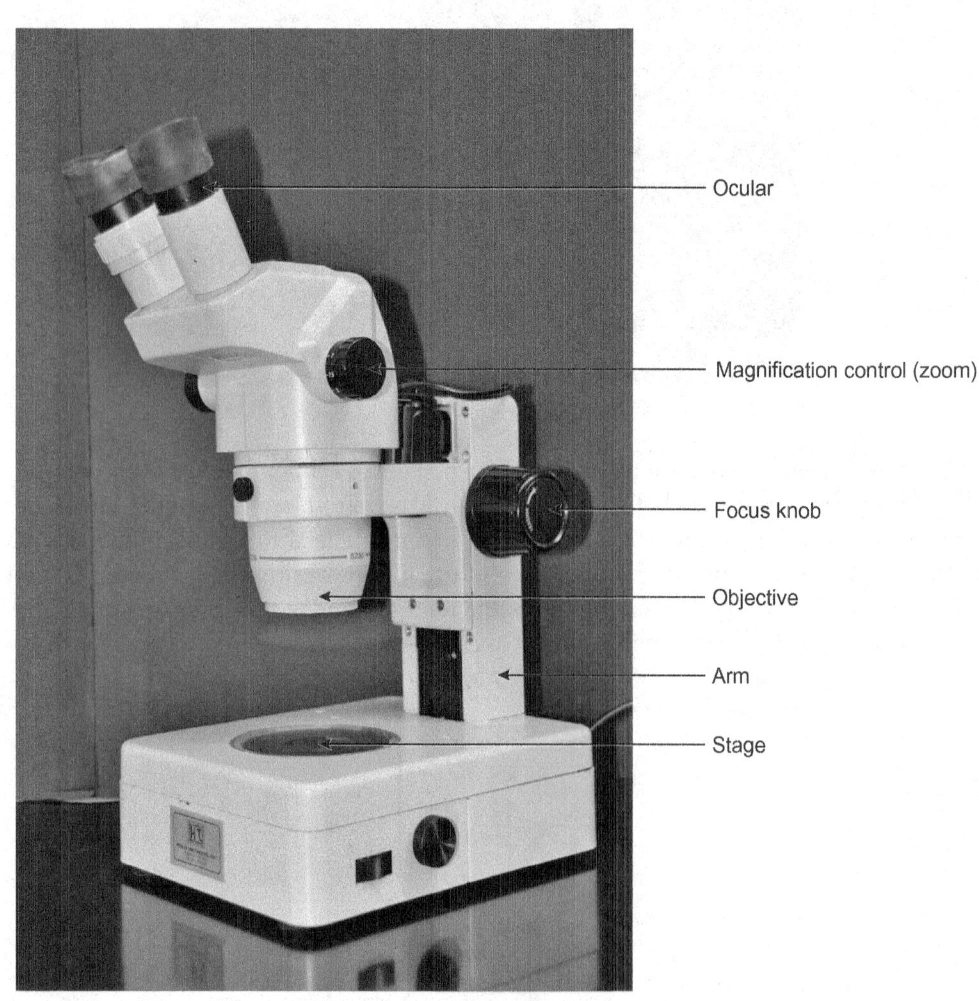

Photo courtesy of Holly J. Morris

CARE OF A STEREOSCOPIC MICROSCOPE

A stereoscope is an expensive and fragile piece of equipment. Follow these directions whenever you use a stereoscope:

- Carry the microscope with two hands—one around the arm and the other supporting the base.
- Keep the microscope flat on the lab bench and away from the edge.
- Keep the power cord from dangling or getting in the way.
- Clean the objective and ocular lenses only when they are dirty and then ONLY with the proper cleaning materials as demonstrated.

When putting the microscope away:

- Turn off power.
- Place the magnification control at the lowest objective (0.9×).
- Wrap the electrical cord on itself, and hang it from the ocular.
- Return the microscope to the microscope storage cabinet.

USING A STEREOSCOPIC MICROSCOPE

Follow these directions whenever you use a stereoscopic microscope:

- Make sure the light intensity setting is on "low."
- Turn on the light.
- Place the object to be viewed on the stage.
- Start viewing a specimen with the lowest power objective.
- Use the focus knob to bring the object being viewed into focus.
- Close your left eye when the object is in focus and using the focus knob, bring the image through the right ocular into the sharpest image. Close your right eye, open your left eye, and adjust the left ocular to bring the left ocular into the sharpest focus. At this point, both eyes will be viewing a focused image.
- Increase the magnification by moving the zoom control to a higher magnification.

COMPOUND LIGHT MICROSCOPE

A compound light microscope is the most commonly used microscope in the biology laboratory. It is used to look at small cells or thin sections of tissue fixed on a glass slide and will usually have a coverslip. Objects need to be thin enough that light can pass through them. The microscope has four objective lenses:

- 4× (low power lens) - used for focusing and scanning the slide
- 10× (medium power lens) - used for focusing and searching
- 40× (high/dry power lens) - used for viewing eukaryotic cells
- 100× (oil immersion lens) - used for very small objects such as bacteria

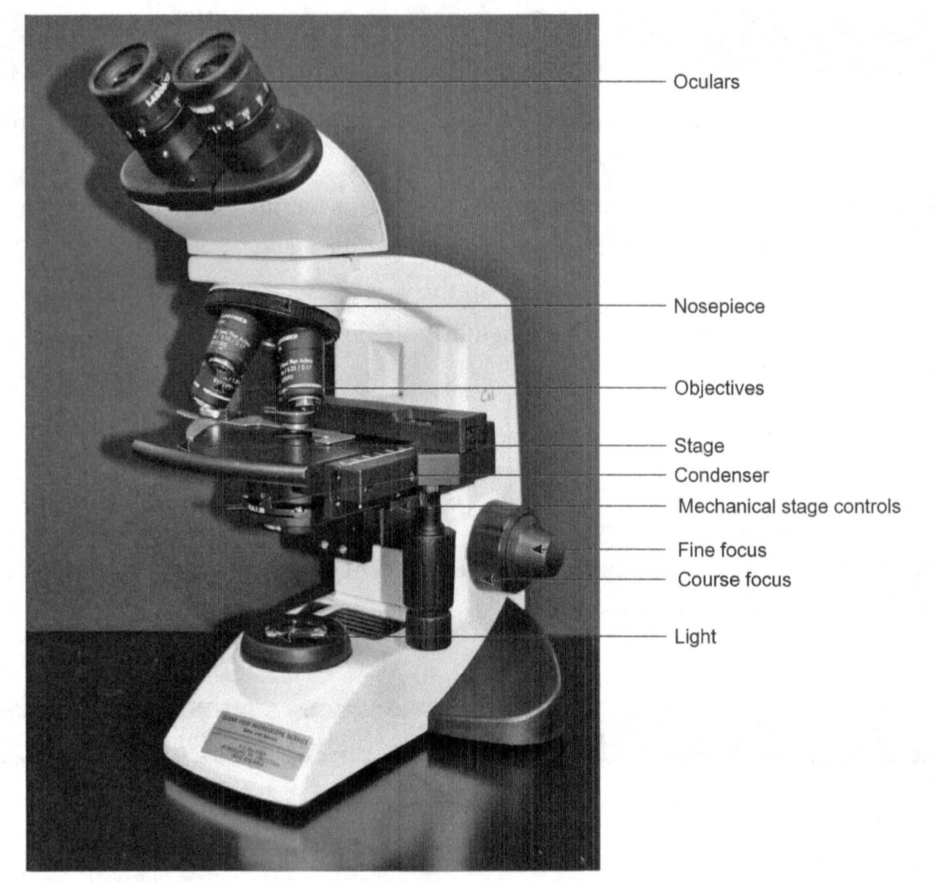

Photo courtesy of Holly J. Morris

CARE OF A COMPOUND LIGHT MICROSCOPE

A compound light microscope is an expensive and fragile piece of equipment. Follow these directions whenever you use a compound light microscope:

- Carry the microscope with two hands - one around the arm and the other supporting the base.
- Keep the microscope flat on the lab bench and away from the edge.
- Keep the power cord from dangling or getting in the way.
- Do not allow immersion oil on any part of the microscope other than the oil immersion (100×) objective lens. If, inadvertently, oil drips on the scope, remove it immediately with tissue paper and lens cleaner.
- Do not get oil on the 40× objective. The 40× objective is almost the same length as the 100×, so if you are using oil immersion and you turn the nosepiece, moving the 40× objective into place, the tip of the objective will touch the oil. This destroys the 40× objective which is not sealed to prevent oil from entering around the lens, as is the oil immersion lens.
- Clean the objective and ocular lenses only when they are dirty and then ONLY with the proper cleaning materials, as demonstrated.

USING A COMPOUND LIGHT MICROSCOPE

Follow these important directions whenever you use a compound light microscope:

- Set the low power objective (4×) in place over the hole in the center of the stage. (Never place a slide on the stage or remove a slide unless the low power lens is in place.)

- If not already there, lower the stage to its bottommost position and place the slide to be viewed in the slide clip.
- Set the light intensity on "low" and turn on the light.
- Make sure that the condenser is in its highest-most position,
- Center the object on the slide over the center of the hole in the stage. Then, while looking at the stage and objective, raise the stage to its highest-most position.
- For initial focusing, select either the 4× or the 10× objective, depending on the size of the object on the slide. Large objects, such as protists, should initially be focused using the 4× objective. Small objects, such as red blood cells, should initially be focused using the 10× objective.
- Look through the oculars (eyepieces) and use the course adjustment to bring the object into near focus. The stage lowers as you do this. Always focus by moving the slide AWAY from the objective lens using the coarse adjustment knob; never move the slide toward the objective lens, unless you are looking at the objective and the stage.
- The oculars can be adjusted to accommodate vision differences in each eye. Close your left eye when the object is in focus and using the fine focus knob, bring the image through the right ocular into the sharpest image. Close your right eye, open your left eye, and rotate the diopter ring on the left ocular to bring the left eye into the sharpest focus. At this point, both eyes will be viewing a focused image.
- Increase the magnification by rotating the nosepiece and clicking the objective into place. Always look at the objectives from the side when you rotate the nosepiece, to prevent rotating an objective into the slide. When focusing, the coarse focusing knob should NEVER be used when the 40× or 100× objective is in place.
- To view an object at high magnification, position the object directly in the middle of the field, then move the high power objective into place. Use only the fine focus knob. These microscopes are parfocal, which means that an object in focus at a lower magnification should only need minor adjustment when going to a higher magnification. If you have trouble focusing with the fine focus, go back down in magnification and refocus. Do not use the course focus at higher magnifications. Doing so is likely to jam the objective into the slide causing damage to both.
- Use proper cleaning materials as demonstrated.

When putting the microscope away:

- Turn off power.
- Lower the stage to its bottommost position and place the low power objective lens in the view position.
- Remove the slide from the stage, and gently blot any excess oil from the slide (if using the oil immersion lens).
- Wrap the electrical cord on itself and secure the plug.
- Return the microscope to the microscope storage cabinet.

In the following exercises, you will be introduced to the use of the stereoscopic microscope and the compound light microscope.

EXERCISE 1 Practicing with a Stereoscopic Microscope

Stereoscopic microscopes are used in many fields of biology and as such, it is important that you are able to correctly and proficiently use a stereoscopic microscope.

Purpose

To master the skills necessary to correctly use a stereoscopic microscope.

Materials

- Stereoscope
- A personal object (such as a coin, pencil etc.)
- Prepared Daphnia slide

Methods

- Members of the lab are separated into groups of three.
- Each group:
  - Obtains a stereoscopic microscope.
  - Selects a personal item to be viewed with a stereoscopic microscope and then each group member draws the object as viewed with their unaided eye in the space provided below.
  - Focuses the object with the lowest power magnification of a stereoscopic microscope and each group member draws the object in the space provided below.
  - Selects a prepared Daphnia slide to be viewed with a stereoscopic microscope and each group member draws the object as viewed with their unaided eye in the space provided below.
  - Focuses the prepared Daphnia slide with a stereoscopic microscope objective magnification of __________ and each group member draws the Daphnia as viewed in the space provided.

EXERCISE 2 Practicing with a Compound Light Microscope

Compound light microscopes are used in many fields of biology and as such, it is important that you are able to correctly and proficiently use a compound light microscope.

Purpose

To master the skills necessary to correctly use a compound light microscope.

Materials

- Compound light microscope
- Prepared Daphnia slide
- Graph paper slide

Methods

- Obtain a compound light microscope and a prepared Daphnia slide, focus the slide with the low power objective (4×), and then draw the Daphnia as you viewed them in the space provided below.
- Focus the Daphnia slide with the medium power objective (10×), then focus the slide with the high power objective (40×), and draw the Daphnia as you viewed them in the space provided below.
- Focus the graph paper slide at the low power objective (4×) and high power objective (40×), draw the graph paper as you viewed them in the space provided, and then complete the **Diameter of the Field of View** table.

EXERCISE 3 Making Wet Mounts

Many times, you will need to examine fresh specimens with a microscope. By making a wet mount, you can improve the image quality by suspending the specimen in a drop of water between a microscope slide and a cover slip. Generally, wet mounts cannot be stored for extended periods of time and as a result, are referred to as a "temporary mount." To become proficient, we will make wet mounts of various types of cells and then examine them using a compound light microscope.

Cell theory states that:

- Cells are the smallest functional biological units.
- All living organisms are comprised of cells.
- All cells come from pre-existing cells.

It is important to understand the structure and function of cells. The earliest known cells were prokaryotes and first appeared in fossils formed about 3.5 billion years ago. Prokaryote means "pre-nucleus." These are cells without a true nucleus or membrane bound organelles. About 1.5 billion years ago cells containing a true nucleus and membrane bound organelles appeared. These were the first eukaryotic cells. Eukaryotic means "true nucleus." Because of organelles with specialized functions and the development of communication pathways (endoplasmic reticulum), cells were able to grow larger. The typical prokaryote is about 0.2–2 micrometers (μm) in diameter, whereas eukaryotes are typically about 10–100 μm in diameter.

An exception to the small size of prokaryotes is cyanobacteria (previously called blue green algae). Cyanobacteria are a diverse phylum within the kingdom Bacteria. They obtain energy through photosynthesis and can exist as single cells or in colonies. They range in size from 0.1 μm to 40 μm, which means that larger cyanobacteria can easily be seen with the 40× objective.

Initially, organisms, whether prokaryotic or eukaryotic were single cell organisms. Some eukaryotic cells formed colonies, which provided some protection and specialization of functions. Eventually, multicellular colonies evolved into multicellular organisms.

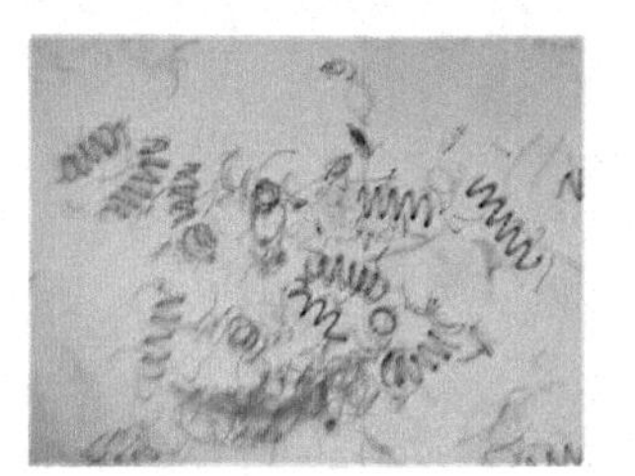

Anabaena 100x

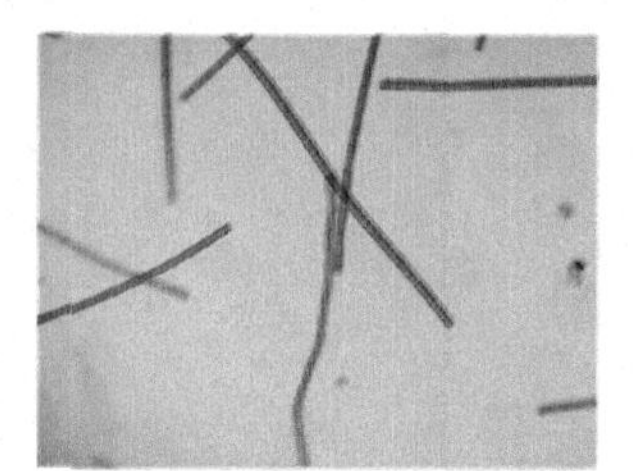

Oscillatoria 100x

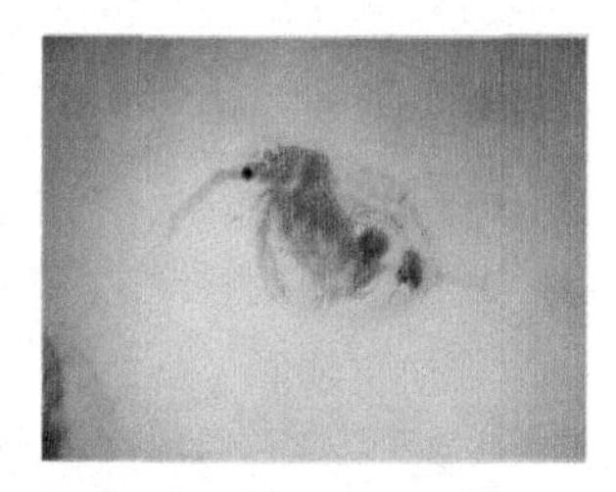

Daphnia 40x (Photos courtesy of Holly J. Morris and David T. Moat)

Purpose

To gain proficiency in making wet mounts to become familiar with various types of cells and cell structures.

Materials

- Compound light microscope
- Slides and cover slips
- Disposable pipettes
- Cultures of prokaryotes and eukaryotes
- Toothpicks
- Elodea leaves
- Methylene blue

Methods

- Make wet mounts of two prokaryotic cells (cyanobacteria), focus the slides using the medium or high power objective, and then draw the specimens as you viewed them in the space provided.

- Make wet mounts of two eukaryotic cells, focus the slides using the medium or high power objective, and then draw the specimens as you viewed them in the space provided.
- Obtain an Elodea leaf, make a wet mount of a thin section of the leaf, focus the slide using the high power objective, and then draw several of the cells in the leaf as you viewed them in the space provided.
- Scape the inside of your cheek gently with a toothpick and roll the scraping onto a slide and add a drop of methylene blue and then a cover slip.
- Focus the slide containing your cheek cells with the high power objective, and then draw several of your cheek cells as you viewed them in the space provided.
- Dispose of your cheek cell slide in the specially marked container and dispose of the other slides in the "used" slide container.

Name:

Exercise 1 Results

1. Make **detailed drawings** of a personal object as you observed them with your unaided eye and the lowest objective lens with a **stereoscopic microscope** in the space below. Include the total magnification at which you viewed the object.

Unaided Eye	Stereoscope
Total magnification: 1×	Total magnification:

2. Make **detailed drawings** of a prepared Daphnia slide as you observed it with your unaided eye and the ___________ objective lens with a **stereoscopic microscope** in the space below. Include the total magnification at which you viewed the object.

Unaided Eye	Stereoscope
Total magnification: 1×	Total magnification:

Exercise 2 Results

1. Make **detailed drawings** of the prepared *Daphnia* slide as you observed it with the low and high power objective lenses in the space below. Include the total magnification at which you viewed the object.

Low Power Lens	High Power Lens
Total magnification:	Total magnification:

2. Make **detailed drawings** of the graph paper slides as you observed them with the low power and high power objectives lenses in the space below.

Low Power Lens	**High Power Lens**
Total magnification:	Total magnification:

3. Complete the table below with the results of your graph paper slide observations.

Diameter of the Field of View	
Low power objective lens	
High power objective lens	

Exercise 3 Results

1. Make **detailed drawings**, in the space below of two types of prokaryotes as you observed them under high total magnification. Include the name of the organism and the total magnification at which you viewed the cells.

Name: ______________	**Name:** ______________
Total magnification:	Total magnification:

2. Make **detailed drawings**, in the space below, of two types of eukaryotes as you observed them under high total magnification. Include the name of the organism and the total magnification at which you viewed the cells.

Name: ______________	**Name:** ______________
Total magnification:	Total magnification:

3. Make a **detailed drawing**, in the space below, of several cells on an *Elodea* leaf wet mount as you observed it under high total magnification in the space below. Label the cell wall, cell membrane, central vacuole, chloroplast, cytoplasm, and nucleus (as appropriate). Include the total magnification at which you viewed the cells.

***Elodea* Leaf**
Total magnification:

4. Make a **detailed drawing**, in the space below, of several cheek epithelial cells that you prepared with methylene blue stain as you observed them under high total magnification in the space below. Label the cell membrane, cytoplasm, and nucleus (as appropriate). Include the total magnification at which you viewed the cells.

Cheek Cells
Total magnification:

5. What is the purpose of making a wet mount?

____

____

Additional Questions

1. What conclusions can you draw regarding:
   - Most appropriate specimens to observe with a stereoscope?

     ____

   - Most appropriate specimens to observe with a compound light microscope?

     ____

   - Best type of microscope (of the types we used in class) to see the fine details in a specimen?

     ____

2. Answer these questions based on your observations using a compound light microscope with the following total magnifications: "low" (40×), "medium" (100×), and "high" (400×):
   - Which total magnification has the largest field of view? ________________.
   - Which total magnification allows you to see the most detail in the specimen? ________________.
   - Which total magnification gives you the darkest image (assuming that you do not change any of the light settings)? ________________.
3. Answer these questions based on your observations using a compound light microscope with the following total magnifications: "low" (40×), "medium" (100×), and "high" (400×):
   - As total magnification increases, the amount of the slide that you can see in a single field of view ________________.
   - As total magnification increases, the brightness of the image ________________.
   - As total magnification increases, the apparent size of the specimen ________________.
   - As total magnification increases, the amount of fine detail that you can see in a specimen ________________.
   - As total magnification increases, the working distance ________________.
   - As you use the stage controls to move the slide toward you, the image you view in the ocular appears to move ________________.
   - As you use the stage controls to move the slide to the left, the image you view in the ocular appears to move to the ________________.
4. You have two compound light microscopes available to use. The first microscope has a resolving power of 1.2 µm; the second has a resolving power of 0.8 µm. Which microscope would allow you to see the most detail in a specimen that you are viewing? Explain your answer.

____

____

The Metric System

Photo courtesy of Holly J. Morris

OBJECTIVES

After completing these exercises, you will:

- Understand how to apply and use the metric system to measure:
  - Length
  - Mass
  - Volume
- Understand the Celsius temperature scale and how to convert between the Fahrenheit and Celsius scales.
- Be able to use the factor label method of conversion.

INTRODUCTION

The universal language of science is the metric system. A standardized system of measurement based on the decimal was first proposed in France about 1670 but it was not until 1791 that the system was actually developed. It was called the "metric" system, based on the French word for measure. At that time, every country had their own system of weights and measures. England actually had three different systems. It was not until 1875 that 17 countries signed the Metric Convention.

What is the metric system? It is a system of units of measurement devised around the convenience of the number ten. The International System of Units (SI) is the modern form of the metric system. SI was developed in 1960 from the old meter-kilogram-second system.

What is the difference between the metric and English systems of measurement? The English system uses a base 12 while the metric system uses a base 10. Because the metric system is the language of science, in this lab, all of our measurements will use the metric system.

The metric system is based on a base unit that corresponds to a certain kind of measurement. For example, length measurements are measured in terms of a meter, mass measurements are measured in terms of kilograms, and volume measurements are measured in terms of liters. The metric system uses prefixes that are appended to the base unit, to make bigger number look smaller and smaller number look bigger. For example, suppose you live 15 miles from school. That's 24,140 meters. Because that's big number, we can add a prefix onto the base unit to make it easier to manage. Therefore, we can say 24,140 meters = 24.14 kilometers.

Here is the list of the metric system prefixes we will use in this lab:

Prefix	Abbreviation	Meaning	Scientific Notation
kilo-	k	1000	10^3
hecto-	h	100	10^2
deca-	d	10	10^1
base		1	10^0
deci-	d	0.1	10^{-1}
centi-	c	0.01	10^{-2}
milli-	m	0.001	10^{-3}
micro-	μ	0.000001	10^{-6}
nano-	n	0.000000001	10^{-9}

METER

The meter is the base unit of length. The meter is the length of the path traveled by light in vacuum during a time interval of 1/299,792,458 of a second. The meter was intended to equal 10^{-7} or one ten-millionth of the length of the meridian through Paris from pole to the equator.

However, the first prototype was short by 0.2 millimeters because researchers miscalculated the flattening of the earth due to its rotation. A platinum iridium bar was cast to the exact length of a meter.

If you think of a millimeter (mm), it's about the thickness of a dime. If you think of a centimeter (cm), it's about the width of your pinky finger. If you think of a meter (m), it's about the height of a doorknob.

Here are some useful length relationships to understand:

1 km	1000 m
1 m	100 cm
1 cm	10 mm
1 m	1000 mm
1 mm	1000 µm
1µm	1000 nm

...and here's a very useful conversion factor:

2.54 cm = 1 inch

Shown below is a combination metric and customary American ruler. Note that on the metric scale, the large numbered vertical lines (1, 2, 3, etc.) represent centimeters. Each of the shorter vertical lines between centimeters represents a millimeter.

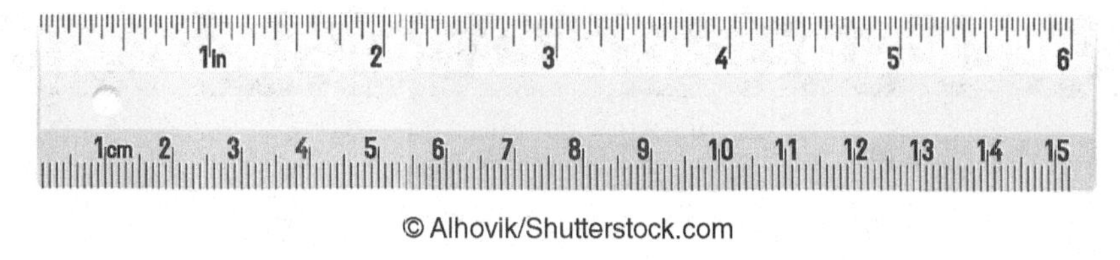

© Alhovik/Shutterstock.com

You can see that 1 inch is about 2.5 cm - 2.54 cm, to be exact, or 25.4 mm.

Approximately how many centimeters is the pencil?

How many meters?

How many millimeters?

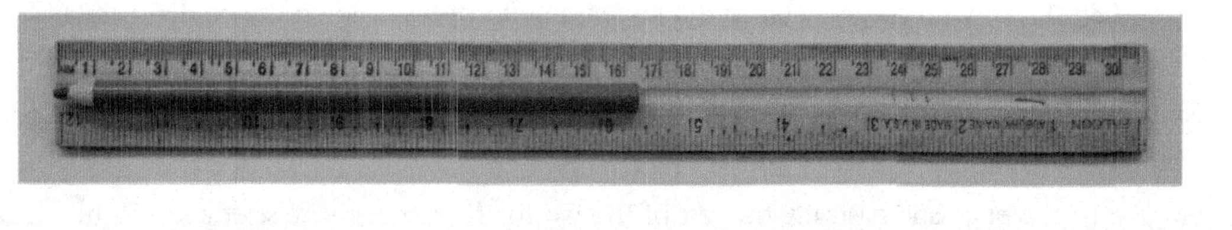

KILOGRAM

The kilogram is the base unit of mass (weight). The kilogram is defined as being equal to the mass of the International Prototype Kilogram which is almost exactly equal to the mass of one liter of water.

If you think of a gram (g), it's about the weight of a paper clip. If you think of a kilogram (kg), it's about 2 pounds.

Here are some useful mass relationships to understand:

1 kg	1000 g
1 g	100 cg
1 cg	10 mg
1 g	1000 mg
1 mg	1000 µg
1 µg	1000 ng

...and here's a very useful conversion factor:

454 g = 1 pound (lb)

LITER

A liter is a volume unit and as such, can be a liquid or a gas. A liter is 1000 cm^3 (or approximately 1.75 pints).

If you think of 5 milliliters (mL), it's about the size of a teaspoon. If you think of a liter (L), it's the size of a bottle of soda. If you think of 2 liters (L), it's the size of a large bottle of soda.

Here are some useful volume relationships to understand:

1 kL	1000 L
1 L	100 cL
1 cL	10 mL
1 L	1000 mL
1 mL	1000 µL
1 µL	1000 nL

...and here's a very useful conversion factor:

29.6 mL = 1 ounce (oz)

A VERY USEFUL METRIC RELATIONSHIP

For water: 1 mL = 1 g = 1 cm^3

A mL of water has a mass of about 1 g and a volume of 1 cm^3.

TEMPERATURE

The temperature scale most commonly used is the Fahrenheit scale. However, the Fahrenheit scale is an outdated and pretty useless scale for measuring temperature. This temperature scale was named after Daniel Gabriel Fahrenheit (1686–1736) who allegedly based the low end point of his scale on the coldest substance he could make in a laboratory (a brine solution), and the upper end point of his scale on human body temperature.

The Kelvin scale is more useful in science because there are no negative temperatures. The scale is based on the temperature at which all molecular motion stops (–273.15 °C).

That being said, in science and basically the rest of the world, the temperature scale used is the Celsius scale, named after Anders Celsius (1701–1744). The scale is based on the melting and boiling points of water: 0° and 100°.

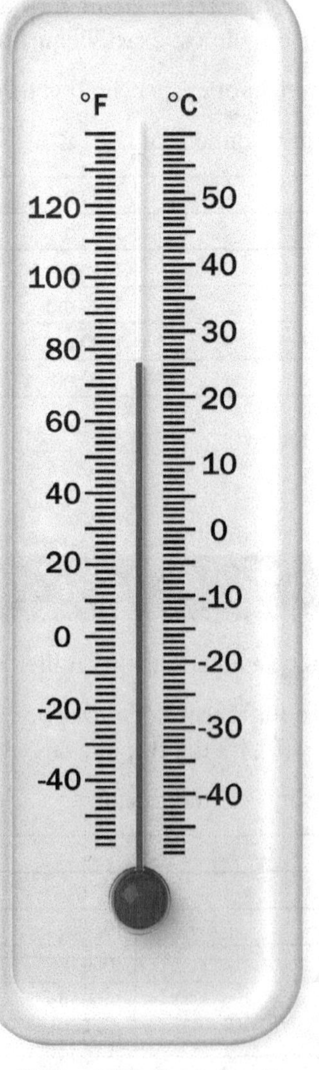

© Pincarel/Shutterstock.com

...and here's another very useful conversion factor:

$$F = 1.8C + 32$$

SCIENTIFIC NOTATION

Scientists often need to work with very large and very small numbers. Writing and manipulating very large and very small numbers can become very boring very quickly. To remedy this situation, scientists have developed a shorthand method of writing very large and very small numbers. This method is called Scientific Notation.

Scientific Notation could be defined as a method of writing or displaying numbers in terms of a decimal number between 1 and 10 multiplied by a power of 10. For example, Avogadro's Number, the calculated number of units in one mole (molecular weight in grams) of a substance is 602,300,000,000,000,000,000,000 atoms. Therefore, one mole of carbon, which weighs 12 grams, would contain 602,300,000,000,000,000,000,000 atoms. In scientific notation, this would be written as 6.023×10^{23}.

A number written in scientific notation contains 2 elements a coefficient and a base raised to an exponent. In the Avogadro's Number example, 6.023 is the coefficient and 10^{23} is the base raised to an exponent. The base is the number that is multiplied (10) and the power to which it is raised is the exponent (23).

To convert a number greater than 1:

- Move the decimal point to the left so there is one non-zero digit to the left of the decimal point. This is the coefficient.
- Count how many places the decimal was moved. This is the exponent. It is positive (in this case).
- For example, $8453 = 8.453 \times 10^3$

To convert a number less than 1:

- Move the decimal point to the right so there is one non-zero digit to the left of the decimal point. This is the coefficient.
- Count how many places the decimal was moved. This is the exponent. It is negative (in this case).
- For example, $0.02753 = 2.753 \times 10^{-2}$

To multiply in Scientific Notation:

- Multiply the coefficients.
- Add the exponents.
- $(2.5 \times 10^2) \times (3.0 \times 10^3) = 7.5 \times 10^5$

To divide in Scientific Notation:

- Divide the coefficients.
- Subtract the exponents.
- $(6.0 \times 10^6)/(3.0 \times 10^2) = 2.0 \times 10^4$

To add/subtract in Scientific Notation

- If the numbers have the same exponent, just add or subtract the coefficients.
- For example,

$$\begin{array}{r} 3.0 \times 10^4 \\ +\ 4.5 \times 10^4 \\ \hline 7.5 \times 10^4 \end{array}$$

- If the numbers have different exponents, convert one number so that its exponent is raised to the same power as the other and then perform the calculation.

- For example,

$$\begin{array}{r} 2.05 \times 10^{2} \\ -\ 9.05 \times 10^{-1} \\ \hline \end{array} \text{ is equivalent to } \begin{array}{r} 2.05 \quad\quad \times 10^{2} \\ -\ 0.00905 \times 10^{2} \\ \hline 2.04095 \times 10^{2} \end{array}$$

FACTOR LABEL METHOD OF CONVERSION

There are many instances in which scientists need to convert a value from the English system into the metric system and vice versa, or to convert values within the English system or within the metric systems.

A method you can use to perform these types of calculations is called the Factor Label Method. To use this method, multiply the original measurement by a conversion factor (i.e., an equality) which is a fraction that relates the original unit and the desired unit. For example, an equality that relates meters to millimeters could be expressed as:

$$\frac{1 \text{ m}}{1000 \text{ mm}} \text{ or } \frac{1000 \text{ mm}}{1 \text{ mm}}$$

Suppose you want to convert 6.5 km to m. First, we need to find a conversion factor that relates km and m. You know that 1 km and 1000 m are equivalent (there are 1000 m in 1 km). Therefore, we can use 1 km/1000 m as our conversion factor (i.e., an equality).

Your solution is: $6.5 \text{ km} \times \frac{1000 \text{ m}}{1 \text{ km}} = 6500 \text{ m}$

Suppose you want to convert 3.5 hours to seconds. Since you probably do not know how many seconds are in an hour, you will need more than one conversion factor in this problem.

Your solution is: $3.5 \text{ hours} \times \frac{60 \text{ minutes}}{1 \text{ hour}} \times \frac{60 \text{ seconds}}{1 \text{ minute}} = 12{,}600 \text{ seconds}$

EXERCISE 1 Using the Metric System and Scientific Notation

The metric system is the language of science. To successfully complete this course you need to be able to utilize the metric system and scientific notation.

Purpose

To become familiar with the use of the metric system and scientific notation which are necessary to work in a biology laboratory.

Materials

- Metric rulers
- Meter sticks
- Thermometers
- Various objects to measure and observe.

Methods

- Complete the following exercises which will help you become familiar with the use of the metric system and scientific notation.

Name:

Exercise 1 Results

1. Measure the following using the most appropriate metric measure of length:
   (a) Width of the lab table: ____________
   (b) Length of the lab table: ____________
   (c) Width of this page: ____________
   (d) Length of this page: ____________
   (e) Item #1: ________________ length: ____________
   (f) Item #2: ________________ length: ____________
2. Determine the temperatures of the 4 Centigrade thermometers located around the lab and then convert this temperature to its Fahrenheit equivalent.

Thermometer	Temperature (°C)	Temperature (°F)
1. Room temperature water		
2. Boiling water		
3. Ice water		
4. Lab refrigerator		

3. (a) The sun converts approximately 6.0×10^8 kg of hydrogen per second to energy. Express this number in standard notation.

   (b) The moon weighs 7.35×10^{22} kg. Convert this number to standard notation.

   (c) Alpha Centauri is 4.37×10^{12} km from Earth. How long does it take light from Alpha Centauri to strike the Earth when traveling at 3.0×10^5 km/sec? Show your work.

4. Perform the following metric-to-metric conversions. Show your work.
   (a) 3.5 m = ____________ cm
   (b) 9.7 kg = ____________ g
   (c) 2 L = ____________ mL
   (d) 59.04 g = ____________ kg
   (e) 767 nm = ____________ km
   (f) Adult humans average 5500 mL of blood. This is equivalent to __________ liters.
   (g) A Seismosaurus weighs 9.072×10^4 kg. What is its weight in ng? ____________
5. Perform the following between the metric and English systems. Show your work.
   (a) 10 miles = ___________ mm
   (b) 150 lb (pounds) = ___________ kg

(c) 65 mL = ____________ ounces

(d) 254,000 mm = ____________ inches

(e) The dinosaur Diplodocus was 37 m in length. Convert this value to inches: ____________

(f) Some redwood trees have a diameter of 11 m. What is this diameter in feet (ft)? ____________

6. Convert the following numbers to scientific notation:

(a) 124.95000000 =

(b) 0.000000000567 =

(c) 12 =

7. Convert the following numbers to standard notation:

(a) 2.94×10^{7} =

(b) 5.43×10^{-8} =

(c) 6.55×10^{-2} =

8. Perform the indicated mathematical operations.

(a) $\begin{array}{r} 3.1 \times 10^{5} \\ + \underline{6.4 \times 10^{5}} \end{array}$

(b) $\begin{array}{r} 5.2 \times 10^{9} \\ + \underline{3.2 \times 10^{5}} \end{array}$

(c) $\begin{array}{r} 8.5 \times 10^{5} \\ - \underline{3.7 \times 10^{5}} \end{array}$

(d) $\begin{array}{r} 7.3 \times 10^{5} \\ - \underline{2.1 \times 10^{2}} \end{array}$

(e) $\begin{array}{r} 3.6 \times 10^{5} \\ \times \underline{2.1 \times 10^{6}} \end{array}$

(f) $\begin{array}{r} 5.8 \times 10^{8} \\ \times \underline{3.3 \times 10^{-5}} \end{array}$

(g) $(4.2 \times 10^{10})/(1.9 \times 10^{5})$ =

(h) $(6.8 \times 10^{-4})/(3.4 \times 10^{-2})$ =

Basic Laboratory Techniques

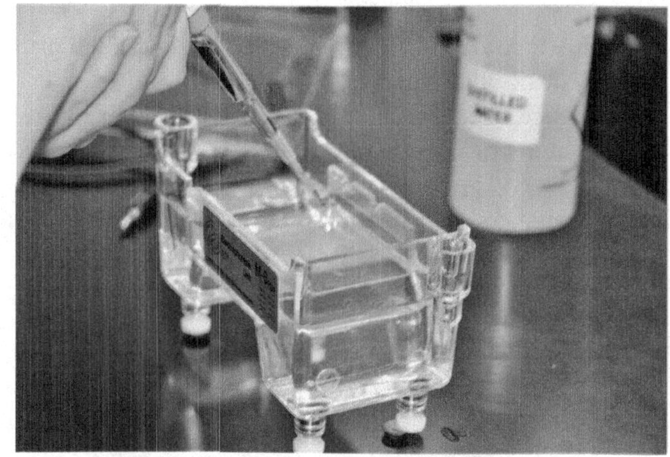

Photo courtesy of Holly J. Morris

OBJECTIVES

After completing this exercise, you will be able to:

- Correctly use laboratory apparatuses such as scales, graduated cylinders, and pipettes.

INTRODUCTION

Scales, graduated cylinders, and pipettes are basic tools in a biology lab. They are used routinely to perform a wide variety of common laboratory tasks.

SCALES

In a laboratory, scales are very useful because they can be used for many different tasks. For example, you can weigh the amount of solute you need to make a solution or you can weigh "cells" to determine whether they are gaining or losing water. There are two types of scales used in this lab:

- Triple beam balance.

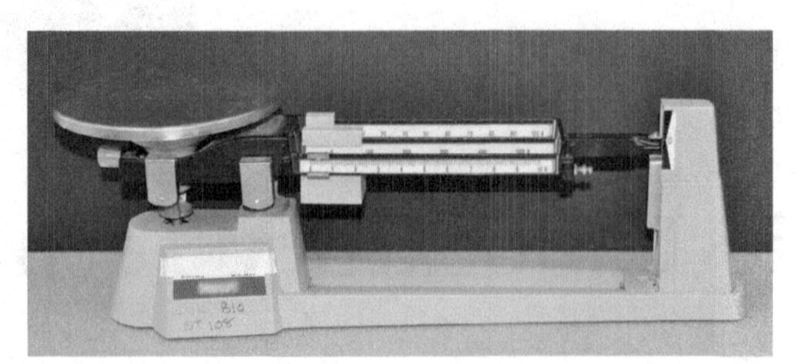

(Photos courtesy of Holly J. Morris and David T. Moat)

To weigh an item with the triple beam balance:

- With the pan empty, ensure that the indicator mark is pointing at the Balance Mark. If it is not, use the Zero Adjust Knob to balance the scale.
- Place the object to be weighed directly on the pan.
- Move the various weights along each of the beams until the indicator mark aligns with the Balance Mark.
- Add up the weights indicated by each of the three beams to determine the total weight of the object.
- A triple beam balance can weigh objects up to about 610 g to the nearest 0.1 gram.

- Electronic balance.

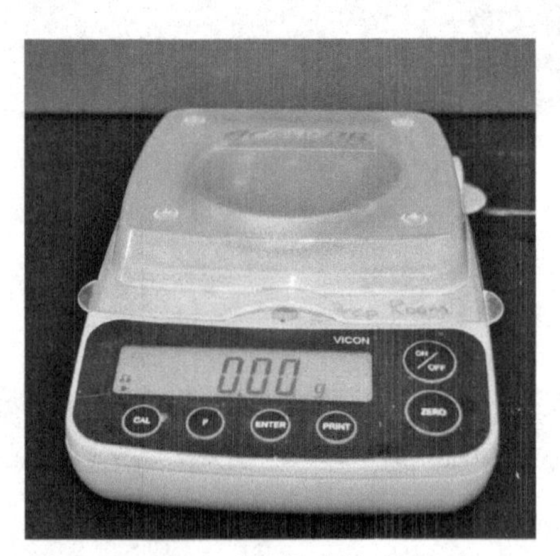

(Photos courtesy of Holly J. Morris and David T. Moat)

To weigh an item with the top-loading electronic balance:

- Turn on the balance and allow it to warm up for a minute.
- Lift the protective plastic cover to expose the pan.
- Press the "Zero" button. The balance should read "0.00 g."
- If the object to be weighed is not a liquid or powder, place it on the pan. Wait several seconds for the weight to stabilize. Record the weight.
- If the object to be weighed is powder or liquid, you need to place it in a container. Follow these steps:
  - Place the empty container that you are going to use to weigh the powder or liquid on the pan.
  - Press the "Zero" button, which will zero the weight of the container. (This is also referred to as "taring" the container.)
  - Put the substance to be weighed in the container and the balance will display the weight of the substance.
  - With a scale that displays 0.00, substances can be weighed to the nearest 0.01 g. (If the display were 0.000, substances could be weighed to the nearest 0.001g.)

GRADUATED CYLINDERS

When using liquids, you need to be able to measure them accurately. Graduated cylinders can be used to measure liquid amounts ranging from 10 mL to 2000 mL.

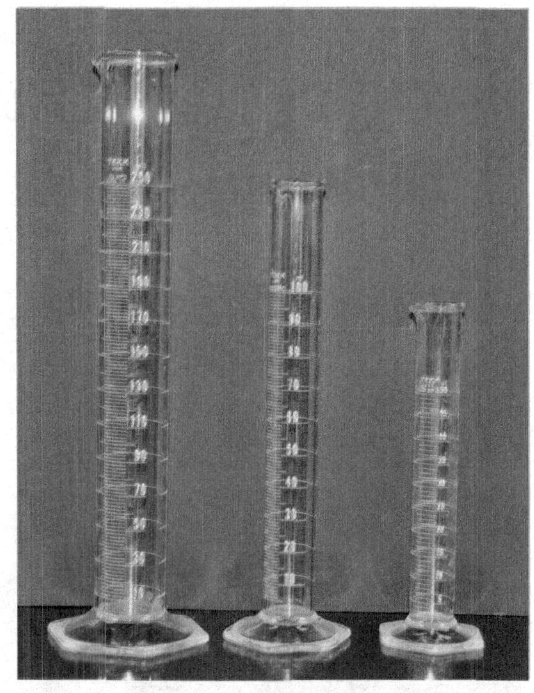

(Photos courtesy of Holly J. Morris and David T. Moat)

To use a graduated cylinder:

- Pour the liquid you want to measure into the graduated cylinder.
- If the graduated cylinder is glass, liquid adheres slightly to the wall, creating a concave appearance. This is called a meniscus (see the picture below). Glassware is calibrated so that the bottom of the meniscus is the correct location to read.

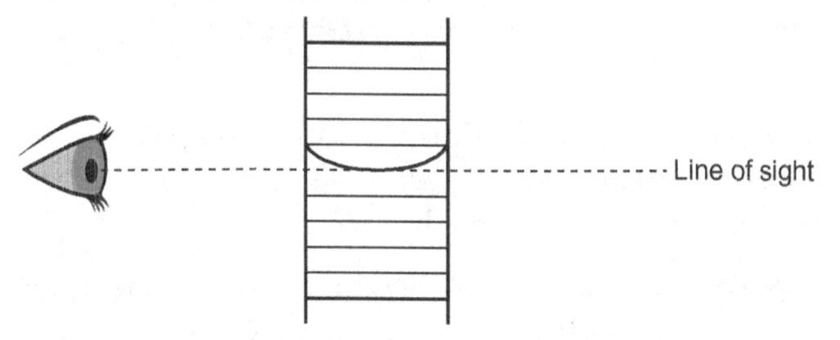

- In a plastic graduated cylinder, a meniscus may not form, depending on the type of plastic. In this case you can read the liquid at the level of the gradation.

Graduated cylinders are not precise. If precision is necessary, a volumetric flask should be used instead of a graduated cylinder. Beakers should not be used for liquid measures unless the instruction reads something like, "Pour about 30 mL of distilled water into a beaker."

PIPETTES

Pipettes are devices that can be used to precisely measure small volumes of liquid. There are two types of pipettes that we will use in this lab. Larger volumes (1–10 mL) will be delivered by either serological or Mohr pipettes, and small volumes (less than 1 mL) will be delivered by micropipettes.

There are two variations of pipettes that we use for larger volumes. We typically have 1 mL, 5 mL, and 10 mL serological pipettes, and 10 mL Mohr pipettes. Look at the photo below. Both pipettes hold a maximum volume of 10 mL, but the serological pipette ends with the number 9 and the Mohr pipette ends with the number 10.

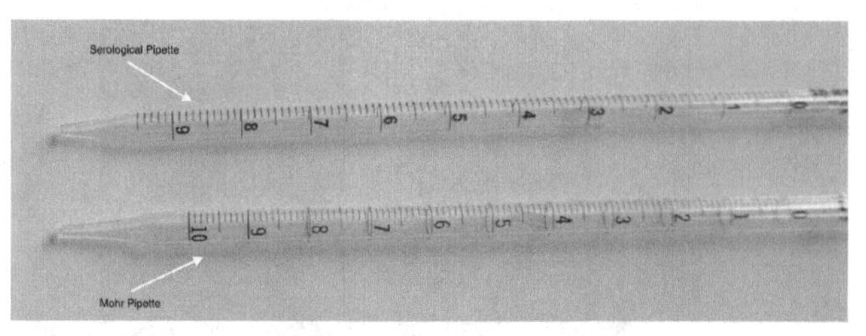

(Photos courtesy of Holly J. Morris and David T. Moat)

You need to make sure you know which type of pipette you are using. The serological pipette has a double stripe at the top, and should have the letters "TD," which means "to deliver." This is also referred to as a "blowout" pipette. That means that the last bit of liquid needs to be blown out in order to deliver the correct volume. The volume in a TD (serological) pipette is from the "0" to the tip of the pipette.

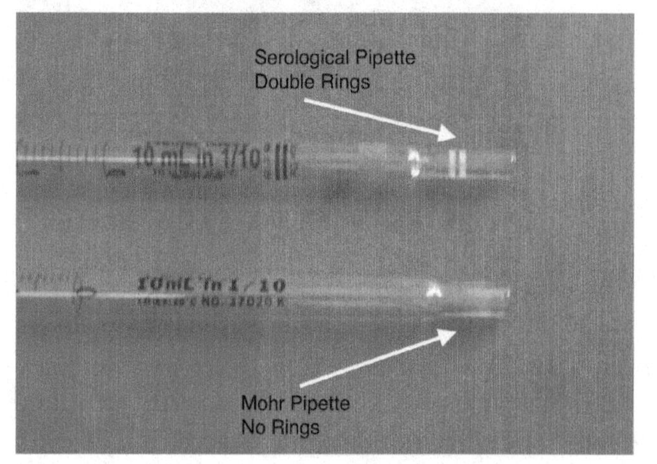

(Photos courtesy of Holly J. Morris and David T. Moat)

A Mohr pipette is calibrated so that the total volume is within the pipette. For example, with the Mohr pipette shown in the photo above, 10 mL would be the volume from the "0" to the "10." Additional liquid will remain in the pipette after the correct volume has been delivered.

A pipette aid, also called a pipetter, can be a bulb or a pump, and is used to draw up and deliver the liquid. Drawing liquid into a pipette should never be done by mouth. We will be using a pump.

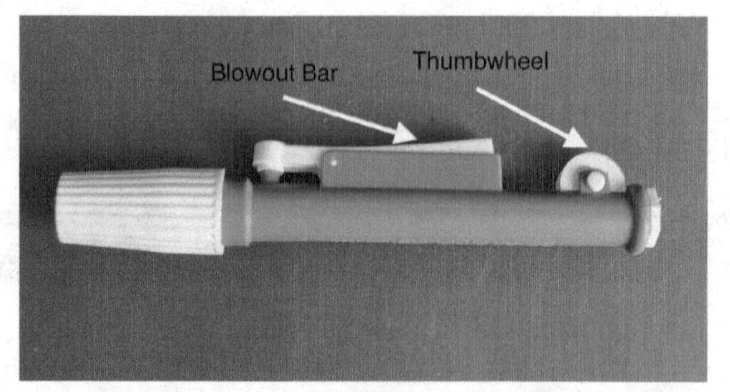

Pipette Pump (Photos courtesy of Holly J. Morris and David T. Moat)

Disposable pipettes usually come from the manufacturer with a cotton plug at the top. This is to help prevent overfilling. Do not remove the cotton. If the pipette does not have a cotton plug, be very careful not to draw liquid into the pump. Liquid can damage the pipetter, and can contaminate subsequent samples as it dribbles back into the pipette.

To use a pipette:

- Select an appropriate size, based on the volume of liquid you need to transfer.
- Attach the pipette to the pump, as shown in the photo below.

(Photos courtesy of Holly J. Morris and David T. Moat)

- Place the tip of the pipette into the liquid to be transferred. Submerge the tip deep enough to draw up liquid without drawing air into the pipette, but not so deep that the tip hits the bottom of the container. Use one hand on the pump and the other hand to steady the pipette.
- Slowly draw the liquid up using the thumbwheel. Draw the liquid just past the desired volume (make sure you have calculated the volume correctly), then carefully reverse the thumbwheel to line the liquid up at the desired volume. Do not draw the liquid up so far that it reaches the cotton plug.
- As you pull the pipette out of the liquid, touch the tip to the side of the container in order to remove any hanging drops on the outside of the pipette.
- To dispense the liquid, turn the thumbwheel in the reverse direction until the correct volume has been dispensed. If you are using a blowout pipette and dispensing the full volume, press the "blowout" bar on the side of the pump, to eject the last drop of liquid.

To use a micropipette:

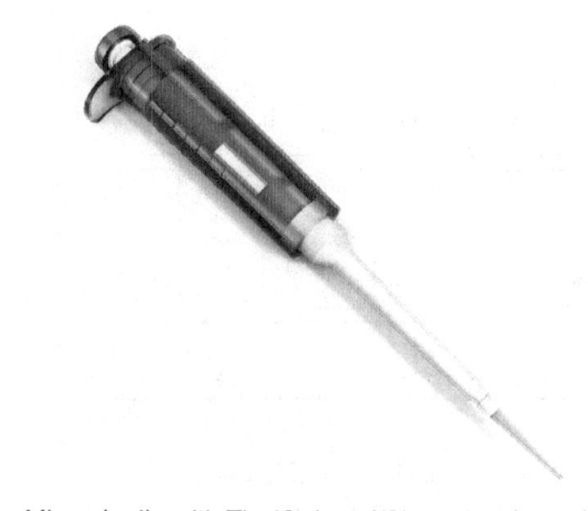

Micropipette with Tip (© dny3d/Shutterstock.com)

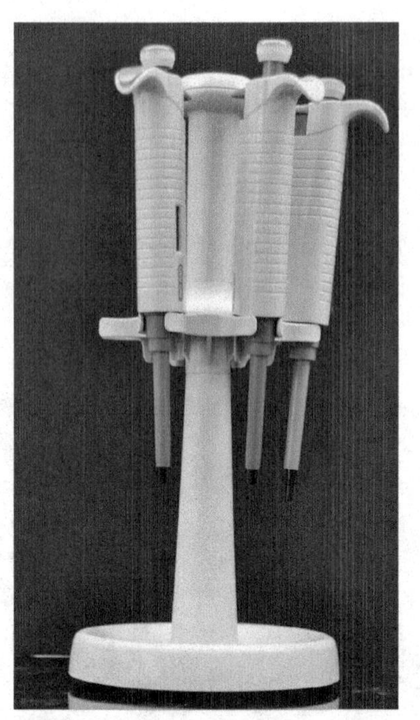

Micropipettes in Stand (Photos courtesy of Holly J. Morris and David T. Moat)

- Select the appropriate size micropipette. There are several sized micropipettes in this lab: P20, P200, and P1000. Their capacities are shown below:

Pipette	Smallest volume (µL)	Largest volume (µL)
P20	0.5	20
P200	20	200
P1000	100	1000

- Set the volume of liquid to be transferred. Our pipettes contain a rotating knob (thumbwheel) which allows you to set the desired volume. Examples of how to set the volume for the various micropipettes follow. When setting the correct volume, turn the thumbwheel slightly past the desired volume, then correct back to the volume.

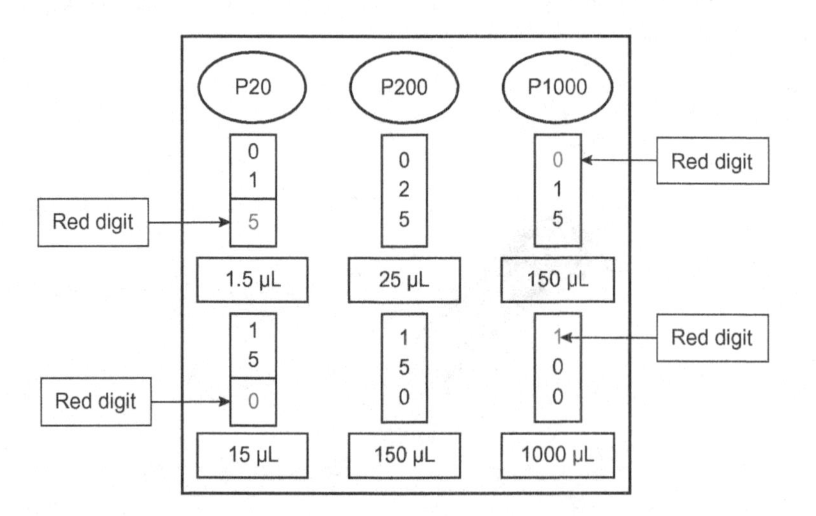

- Attach a micropipette tip to the micropipette, like shown in the picture, below. Make sure you select the correct size tip. Do not touch the end of the tip with your fingers.

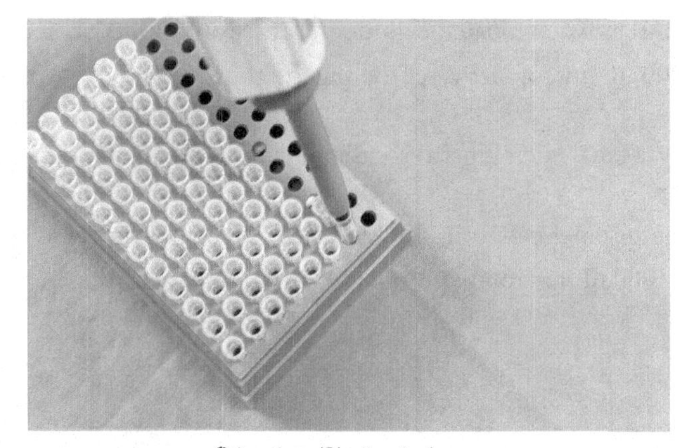

© toeytoey/Shutterstock.com

- Push the plunger until you feel resistance (the first stopping point) and then put the micropipette tip into the container with the liquid to be transferred. Slowly release the plunger to draw the desired volume of liquid into the micropipette. (If you release the plunger too quickly, you will not draw the correct volume into the tip.) Touch the tip to the side of the container to remove any hanging drops.
- To dispense the liquid, push the plunger to the first stopping point and then continue to push the plunger until the second stopping point. Just like blowing out the liquid on the serological pipette, this forces all of the liquid to be pushed out of the micropipette.
- Eject the micropipette tip into a waste beaker. Each tip is only used once to help prevent contamination of the liquids.

EXERCISE 1 Performing Basic Lab Techniques

This exercise will provide you with some experience using basic lab techniques, such as pipetting, using graduated cylinders, and using scales, that are useful and applicable in a laboratory setting.

Purpose

To be able to correctly perform basic lab techniques, such as using pipettes, graduated cylinders, and scales.

Materials

- Pipettes
- Pipette pump
- Micropipettes
- Micropipette tips
- Various graduated cylinders
- Triple beam balance
- Electronic balance
- Test tubes
- Various objects to be weighed

Methods

- Weigh any two of the various items on display using a triple beam balance and record your results.
- Weigh the same two items, that you used previously, using an electronic balance and record your results.
- Practice using graduated cylinders to ensure that you can accurately measure liquids using them.

- Read the volume of the two graduated cylinders at the front of the lab and record their volumes.
- Using a 20 μL to 200 μL micropipette, set the pipette to measure _____ μL of water and show it to the lab instructor.
- Using a 100 μL to 1000 μL micropipette, set the pipette to measure _____ μL of water and show it to the lab instructor.
- Using a manual pipette, pipette 4.3 mL of water into a test tube and show it to the lab instructor.
- Using a 100 μL to 1000 μL micropipette, pipette 350 μL of water into a test tube and show it to the lab instructor.

Name:

Exercise 1 Results

1. Determine the mass of the following using a triple balance scale:
   (a) Item #1: ________________ mass: ________________
   (b) Item #2: ________________ mass: ________________
2. Determine the mass of the following using an electronic scale:
   (a) Item #1: ________________ mass: ________________
   (b) Item #2: ________________ mass: ________________
3. Read the volume of the graduated cylinder 1 at the front of the room and record its volume.

   Volume: ________________
4. Read the volume of the graduated cylinder 2 at the front of the room and record its volume.

   Volume: ________________
5. Using a 20 μL to 200 μL micropipette, set the pipette to measure ________________ μL of water and show it to lab instructor.

   OK / not OK.
6. Using a 100 μL to 1000 μL micropipette, set the pipette to measure ________________ μL of water and show it to lab instructor.

   OK / not OK.
7. Using a manual pipette, pipette 4.3 mL of water into a test tube and show it to lab instructor.

   OK / not OK.
8. Using a 100 μL to 1000 μL micropipette, pipette 350 μL of water into a test tube and show it to the lab instructor.

   OK / not OK.

CHAPTER 5

Acids, Bases, and pH

Photo courtesy of Holly J. Morris

OBJECTIVES

After completing these exercises, you will be able to:

- Understand the difference between an acid and a base.
- Understand the pH scale and what it measures.
- Understand the use of chemical indicators.
- Understand the role of buffers and why they are important to living systems.

INTRODUCTION

Acids and bases play an essential role in biology. Some biologically important molecules are acids including amino acids (protein building block), nucleic acids (DNA and RNA building block), and pyruvic acid (metabolic pathway intermediary). Additionally, acids and bases are biologically important because enzymes only work in specific acid/base environments.

An acid is any substance that can donate a proton (hydrogen ion, H^+). For example, carboxyl groups (COOH) are common proton donors. The carboxyl group dissociates to yield a proton (H^+) and a negatively charged COO^- ion carboxyl group. Acetic acid (the acid that gives vinegar is characteristic taste) disassociates into the acetate ion and a hydrogen ion as shown below:

$$CH_3COOH \rightarrow CH_3COO^- + H^+$$

A base is any substance that accepts protons (hydrogen ions, H^+). Sodium hydroxide is an example of a base. Upon ionization in water, sodium hydroxide releases a positively charged metallic ion and a negatively charge hydroxide ion:

$$NaOH \rightarrow Na^+ + OH^-$$

Hydroxide ions are effective bases because they very strongly attract protons. However, there are many biological compounds that act as bases even though they do not contain hydroxide ions. For example, a water molecule (a polar molecule with two unshared pairs of electrons) can accept a proton and therefore, act as a base. For example, water acts as a base by accepting protons from hydrochloric acid:

$$HCl + H_2O \rightarrow H_3O^+ + Cl^-$$

On the other hand, H_3O^+ (hydronium ion) can act as an acid by donating a proton to another substance.

Some acids are called strong acids while other acids are called weak acids. The strength of an acid is determined by how readily it gives up protons (how readily it ionizes). For example HCl (hydrochloric acid) is a strong acid and CH_3COOH (acetic acid) is a weak acid.

Similarly, some bases are called strong bases while other bases are called weak bases. The strength of a base is determined by how readily it accepts protons. For example, OH^- (hydroxide ion) is a strong base while H_2O (water) is a weak base.

The pH of a solution is a measure of the concentration of hydrogen ions in a solution and as such is a measure of the acidity or basicity of the solution. The pH scale, which is determined by the concentration of hydrogen ions, measures this concentration and runs from 0 to 14. The lower numbers refer to acid solutions. The higher numbers refer to basic solutions. Any solution with a pH of less than 7 has more hydrogen ions than hydroxide ions in solution. Any solution with a pH of more than 7 has fewer hydrogen ions than hydroxide ions in solution. The midpoint in the scale is 7, where the concentration of hydrogen ions equals the concentration of hydroxide ions. The pH of distilled water is 7. The diagram below illustrates the pH scale:

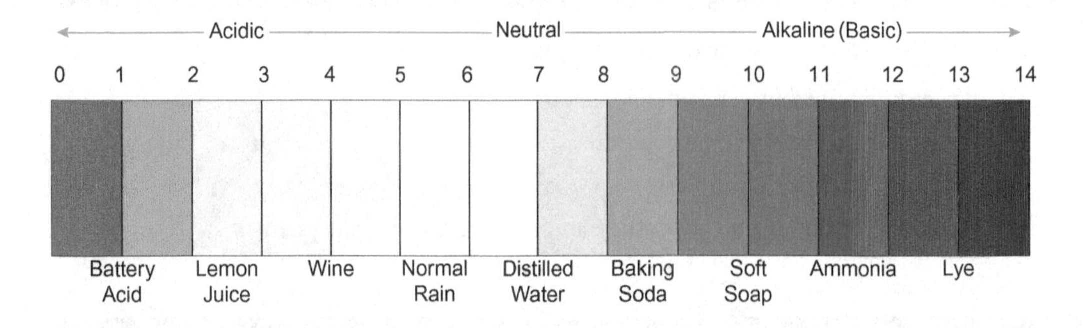

More specifically, pH is a measure of the concentration of hydrogen ions, represented as the negative log of the concentration. Thus, an aqueous solution that contains 1×10^{-3} moles/L of hydrogen ions has a pH of 3. A solution with a concentration of hydrogen ions equal to 1×10^{-8} moles/L has a pH of 8. A change of one pH number represents a 10-fold difference in hydrogen ion concentration. That is, a solution with a pH of 2 has 10 times more hydrogen ions than a solution with a pH of 3, and 100 times more hydrogen ions than a solution with a pH of 4.

The concentrations of H^+ and OH^- are reciprocal, and together add up to 10^{-14} moles/L. As one concentration goes up, the other comes down by a proportional amount, thus if the concentration of H^+ is 10^{-2}, the concentration of OH^- is 10^{-12}.

pH indicators are weak acids or weak bases that can be used to indicate the approximate concentration of H^+ ions in a solution by causing the color of the solution to change. The table below shows some sample pH indicators.

pH INDICATORS

pH Indicator	Color in Acid	Color in Base	pH Range Where Color Changes
Phenolphthalein	Colorless	Pink to violet	8.3–10
Phenol red	Yellow	Red	6.8–8.2
Bromothymol blue	Yellow	Blue	6–7.6
Litmus	Red	Blue	5.0–8.0 (color will be purple in this range)
Methyl orange	Red	Yellow	3.1–4.4

BUFFERS

Buffers are molecules that help maintain a relatively stable pH in a solution by either releasing or binding hydrogen ions, as needed. A buffer system usually contains a set of molecules consisting of a weak acid and its conjugate base. Buffers do not neutralize a solution, nor do they turn acids into bases, or bases into acids. Rather, they minimize changes in pH when either an acid or a base is added to a solution, keeping the pH stable. Buffers typically work within a specific pH range.

Enzymes are especially dependent on specific pH ranges which are sometimes very narrow. Either excess H^+ or OH^- can cause structural changes to occur in enzymes, reducing or eliminating their activity. Thus, buffers help organisms maintain a relatively stable pH so that biochemical processes can proceed normally.

One location in humans where a narrow pH range is critical is the blood. Human blood needs to remain in a pH range of 7.35 to 7.45, even though metabolic processes constantly add and remove acids and bases. The primary buffer system that helps maintain that narrow pH range is the carbonic acid/bicarbonate buffer system.

$$CO_2 + H_2O \leftrightarrow \underset{\text{Carbonic Acid}}{H_2CO_3} \leftrightarrow \underset{\text{Bicarbonate}}{H^+ + HCO_3^-}$$

When carbon dioxide levels increase, which results from cell respiration in the tissues, it is combined with water to form carbonic acid. When carbon dioxide levels decrease, which occurs in the lungs, carbonic acid is converted back to carbon dioxide and water, and the CO_2 is exhaled through the lungs.

In the blood, between the tissues and the lungs, if hydrogen ion (H^+) concentration increases, the bicarbonate ions will combine with free hydrogen ions, forming carbonic acid, thus removing H^+ from the solution, and raising the pH. When hydrogen ion concentration decreases, carbonic acid will release H^+ into the solution, forming bicarbonate. The resulting increase in free H^+ decreases the pH of the solution. Without a buffer system, the pH of the blood would fluctuate enough to interfere with enzyme function, and threaten an individual's survival.

In the following exercises, you will be introduced to the use of pH indicators, the measurement of pH, neutralizing acids, and buffering systems.

EXERCISE 1 Using Phenolphthalein

Phenolphthalein is a chemical compound that is often used as an acid–base indicator. It is colorless below pH 8, light pink between pH 8 and pH 10, and pink above pH 10.

Purpose

To demonstrate the use of pH indicators as a tool to determine the state of a chemical reaction.

Materials

- Sodium carbonate
- Small blue scoop
- Glass stirring rod
- Straw
- Distilled water
- 400 mL beaker
- Phenolphthalein solution
- 100 mL graduated cylinder
- Disposable pipettes

Methods

- Get one well rounded spoonful of sodium carbonate using the small blue spoon.
- Place the sodium carbonate into a 400 mL beaker and fill with distilled water up to 100 mL.
- Mix the solution with a glass stirring rod until the sodium carbonate is completely dissolved and distributed evenly throughout the solution.
- Place two drops of phenolphthalein into the solution. The solution should turn and remain pink.
- Place the straw in the solution and blow into the solution until it becomes colorless. Do not exhale too hard into the straw to the point where the solution will bubble out of the beaker. It should take about three to four breaths to make the solution colorless.

EXERCISE 2 Measuring pH

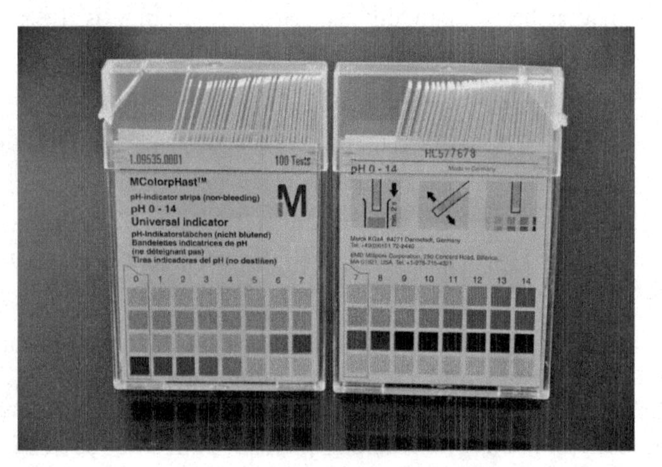

Photo courtesy of Holly J. Morris and David T. Moat

The two main ways to measure pH are pH strips and pH meters. pH strips are impregnated with dyes that change from one color to another as the pH of their environment is altered. The pH is determined by comparing the colors on the strip with those of a standard. pH meters directly measure the pH of a substance.

Purpose

To become familiar with the methods used to measure pH.

Materials

- pH strips
- Vernier Lab Quest 2 Data Collector
- Vernier pH probe
- Substances to be measured for their pH

Methods

- Measure five substances with pH strips and then with the Vernier pH probe (Instructions for the data collector and probe will be handed out separately.) Record your results in the **pH Results** table.

EXERCISE 3 Neutralizing Antacids

Antacids contain alkaline ions that directly neutralize stomach gastric acid. Phenol red, another chemical compound used as an acid-base indicator, can be used to indicate how effectively antacids can neutralize acids. Phenol red is yellow below pH 6.8, orange between pH 6.8 and pH 8.2, and red above pH 8.2.

Purpose

To determine which antacid is the most effective acid reducing agent.

Materials

- 0.1 M Hydrochloric acid solution
- 0.02% Phenol Red indicator
- Distilled water
- Antacids - Assorted solid and liquid samples
    - Use 1 tablet (solid sample)
    - Use 10 drops (liquid sample)
- Three 150 mL beakers
- 100 mL graduated cylinder
- Three clean test tubes
- Mortar and pestle
- Glass stirring rod
- Disposable pipettes

Methods

- *For antacid tablets*: Use the mortar and pestle to crush the solid antacid tablets to a fine powder and place into a beaker.

  For antacid liquids: Place 10 drops of the liquid antacid into a beaker.
- Add 100 mL of distilled water to each of the beakers containing antacids and mix the solutions.
- After dissolution, place 10 mL of each solution in a test tube.
- Add 3 drops of 0.02% Phenol Red and mix using a glass stirring rod. Note the color of the solution.

- Carefully add 0.1 M hydrochloric acid (HCl) solution dropwise to the test tubes making sure to count each drop.
- Continue adding HCl to your solutions in the test tubes until the solution turns yellow.
- Record the number of drops of 0.1 M HCl was used to neutralize the antacids in the **Antacid Results** table.

EXERCISE 4 Buffering Systems

A buffer is any molecule that releases or binds hydrogen ions in order to maintain a relatively stable pH in a solution. The bicarbonate ion is this kind of molecule.

Purpose

To demonstrate the pH regulating capacity of the bicarbonate buffer.

Materials

- Four 100 mL beakers
- Label tape
- Distilled water
- 0.1 M Hydrochloric acid solution
- 0.1 M Sodium hydroxide solution
- 0.2% Sodium bicarbonate
- Vernier Lab Quest 2 Data Collector
- Vernier pH probe

Methods

- Using label tape, label one beaker "HCl-Water" and a second beaker "NaOH-Water."
- Add approximately 50 mL of distilled water to the two beakers.
- Label the third beaker "HCl-Bicarb" and the fourth beaker "NaOH-Bicarb."
- Add approximately 50 mL of sodium bicarbonate to beakers three and four.
- Determine the pH of the solution in each beaker with the Vernier pH probe. (Instructions for the data collector and probe will be handed out separately.) Record the pH for each solution in the **Buffer Experiment Results** table.
- Add 1 drop of HCl to the "HCl-Water" and the "HCl-Bicarb" beakers. Determine the pH of each solution and record it in the **Buffer Experiment Results** table. Repeat this step 4 more times.
- Add 1 drop of NaOH to the "NaOH-Water" and the "NaOH-Bicarb" beakers. Determine the pH of each solution and record it in the **Buffer Experiment Results** table. Repeat this step 4 more times.

Name:

Exercise 1 Results

1. What color was the solution after you added and stirred the phenolphthalein?

2. Describe the color changes that occur after blowing through the straw.

3. Explain what happened as the CO_2 from your breath was blown into the basic solution.

Exercise 2 Results

1. Record your results below.

pH Results		
Sample	**pH Strip**	**pH Probe**
1		
2		
3		
4		
5		

Exercise 3 Results

1. Record your results below.

Antacid Results	
Antacid	**Drops of 0.1 M HCl to Neutralize an Antacid**
1	
2	
3	
4	

2. What is the basic function of an antacid? Explain your answer.

3. Which antacid is potentially the strongest? Explain your answer.

Exercise 4 Results

1. Record your results below.

Buffer Experiment Results				
	HCl-H_2O	HCl-Bicarb	NaOH-H_2O	NaOH-Bicarb
Starting pH				
After 1 drop				
After 2 drops				
After 3 drops				
After 4 drops				
After 5 drops				

2. Discuss the pH regulating capacity of the bicarbonate ion.

Biomolecules

Photo courtesy of Holly J. Morris

OBJECTIVES

After completing these exercises, you will be able to:

- Understand the four categories of biomolecules and their functions.
- Understand the concept of positive and negative controls and their use.
- Understand how to perform the Benedict's test to identify reducing sugars, interpret the results and know their significance.
- Understand how to perform the Iodine test to identify starch, interpret the results and know their significance.
- Understand how to perform the Biuret's test to identify proteins, interpret the results and know their significance.
- Understand how to perform the Sudan IV test to identify lipids, interpret the results and know their significance.

INTRODUCTION

The cell is the fundamental unit of organization of life. It is not known how the first cells were formed but probably many different types of chemical and physical processes led to the formation of simple molecules. Additional and more complex interactions between these molecules resulted in the formation of more complex structures and groupings that eventually led to the formation of a living cell.

These biomolecules form the basic components of a living cell and are in all living cells. Biomolecules consist mostly of carbon, hydrogen, nitrogen, oxygen, sulfur, and phosphorus. They can be grouped into four main categories:

- Nucleic acids
- Carbohydrates
- Lipids
- Proteins

Nucleic acids

Nucleic acids are long chains of nucleotides, the building blocks of nucleic acids. A nucleotide consists of a nitrogenous base (adenine, guanine, thymine, cytosine, and uracil), a pentose sugar and a phosphate group. Polymerized nucleotides form deoxyribonucleic acid (DNA) and ribonucleic acid (RNA).

DNA is the molecule that contains a cell's genetic information. RNA's principal role is to act as a messenger carrying instructions from DNA, controlling the synthesis of proteins, although in some organisms, RNA, rather than DNA, contains the cell's genetic information. Thus, DNA and RNA function in encoding, transmitting and expressing genetic information conveyed through the nucleic acid sequence.

Carbohydrates

Carbohydrates typically contain carbon, hydrogen, and oxygen. They can be grouped into three categories including:

- Simple sugars
- Complex carbohydrates
- Structural polysaccharides

Simple sugars commonly include monosaccharides (single sugars such as glucose) and disaccharides (two single sugars linked together such as maltose). Simple sugars, such as glucose, fructose, maltose, and sucrose, are the primary energy sources for cells. Monosaccharides are the building blocks of disaccharides and polysaccharides (such as cellulose and starch)

Complex carbohydrates are polysaccharides, long strings of glucose, which are stored in plant and animal tissues as a reserve energy source when other energy sources are not available. Plants make and store a complex carbohydrate called starch, while animals store a complex carbohydrate called glycogen.

There are two kind of starch - the helical amylose and the branched amylopectin. Plants will vary, but it is typical for the starch to contain 20% to 25% amylose and 75% to 80% amylopectin. Glycogen, the glucose store of animals, is a more branched version of amylopectin.

Structural polysaccharides are also long strings of glucose and are used as building materials for the organism to give them structure and strength. Cellulose, the plant version of structural polysaccharides, is the main component of plant cell walls. Chitin, the animal version of structural polysaccharides, is the main component of structures such as the cell walls of fungi and the exoskeletons of arthropods such as crustaceans and insects.

Lipids

Lipids are a class of organic compounds composed of long hydrocarbon chains. They contain a large amount of stored energy and form a reserve energy source for the cell. Lipids are the building blocks of biologic membranes including the membranes surrounding the cell and surrounding its organelles. Cholesterol, another lipid, is an essential structural component of animal cell membranes and additionally functions as the building block for the biosynthesis of steroid hormones such as testosterone and estradiol.

The common feature of lipids is that they are all nonpolar. As such, they are hydrophobic, insoluble in water, and soluble in nonpolar solvents.

Proteins

Proteins are large complex molecules that play many critical roles in the body. There are more than 100,000 different proteins in a human being, all of which are composed of chains of the 20 different amino acids joined together by peptide bonds. All amino acids have the same basic structure: an amino group, a carboxyl group and a central carbon bound to a hydrogen, and a side chain (commonly known as the R Group). The sequence of amino acids in a protein molecule is determined by the cell's genetic material. The amino acid sequence determines the protein's three-dimensional structure and as a result, its specific function.

Proteins function as antibodies (to help protect the organism), enzymes (to regulate where and when cellular reactions occur), messengers (to transmit signals to coordinate biologic processes), structural components (to provide structure and support for the cell), and carriers (to transport and store molecules).

Biomolecular Themes

Although the biomolecules are varied groups, there are some commonalities that they all share:

- All of the biomolecules have 'backbones" or "skeletons" made of chains or rings of carbon. The "backbones" or "skeletons" give each biomolecule its characteristic shape. This characteristic results from carbon's capability to form four covalent bonds which enables it to form very complex structures.
- Most biomolecules are composed of smaller subunits connected together in a repeating fashion like the cars in a train. The larger multi subunit molecule is called a polymer and the smaller subunits that make it up are called monomers.
- Monomers can be assembled into polymers by dehydration synthesis (also known as condensation). Dehydration synthesis removes a water molecule in the process of joining two monomers together. This reaction is shown below:

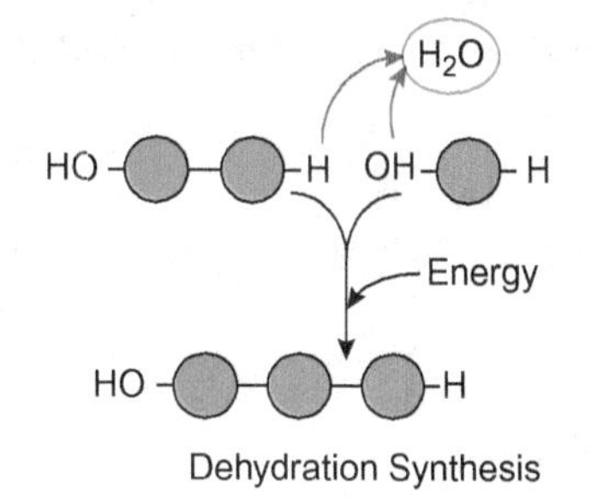

Dehydration Synthesis

- Polymers can be broken down into their component subunits by hydrolysis. Hydrolysis uses a water molecule to break the bond between two adjacent monomers. This reaction is shown below:

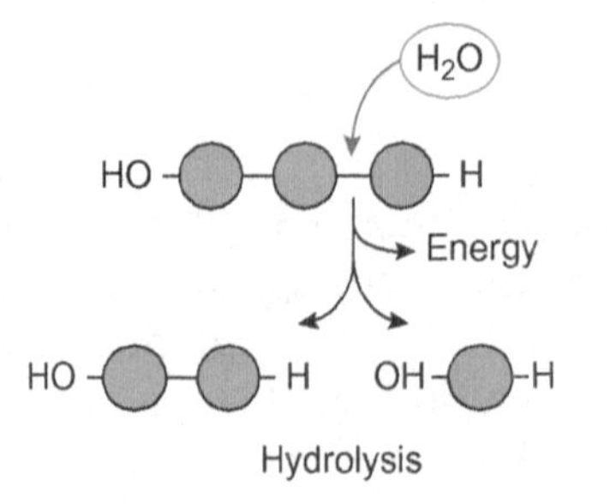

In the following exercises, you will be introduced to the use of chemical tests to identify the presence of the various biomolecules.

EXERCISE 1 Benedict's Test

All monosaccharides as well as some disaccharides and polysaccharides are reducing sugars. A reducing sugar is a sugar that can convert between a ring configuration and an open chain configuration. The result of this conversion into an open chain configuration is the formation of a reactive aldehyde which allows the sugar to act as a reducing agent. Benedict's reagent can be used to determine if a reducing sugar is present.

Not all sugars are reducing sugars. For example, sucrose cannot exist in an open chain configuration. As such, it does not react with Benedict's reagent and is not a reducing sugar.

When Benedict's reagent is mixed with a reducing sugar and heated, copper (II) is reduced to copper (I) forming a precipitate. Benedict's reagent is blue and when the precipitate forms, depending on the concentration of reducing sugars, various colors develop from green to yellow to orange to red. A yellowish-green color indicates a low concentration, whereas an orange-reddish color indicates a high concentration of reducing sugars.

The test tube on the left shows a positive Benedict's test and the test tube on the right shows a negative Benedict's test.

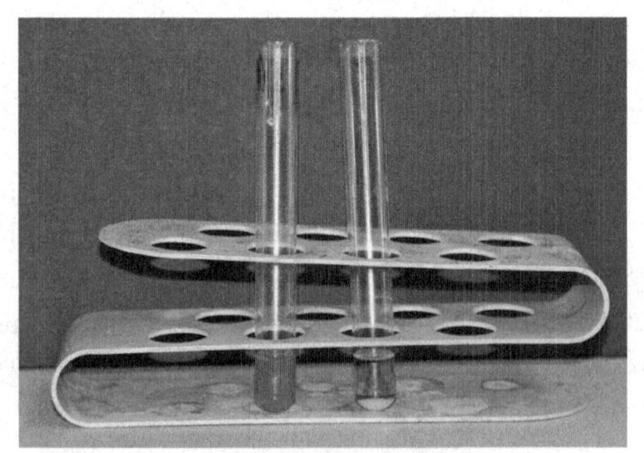

(Photos courtesy of Holly J. Morris and David T. Moat)

Purpose

To demonstrate the use of the Benedict's Test to identify the presence of reducing sugars.

Materials

- Positive control - 1% glucose solution
- Negative control - water

- 1% sucrose solution
- Samples to be tested
- Benedict's reagent
- 6 clean test tubes
- Disposable pipettes
- Large beaker
- Hot plate
- Boiling chips

Methods

- Obtain the positive control, negative control, 1% sucrose solution, and any 3 other samples to be tested.
- Place about 1 mL of each substance to be tested in to a separate labeled test tube.
- Place about 1 ml of Benedict's Reagent into each test tube and mix the solutions.
- Begin heating a large beaker that is filled about 1/3 full of water. Add a few boiling chips to help the water boil smoothly.
- When the water in the beaker is boiling, place each test tube in the boiling water bath and let it boil for 2 minutes.
- Analyze the results. If the solution is blue, no reducing sugar is present. If the solution is any color other than blue, a reducing sugar is present. The color of the resultant solution depends on the concentration of the reducing sugar present in the sample.
- Record the results in the **Benedict's Test Results** table.

EXERCISE 2 Iodine Test

The Iodine Test indicates the presence of starch on the basis of color changes occurring when the starch contacts the iodine solution. Amylose, one of the components of starch, turns deep blue-black color when it reacts with iodine as a result of the iodine becoming intertwined in the helical amylose molecules.

The test tube on the left shows a negative iodine result and the test tube on the right shows a positive iodine test result.

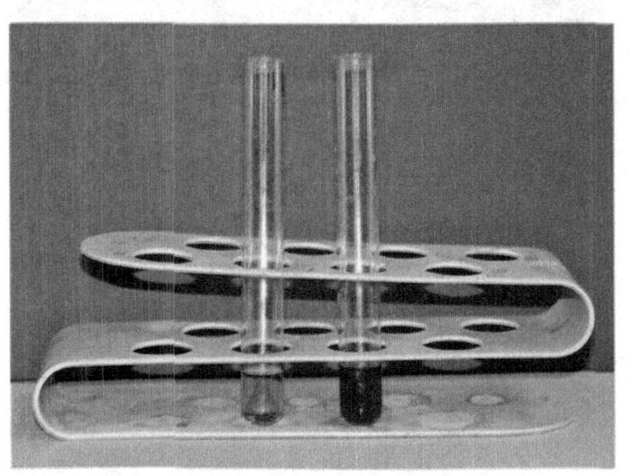

(Photos courtesy of Holly J. Morris and David T. Moat)

Purpose

To demonstrate the use of the Iodine Test to identify the presence of starch.

Materials

- Positive control - 1% starch solution
- Negative control - water
- Samples to be tested
- I_2KI solution (Iodine potassium iodide)
- Six clean test tubes
- Disposable pipettes

Methods

- Obtain the positive control, negative control, and any four other samples to be tested.
- Place about 1 mL of each substance to be tested in to a separate labeled test tube.
- Place 2 or 3 drops of I_2KI into each test tube and mix the solutions.
- Record the results in the **Iodine Test Results** table.

EXERCISE 3 Sudan IV Test

This test indicates the presence of lipids. It is based on the capability of lipids to combine with the pigments in fat soluble dyes such as Sudan IV. Sudan IV is a red dye that will combine with other lipids turning them bright red and forming reddish droplets or reddish bands. Non-lipid solutions will mix with the Sudan IV dye and cause the red dye to be dispersed throughout the solution. However, the Sudan IV dye does not actually combine with the molecules in the solution and produce the bright red droplets or bands of a positive test.

The test tube on the left shows a positive Sudan IV test and the test tube on the right shows a negative Sudan IV test.

(Photos courtesy of Holly J. Morris and David T. Moat)

Purpose

To demonstrate the use of the Sudan IV to identify the presence of lipids.

Materials

- Positive control - vegetable oil
- Negative control - water
- Samples to be tested
- Sudan IV dye

- Water
- Six clean test tubes
- Disposable pipettes

Methods

- Obtain the positive control, negative control, and any four other samples to be tested.
- Place about 1 mL of each substance to be tested in to a separate labeled test tube.
- Place 2 or 3 drops of Sudan IV dye into each test tube and mix the solutions.
- Wait 2 minutes, add 1 mL of water and mix the solutions.
- Wait a few minutes and then record the results in the **Sudan IV Test Results** table.

EXERCISE 4 Biuret Test

The Biuret test is a test that indicates the presence of peptide bonds, the bonds that link the amino acids forming a protein. The presence of peptide bonds causes the blue Biuret reagent to become purple.

The test tube on the left shows a negative Biuret test and the test tube on the right shows a positive Biuret test.

(Photos courtesy of Holly J. Morris and David T. Moat)

Purpose

To demonstrate the use of the Biuret's Test to identify the presence of proteins.

Materials

- Positive control - albumin
- Negative control - water
- 1% amino acid solution
- Samples to be tested
- Biuret reagent
- Benedict's reagent
- Six clean test tubes
- Disposable pipettes

Methods

- Obtain the positive control, negative control, 1% amino acid solution, and any three other samples to be tested.
- Place about 1 mL of each substance to be tested in to a separate labeled test tube.
- Place ½ mL of Biuret reagent into each test tube test tube.
- Place ½ mL of Benedict's reagent into each test tube test and mix the solutions.
- Record the results in the **Biuret's Test Results** table.

Name:

Exercise 1 Results

1. Complete the table below.

Benedict's Test Results		
Item Tested	**Test Result**	**Reducing Sugar Present?**
Positive control (glucose)		
Negative control (water)		
Sucrose		

Exercise 2 Results

1. Complete the table below.

Iodine Test Results		
Item Tested	**Test Result**	**Starch Present?**
Positive control (starch)		
Negative control (water)		

Exercise 3 Results

1. Complete the table below.

Sudan IV Test Results		
Item Tested	**Test Result**	**Lipid Present?**
Positive control (vegetable oil)		
Negative control (water)		

Exercise 4 Results

1. Complete the table below.

Biuret's Test Results		
Item Tested	**Test Result**	**Protein Present?**
Positive control (albumin)		
Negative control (water)		
Amino acid solution		

Additional Exercises

1. What is a positive control? What is a negative control? Why is it important to have a positive and negative control when identifying an unknown?

2. Suppose you test a food sample using the Benedict's Test and get a negative result. Can you say for sure that there are no sugars in the sample? Why or why not?

3. Sudan IV dye is used to test for the presence of lipids. The molecules of this dye are nonpolar. When lipids are present, the Sudan IV molecules will combine with the lipid and stain them dark red. When lipids are not present, the Sudan IV molecules will not form any kind of recognizable dark red droplets or bands. What does this tell you about the polarity of lipids?

4. Did the amino acid solution yield a positive or negative result when tested with Biuret reagent? What does this tell you about the amino acids in the amino acid solution?

5. Complete the table showing the results of performing the Benedict's, Iodine, Sudan IV, and Biuret tests on the foods listed below.

Food Testing Results				
	Benedict's	**Iodine**	**Sudan IV**	**Biuret's**
Banana				
Coconut				
Milk				
Peanut				
Potato				

6. Which foods, in the food testing exercise, confirmed your expectations about the composition of the foods? Which foods surprised you regarding their composition? Why?

___

___

___

Making Solutions and Performing Serial Dilutions

Photo courtesy of Holly J. Morris

OBJECTIVES

After completing these exercises, you will be able to:

- Understand the various types of solutions and the differences between them.
- Be able to make any of the various types of solutions and test for the accuracy of their preparation.
- Understand when serial dilutions need to be performed.
- Be able to perform a serial dilution.

INTRODUCTION

A solution is composed of:

- Solute (component of a solution being dissolved).
- Solvent (dissolving agent of a solution).

In the fields of biology and chemistry, solutions, containing a solute of specific interest, are commonly used for assays and other analytical purposes.

HOW TO MAKE A SOLUTION

If the solution to be prepared contains a solid (solute) being dissolved a liquid (solvent), the following procedure should be followed:

1. Determine the amount of solute needed.
2. Weigh the solute.
3. Fill a volumetric flask about halfway with the solvent.
4. Transfer the solute to the volumetric flask.
5. Rinse the weighing dish with the solvent to ensure all of the solute is transferred into the flask.
6. Stir the solution until the solute is dissolved. Heat or more solvent may be needed to completely dissolve the solute.
7. Fill the volumetric flask to the desired level of solution with solvent.

If the solution does not need to be very accurate, a graduated cylinder or an Erlenmeyer flask can be used to prepare the solution instead of a volumetric flask.

If the solution to be prepared contains a liquid (solute) being dissolved in a liquid (solvent), the following procedure should be followed:

1. Determine the amount of solute and solvent needed.
2. Measure the amount of solute.
3. Fill a volumetric flask with less than the amount of solvent needed.
4. Dissolve the solute in the solvent.
5. Fill the volumetric flask to the desired level with solvent.

METHODS TO EXPRESS SOLUTE CONCENTRATION

Several methods can be used to express the concentration of a solute in a solution:

- Known concentration
  - Weight per unit volume (mass/volume)
  - Volume per unit volume (volume/volume)
  - Mass per unit mass (mass/mass)
- Percent concentration
  - Weight/weight percent
  - Weight/volume percent
  - Volume/volume percent
- Molarity (moles/L)
- Normality (equivalents/L)
- Molality (moles/1000 g)
- Parts (part/total number of parts)

The units of concentration depend on the nature of the solute. If the exact form or composition of the solute is not known, then any methods based on molar concentration terms cannot be used (molarity, normality).

Known Concentration

To prepare a solution to a known concentration requires that you know:

- Volume of solution you are going to prepare.
- Desired concentration of the solution.

The following expression can be used to calculate the amount of solute you need to prepare a solution of known concentration:

(amount of solution to be made) × (solution concentration) = amount of solute needed

For example, if you want to prepare 500 mL of a solution to a concentration of 10 mg/L in water, how much solute must be dissolved in 500 mL of solvent to obtain this concentration? Assume that the solute is a solid such as NaCl.

Step 1: Notice that the amount of solution is expressed in mL and the concentration is expressed in mg/L. As such, it is necessary to convert the volume to L. Thus, you need to make 0.500 L of solution.

Step 2: Put the values into the formula and determine how much solute you need to add to make a total volume of 500 mL (0.500 L):

$$0.500 \text{ L} \times 10 \text{ mg/L} = 5.0 \text{ mg}$$

Step 3: Five milligrams of solute in a total volume of 500 mL (0.500 L) of solution is needed. Weigh 5 mg of solute on an analytical balance. Fill a 500 mL volumetric flask with approximately half of the solvent needed and add the solute to it. After dissolving the solute, add enough water so the final volume is 500 mL.

If the target concentration requires the use of a liquid, such as ethanol, and the concentration term is expressed as v/v (rather than m/v as above), then the method of preparation would involve transferring the appropriate solute volume into a volumetric flask rather than weighing a solid (although liquids can be weighed and densities can be used to convert between mass and volume when known).

For example, if the target concentration would be a 5 mL/L concentration of ethanol in water and 100 mL of this concentration needed to be prepared, then the following steps would be performed:

Step 1: Calculate the amount of ethanol needed as a volume:

$$0.100 \text{ L} \times 5.00 \text{ mL/L} = 0.50 \text{ mL}$$

Step 2: Transfer by pipette 0.50 mL of ethanol to a 100 mL volumetric flask. Add a few mL of the solvent to the volumetric flask before adding the solute and then add enough solvent to bring the final volume to 100 mL.

Percent Concentration

Concentrations expressed as a percent indicate the amount of solute in grams per 100 g of total solution (weight/weight percent (w/w)). The formula used to determine the weight percent solute in a sample is:

$$\text{Weight Percent} = [(\text{solute weight})/(\text{total solution weight})] \times 100$$

Percent can also be expressed as a weight/volume percent (w/v):

$$\text{Volume Percent} = [(\text{solute weight})/(\text{total solution volume})] \times 100$$

Finally, percent can also be expressed as a volume/volume (v/v) percent:

$$\text{Volume/Volume Percent} = [(\text{solute volume})/(\text{total solution volume})] \times 100$$

Molarity

Molarity is defined as the number of moles of a substance in 1 L of solution. For example, NaCl has a formula weight of 58.44 g/mole equivalent. A 1 molar (1 M) solution of NaCl in water would contain 58.44 g of NaCl in 1 L of solution volume.

Molality

Molality is based on solution weight rather than volume. A 1 molal (1 m) solution of NaCl (58.44 g/mole) would contain 58.44 g of NaCl in 1 Kg (1000 g) of water.

Normality

Normality provides information about the number of reactive units in 1 L of solution. Normality is expressed in equivalents per liter. An equivalent is the number of moles of hydrogen ions one mole of an acid will donate or one mole of a base will accept. For example, HCl dissociates in solution to give 1 equivalent of H^+ ions per mole of substance. Thus, a 1 M solution of HCl would also be a 1 N (N is the abbreviation for normal) solution. Sulfuric acid (H_2SO_4) dissociates in water to give 2 H^+ ions per mole of substance. Thus, 1 mole of sulfuric acid in 1 L of solution would be a 2 N concentration.

Parts

Parts per notation is a way of expressing very dilute concentrations of substances. Just as percent means "out of a hundred" (such as a 5% NaCl solution), parts per million (ppm) means "out of a million," parts per billion (ppb) means "out of a billion," etc. One ppm is equivalent to 1 mg of something per liter of water (mg/L) or 1 mg of something per kilogram of soil (mg/kg).

For example, 5 ppm chlorine means 5 g of chlorine in 1 million g of solution. This can also be expressed in any other suitable units, such as 5 mg of chlorine in 1 million mg of solution or 5 lbs. of chlorine in 1 million lbs. of solution.

Serial Dilutions

It is often necessary to work at very low concentrations when analyzing a sample to avoid "saturating" or overwhelming the analytical method beyond the capabilities it can accurately measure. If you wanted to prepare a 1 ppm NaCl solution in 100 g of solution, it would require that you weigh 0.0001 g of NaCl. Because most analytical balances have an accuracy of ±0.0002 g, the error in weight could be 100% which is generally considered unacceptable for any scientific measurement. As such, the best way to achieve this concentration would be to make a *stock solution* and then dilute it to the desired concentration.

In this example, start by making a *stock solution* consisting of 0.01 g NaCl (which can be determined with a higher degree of accuracy) in 100 g of water and then prepare a 1/100 dilution by adding 1 ml of the *stock solution* to 99 mL of water.

Serial dilutions can be performed by using the same stock solution for each dilution in the series unless a dilution generally greater than 1/100 is required. If, for example, the final dilution required is 1/1,000,000, the best method would be to do 3 dilutions using the previous dilution as the stock solution for the next. The first dilution would be 1/100 (1:99) which would be the stock solution for the next 1/100 dilution and would yield a total dilution of 1/10,000 (1/100 × 1/100). Using this dilution as the stock solution for the final dilution and then diluting it by 1/100 would give a total dilution of 1/1,000,000 (1/10,000 × 1/100).

Some important terms to remember relating to serial dilutions are:

- **Stock solution:** original solution being diluted.
- **Undiluted solution:** by definition is called 1/1.
- **1:2 dilution:** means there are three parts total volume (1 part stock solution + 2 parts solvent).
- **1/2 dilution:** means there are two parts total volume with one part being the stock solution.

Therefore:

- 1/2 is the same as 1 : 1
- 1 : 2 is the same as 1/3
- 1 : 3 : 5 A:B:C means that there are 9 total parts of which 1 part is A, 3 parts are B and 5 parts are C.

EXERCISE 1 Preparing Solutions

In the laboratory, many different types of solutions are needed for analytical purposes. This exercise will help you become familiar with the various types of solutions that are commonly used.

Purpose

To prepare various types of solutions and determine the accuracy of their preparation.

Materials

- Test tubes
- Beakers
- Hot plates
- Sugar refractometer
- Salinity refractometer
- TDS meter
- Various chemicals to be used as solutes

Methods

- Members of the lab are separated into groups of three or four.
- Each group will be assigned six different solutions that they are to prepare.
- Each group determines the solute and solvent requirements for each of their assigned solutions.
- After reviewing their solutions with the lab instructor, each group will prepare their assigned solutions.
- After preparing their solutions, each group will determine how to test the accuracy of their solution preparation and then will test their solutions for accurate preparation.

Name:

Exercise 1 Results

1. How would you prepare 50 mL of a solution that is 2 mg/mL NaCl? Show details.

2. How would you prepare 10 L of 0.3 M KH_2PO_4? Show details.

3. How would you prepare 500 mL of a 5% (w/v) solution of NaCl? Show details.

4. How would you prepare 100 mL of a 10% (v/v) solution of ethanol in water? Show details.

5. How would you prepare 500 g of a 5% NaCl solution by weight (w/w)? Show details.

6. How would you prepare 1 L of 6 N H_3PO_4? Show details.

7. A solution contains 10 g salt/L. How could you set up a serial dilution to obtain 1 mL of a solution with a concentration of 0.100 mg/L?

Note: 10 g of salt/L is equivalent to 10,000 mg.

Determine the dilution factor for each step:

(1) 1 mL of original stock + 99 mL of water, dilution = ?

(2) 1 mL of first dilution + 99 mL of water, dilution = ?

(3) How would you dilute solution (2) to get to a final concentration of 0.100 mg/L?

8. How would you prepare 10 mL of a 1/10 dilution of blood? Show details.

9. How would you prepare 250 mL of a 1/300 dilution of blood? Show details.

10. You have a recipe that calls for 1 part of salt solution to 3 parts of water. How would you mix 10 mL of this solution? Show details.

11. Convert 3 ppm to milligrams per liter and then convert it to milligrams per milliliter.

12. How would you prepare 1 L of a solution that has 100 ppm fructose in water? Show details.

13. Complete the table below detailing the process to prepare the solutions for the in-class exercise.

Making the Solutions		
Solution	**Solution Description**	**How Did You Make The Solution?**
1		
2		
3		
4		
5		
6		

14. Complete the table below detailing the process to verify the accuracy of the solution preparation for the in-class exercise.

Verifying the Accuracy of Solution Preparation			
Solution	**Verification Method Used**	**Expected Solute Value**	**Actual Solute Value**
1			
2			
3			
4			
5			
6			

Verification Method: Sugar Refractometer, Salinity Refractometer, Titration, or TDS Meter

Enzymes

Photo courtesy of Holly J. Morris

OBJECTIVES

After completing these exercises, you will be able to:

- Understand what an enzyme is.
- Understand the importance of enzymes to living organisms.
- Understand how an enzyme works.
- Understand the properties of an enzyme.
- Understand the effects of temperature and pH on enzyme activity.

INTRODUCTION

Nearly all enzymes are proteins. However, not all proteins are enzymes. Although proteins perform many functions, their most important function is to serve as enzymes controlling the rate of chemical reactions in the cells. As such, enzymes function as biologic catalysts to reduce the amount of activation energy required for a reaction to occur. Although enzymes take part in the reaction, they are not consumed by the reaction and only change the rate of reaction.

For two molecules to react, they must collide with each other, be in the right orientation, and have sufficient activation energy. Many reactions that are thermodynamically possible do not occur very quickly or at all because there is insufficient activation energy. Having sufficient activation energy means that the molecules themselves have enough energy to initiate the reaction. Reducing the amount of required activation energy enables the reaction to occur more quickly without raising the temperature and as a result, increases the rate at which the reaction occurs. However, in an enzyme catalyzed reaction, the enzyme and substrate form a reaction intermediate that has a lower activation energy requirement than an uncatalyzed reaction and as a result, the reaction occurs.

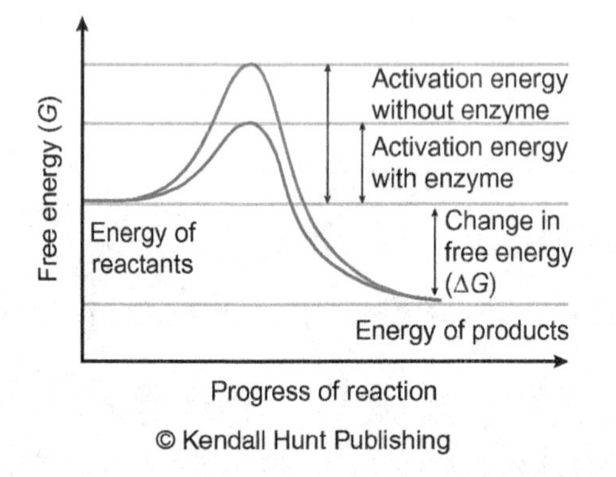

© Kendall Hunt Publishing

CHARACTERISTICS OF AN ENZYME

There are several characteristics of an enzyme that affect its ability to catalyze a reaction:

- **Optimum conditions:** All enzymes have an optimum temperature range and an optimum pH range at which they work best. As the temperature increases, kinetic energy will increase until the enzyme starts to become unstable and denaturation occurs. Likewise, the structure of a specific enzyme will be most stable at a specific pH. That stability is dependent on the specific amino acid composition of the enzyme. Changes in pH will alter the shape of the enzyme and, therefore, alter its activity.

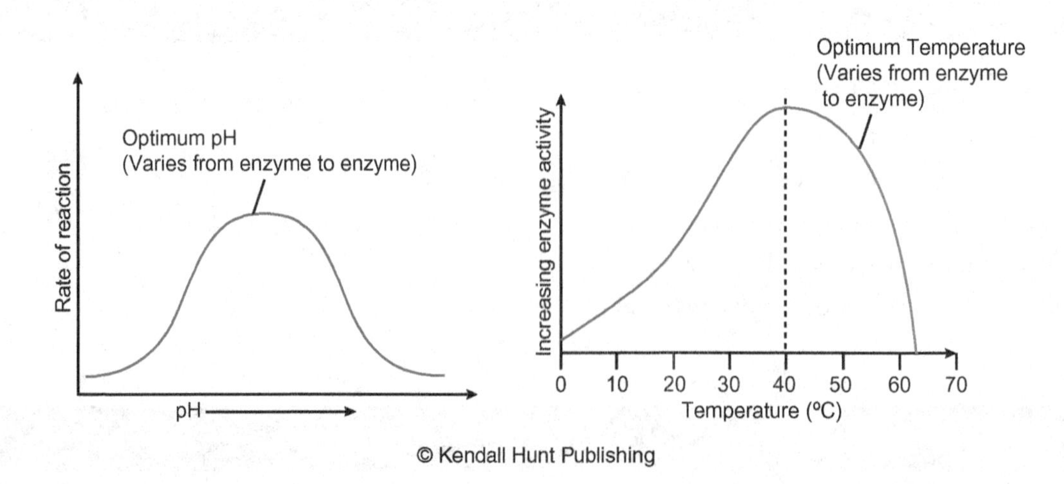

© Kendall Hunt Publishing

- **Specificity:** Each enzyme will catalyze only one specific reaction. This reaction occurs at the active site of the enzyme, where the substrates react to form products. Because the active site is specific for the substrate, almost no other substances will be able to bind there and subsequently react. Only the substrate is able to form the reaction intermediate.

The Lock and Key Mechanism

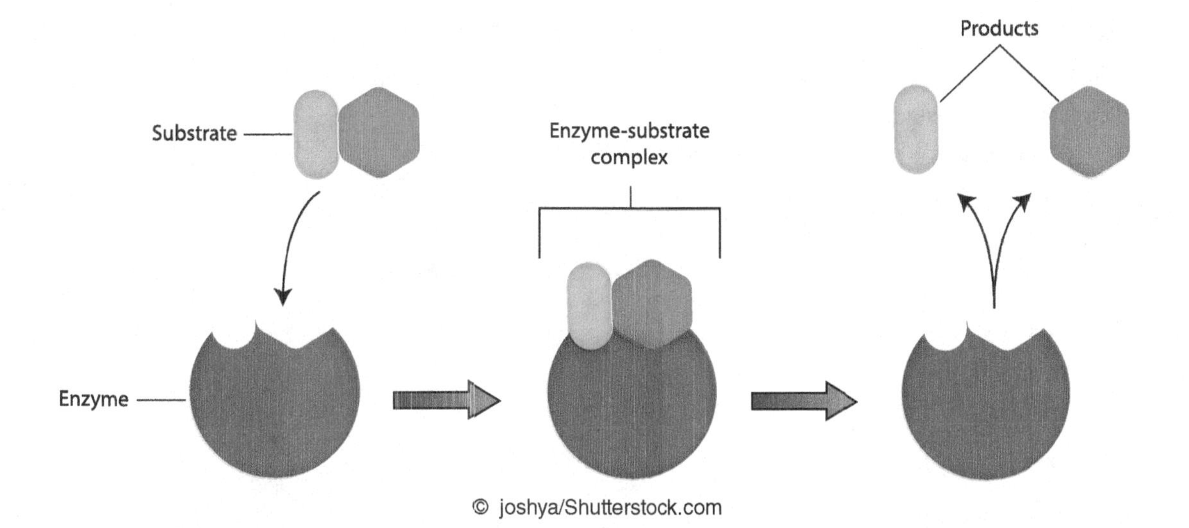

© joshya/Shutterstock.com

- **Denaturation:** Enzymes must be able to maintain their three-dimensional structure to maintain their function. Since enzyme function depends on protein structure, any factor that interferes with the enzyme's structure will affect its function. Factors such as changes in pH and temperature can interfere with the enzyme's structure and cause it to lose its function. When these factors cause an enzyme to lose its three-dimensional structure and as a result, its functionality, the enzyme is said to be denatured.
- **Regulation:** Not all enzymes need to be active all the time, therefore mechanisms exist which can regulate whether or not an enzyme is active. These mechanisms can include regulating the production of the enzyme (controlling the amount), isolating the enzyme (compartmentalization), controlling the concentration of the substrate, controlling the temperature of pH, and so forth.
- **Inhibition:** Enzymes can be inhibited, which can be a type of regulation or it can be a way of interfering with normal function. Enzyme inhibition involves an inhibitor interacting with the enzyme to reduce or stop its catalytic activity. For example, a substance may have a similar structure to the enzyme's real substrate and the enzyme may not be able to distinguish between the real substrate and the imposter. The presence of the imposter molecule at the active site will prevent the enzyme from catalyzing a reaction.

In the following exercises, you will be introduced to enzymes and their properties as well as their uses and how they are affected by temperature and pH.

EXERCISE 1 Making Cheese

The enzyme rennilase converts milk proteins (casein) into insoluble paracasein which precipitates out of the liquid. The paracasein is called the curd (cheese) while the remaining liquid is called the whey. Rennilase interferes with some of the milk proteins' bonds causing it to precipitate out of the liquid but it does not break the peptide bonds of the curd.

Bacterial protease does break the peptide bonds of the curd into its component amino acids. The chemical indicator ninhydrin reacts with the free amino acids to demonstrate that the bacterial protease did break down the curd into free amino acids.

Purpose

To demonstrate the effect of a bacterial protease enzyme on proteins.

Materials

- Milk
- Rennilase
- Bacterial protease
- Two 400 mL beakers
- Two small beakers
- 1 N HCl
- Cheesecloth
- Two clean test tubes
- Boiling chips
- Thermometer
- Hot plate
- Ninhydrin

Methods

- Pour 250 mL of milk into a beaker and add 5 drops of 1 N HCl.
- Heat the milk to 32°C with continuous stirring.
- Add 3 drops of rennilase to the milk while stirring.
- Remove the milk from the hot plate and allow it to stand undisturbed until the paracasein (curd) precipitates out of the milk.
- Filter the whey through the cheesecloth into the other 400 mL beaker.
- Remove the curd from the 400 mL beaker and divide it into 2 equal portions.
- Put 1 portion of the curd into each of the 2 small beakers with 100 mL water.
- Stir to break up the curd.
- Label 1 of the small beakers "Protease" and the other "No Protease."
- Add 1 g of bacterial protease to the beaker marked "Protease". Do not add any bacterial protease to the beaker marked "No Protease."
- Stir each of the beakers briefly and allow them to sit for 5 minutes.
- Label 1 of the test tubes "Protease" and the other "No Protease."
- Filter 5 mL of the solution through the cheesecloth from the "Protease" beaker into the test tube marked "Protease."
- Filter 5 mL of the solution through the cheesecloth from the "No Protease" beaker into the test tube marked "No Protease."
- Add 1 mL of ninhydrin to each of the 2 test tubes and put them in a boiling water bath for 10 minutes. Add a few boiling chips to help the water boil smoothly.
- Remove the 2 test tubes from the water bath.
- Record your observations in the **Breaking Cheese into Amino Acids** table.

EXERCISE 2 Enzyme Reusability

Enzymes speed up chemical reactions without being consumed by the reaction. Hydrogen peroxide is a byproduct of metabolic processes and has been widely regarded as a cytotoxic agent whose levels must be minimized by the action of antioxidant defenses. Catalase is an enzyme that catalyzes the decomposition of hydrogen peroxide to water and oxygen.

Purpose

To demonstrate that although enzymes take part in a reaction, they are not consumed by the reaction.

Materials

- Hydrogen peroxide solution
- Water
- Two clean test tubes
- Test tube rack
- Disposable pipettes
- Beefsteak
- Scalpel

Methods

- Place about 5 mL of water in one of the labeled test tubes.
- Place about 5 mL of hydrogen peroxide solution in the second labeled test tube.
- Cut 2 small pieces of beefsteak and place 1 piece in each of the 2 test tubes.
- Record your observations.
- After a few minutes, pour of the liquid from the 2 test tubes but keep the beefsteak in each test tube.
- Refill each test tube with 5 mL of the same liquid as originally used.
- Record your observations in the **Enzyme Reusability** table.

EXERCISE 3 Effect of Bromelain on the Formation of Gelatin

Gelatin is obtained by boiling the skin, tendons, and ligaments of animals. As a result, it contains a protein called collagen (a primary component of joints, cartilage, and nails).

Pineapples contain a protein digesting enzyme called bromelain. Fresh pineapple juice contains active bromelain whereas canned pineapple juice contains denatured bromelain because the bromelain has been denatured in the pasteurization process.

Purpose

To demonstrate the effect that active and denatured bromelain have on gelatin formation.

Materials

- Fresh pineapple juice
- Canned pineapple juice
- Gelatin
- Two beakers
- Three clean test tubes
- Test tube rack
- Disposable pipettes
- Hot plate
- Ice

Methods

- Prepare gelatin by dissolving 1 packet of gelatin in 250 mL of water in a beaker and place the beaker on a hot plate to keep the gelatin warm.
- In test tube 1, place 2 mL of water.
- In test tube 2, place 2 mL of fresh pineapple juice.
- In test tube 3, place 2 ml of canned pineapple juice.
- Place 3 mL of warm gelatin in each of the 3 labeled test tubes.
- Place all 3 test tubes in ice until test tube 1 (water) solidifies.
- Observe test tubes 2 and 3.
- Record your observations in the **Effect of Bromelain on Gelatin Formation** table.

EXERCISE 4 Effect of Temperature on Bromelain

Since protein functionality depends upon protein structure, any factor that interferes with the enzyme's structure will affect its functionality. Temperature is one of the factors that can affect the functionality of an enzyme.

Purpose

To determine the optimum temperature at which bromelain works.

Materials

- Fresh pineapple juice
- Gelatin
- Two beakers
- Six clean test tubes
- Test tube rack
- Disposable pipettes
- Hot plate
- Water baths (40°C, 60°C, 80°C)
- Ice

Methods

- Prepare gelatin by dissolving 1 packet of gelatin in 250 mL of water in a beaker and place the beaker on a hot plate to keep the gelatin warm.
- In test tube 1, place 2 mL of water.
- In test tubes 2 → 6, place 2 mL of fresh pineapple juice.
- Place test tube 2 in a beaker filled with ice for 10 minutes to let it acclimate to the environment.
- Place test tube 3 in the test tube rack on your desk for 10 minutes to let it acclimate to the environment.
- Place test tube 4 in the 40°C water bath for 10 minutes to let it acclimate to the environment.
- Place test tube 5 in the 60°C water bath for 10 minutes to let it acclimate to the environment.
- Place test tube 6 in the 80°C water bath for 10 minutes to let it acclimate to the environment.
- Place 3 mL of warm gelatin in each of the 6 labeled test tubes.
- Place all 6 test tubes in ice until test tube 1 (water) solidifies.

- Observe all of the test tubes.
- Record your observations in the **Effect of Temperature on Bromelain** table.

EXERCISE 5 Effect of pH on Bromelain

Since protein functionality depends upon protein structure, any factor that interferes with the enzyme's structure will affect its functionality. pH is one of the factors that can affect the functionality of an enzyme.

Purpose

To determine the optimum pH at which bromelain works.

Materials

- Fresh pineapple juice
- Gelatin
- Two beakers
- Seven clean test tubes
- Test tube rack
- Disposable pipettes
- Hot plate
- Buffer solutions (pH 3, pH 5, pH 7, pH 9)
- 1 N NaOH
- 1 N HCl
- Ice
- pH strips

Methods

- Prepare gelatin by dissolving 1 packet of gelatin in 250 mL of water in a beaker and place the beaker on a hot plate to keep the gelatin warm.
- In test tube 1, place 2 mL of water.
- In test tubes 2 → 7, place 1 mL of fresh pineapple juice.
- In test tube 2, place 1 mL of 1 N HCl.
- In test tube 3, place 1 mL of pH 3 buffer solution.
- In test tube 4, place 1 mL of pH 5 buffer solution.
- In test tube 5, place 1 mL of pH 7 buffer solution.
- In test tube 6, place 1 mL of pH 9 buffer solution.
- In test tube 7, place 1 mL of 1 N NaOH.
- Determine the pH of all of the test tubes.
- Place 3 mL of warm gelatin in all of the test tubes.
- Place all 7 test tubes in ice until test tube 1 (water) solidifies.
- Observe all of the test tubes.
- Record your observations in the **Effect of pH on Bromelain** table.

Name:

Exercise 1 Results

1. What causes the milk proteins to coagulate/precipitate? Explain your answer.

2. Complete the table below summarizing the effect of bacterial protease on cheese.

Breaking Cheese into Amino Acids		
Test Tube	**Color at 10 Minutes**	**Amino Acids Present?**
Bacterial protease		
No bacterial protease		

3. Did the intensity of the color change as time elapsed? Explain your answer.

Exercise 2 Results

1. Complete the table below with your observations on enzyme reusability.

Enzyme Reusability	
Test Tube	**Observations**
Water (5 mL)	
Hydrogen peroxide (5 mL)	

2. What is the importance of the reusability of enzymes? Explain your answer.

Exercise 3 Results

1. Complete the table below with your observations on the effect of bromelain on gelatin formation.

Effect of Bromelain on Gelatin Formation		
Test Tube	**Ingredients**	**Results**
1	2 mL water, 3 mL warm gelatin	
2	2 mL fresh pineapple juice, 3 mL warm gelatin	
3	2 mL canned pineapple juice, 3 mL warm gelatin	

2. Construct a hypothesis addressing the effect of bromelain on gelatin formation.

____

____

3. Identify the independent variable.

____

____

4. Identify the dependent variable.

____

____

5. Identify the control.

____

____

6. Explain the results of this exercise.

____

____

Exercise 4 Results

1. Complete the table below with your observations on the effect of temperature on bromelain.

Effect of Temperature on Bromelain		
Test Tube	**Ingredients**	**Results**
1	2 mL water, 3 mL warm gelatin	
2	2 mL fresh pineapple juice (ice), 3 mL warm gelatin	
3	2 mL fresh pineapple juice (room temperature), 3 mL warm gelatin	
4	2 mL fresh pineapple juice (40°C), 3 mL warm gelatin	
5	2 mL fresh pineapple juice (60°C), 3 mL warm gelatin	
6	2 mL fresh pineapple juice (80°C), 3 mL warm gelatin	

2. Construct a hypothesis addressing the effect of temperature on bromelain.

____

____

3. Identify the independent variable.

____

____

4. Identify the dependent variable.

____

____

5. Identify the control.

____

____

6. Explain the results of this exercise.

____

____

7. What do you think is the optimal temperature for bromelain? Explain your answer.

____

____

Exercise 5 Results

1. Complete the table below with your observations on the effect of pH on bromelain.

Effect of pH on Bromelain		
Test Tube	**Ingredients**	**Results**
1	2 mL water, 3 mL warm gelatin	
2	1 mL fresh pineapple juice, 1 mL 1 N HCl, 3 mL warm gelatin	
3	1 mL fresh pineapple juice, 1 mL pH 3 buffer, 3 mL warm gelatin	
4	1 mL fresh pineapple juice, 1 mL pH 5 buffer, 3 mL warm gelatin	
5	1 mL fresh pineapple juice, 1 mL pH 7 buffer, 3 mL warm gelatin	

6	1 mL fresh pineapple juice, 1 mL pH 9 buffer, 3 mL warm gelatin	
7	1 mL fresh pineapple juice, 1 mL 1 N NaOH, 3 mL warm gelatin	

2. Construct a hypothesis addressing the effect of pH on bromelain.

____

____

3. Identify the independent variable.

____

____

4. Identify the dependent variable.

____

____

5. Identify the control.

____

____

6. Explain the results of this exercise.

____

____

7. What do you think is the optimal pH for bromelain? Explain your answer.

____

____

Diffusion and Osmosis

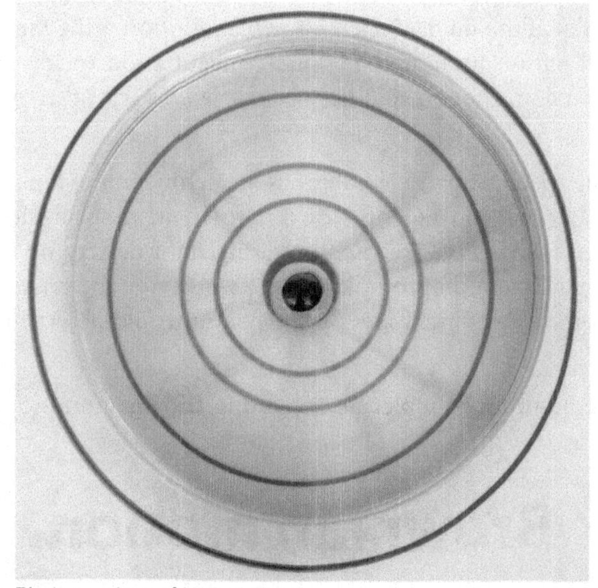

Photo courtesy of Holly J. Morris

OBJECTIVES

After completing these exercises, you should:

- Understand and be able to describe the processes of:
  - Brownian Motion.
  - Diffusion.
  - Osmosis.
- Be able to define and use the terms hypertonic, hypotonic, and, isotonic.

INTRODUCTION

All organisms are composed of one or more cells. A cell is enclosed by a membrane that is comprised of a phospholipid bilayer embedded with a variety of proteins. Cell membranes help maintain a relatively stable environment inside the cell by regulating what can enter and exit the cells. Since some molecules can pass through the cell membrane but others cannot, cell membranes are said to be selectively permeable.

The overall process of maintaining homeostasis in cells will be covered elsewhere. Here, we are going to focus on one component of movement in and out of cells, passive movement by means of **diffusion** or **osmosis**.

In order to understand diffusion, we have to first look at the random movement of particles that are suspended in fluids (liquids or gases). This random movement is called Brownian motion, and is named after the botanist Robert Brown. In 1827, Brown observed pollen grains suspended in water, under a microscope. He observed that the particles moved randomly, by an unknown force. Later, Albert Einstein explained that the motion Brown observed was a result of the particles being moved by individual water molecules. The small particles of matter suspended in the fluid are constantly being moved around by the molecules of the fluid. The direction of the force is constantly changing, and the particle can be hit on many sides leading to the apparent random nature of the motion. Brownian Motion is the inherent random motion of the molecules of the fluid. As such, Brownian motion is the mechanism by which diffusion takes place. Molecules will diffuse from areas of high concentration to low concentration because the molecules are in constant random motion, and they collide with each other more if the molecules are in concentrated areas.

Some factors that affect the speed of particle collisions are temperature, fluid density, and fluid viscosity. As temperature decreases, atomic movement, that is, Brownian motion, slows down. As temperature increases, atomic movement speeds up. Likewise, increasing fluid density or fluid viscosity will decrease Brownian motion. Imagine adding a drop of ink to a glass of water and a glass of honey. Honey is more viscous than water. Predict in which fluid, water or honey, you would observe faster movement (dispersal) of ink particles.

Below is a depiction of the motion of a molecule exhibiting Brownian motion.

Brownian motion

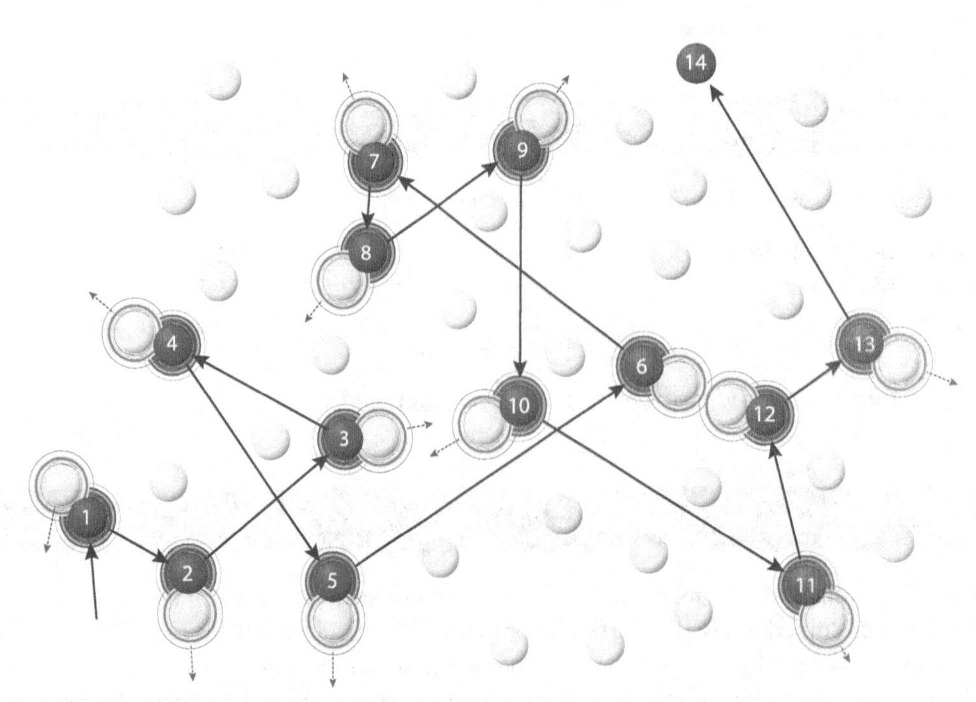

© Designua/Shutterstock.com

Diffusion

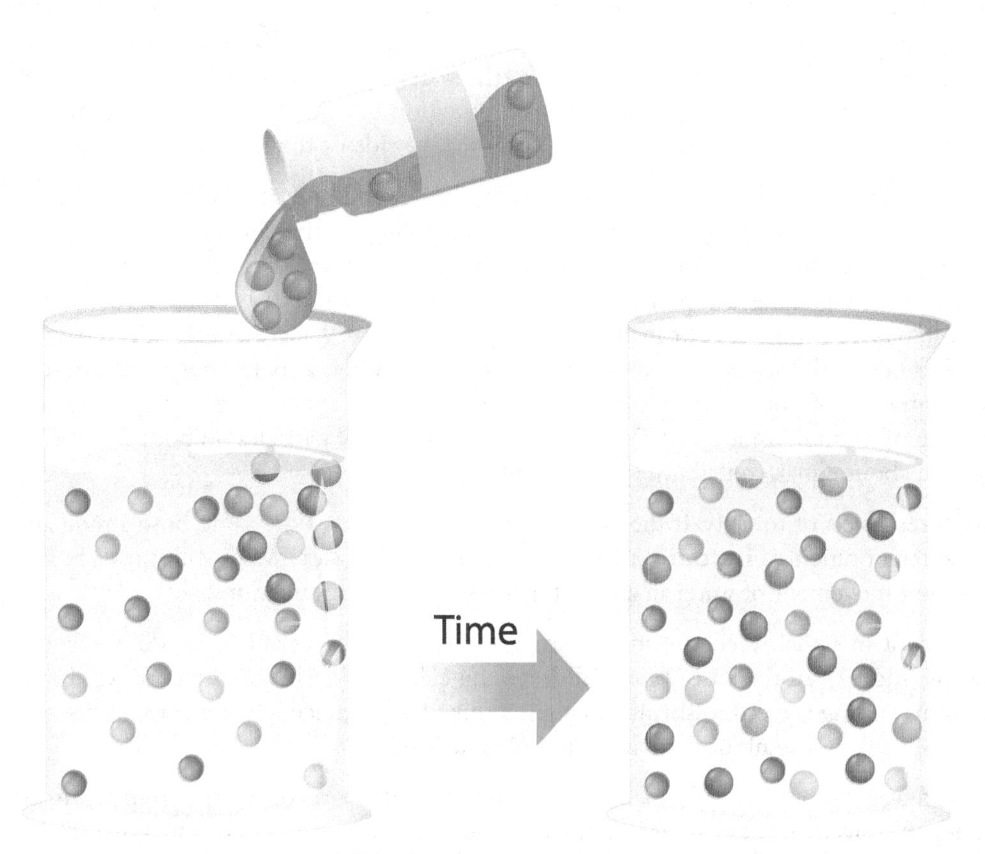

© Designua/Shutterstock.com

Diffusion is the passive movement of particles in a fluid from a region of higher concentration, in the direction of lower concentration, until equilibrium is reached. At the point of equilibrium, particles will continue to move (by Brownian motion), but the net movement is equal in all directions. Factors that affect how fast diffusion proceeds are concentration, temperature, size of the container, and size of the particles. The rate of diffusion is faster when the initial concentration of particles is higher. Likewise, the rate of diffusion is faster when the temperature is higher, the vessel is larger, or the size of the particles is smaller.

Tonicity and Osmosis

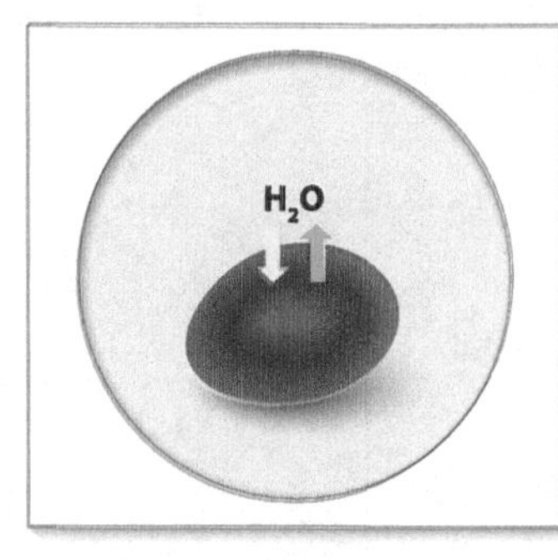

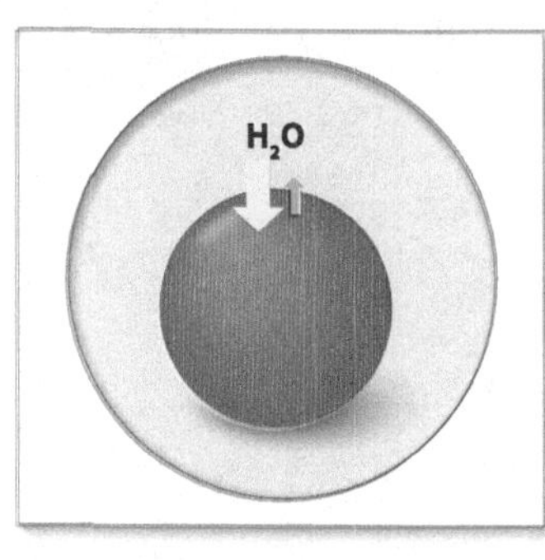

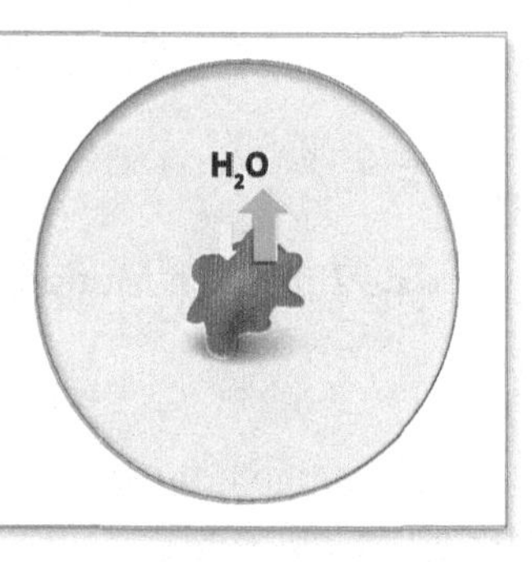

Isotonic **Hypotonic** **Hypertonic**

© Designua/Shutterstock.com

Another type of diffusion in biological systems is called osmosis. Osmosis involves the movement of water across a selectively permeable membrane in the direction of higher concentration of solute. A selectively permeable membrane allows water to pass but restricts the movement of some or all solute particles. The natural condition is for the concentration of a solute to move from higher to lower concentration until reaching equilibrium within its given vessel. If there is the barrier of a selectively permeable membrane, the natural tendency to reach equilibrium still exists, so the water moves in the direction of the higher concentration of solute in an attempt to even out the concentration of solute per water molecules on both sides of the membrane. This increases the volume of water on the side of the membrane with more solute.

The higher the concentration of solute is on one side of the membrane, the more "pull" it has for the water on the other side of the membrane. This pull is called osmotic pressure. The movement of water by means of osmosis is called **tonicity**. Tonicity is determined by the solutes that cannot pass through the selectively permeable membrane, that is, the particles that are responsible for osmotic pressure. An important characteristic of tonicity is that it does not matter what the particles are. What matters is their overall concentration on either side of the selective membrane. For example, there is a higher concentration of sodium and lower concentration of potassium outside of a cell, and a higher concentration of potassium and a lower concentration of sodium inside a cell. The overall concentration of particles on both the outside and inside of the cell are about equal, contributing to the tonicity.

There are three levels of tonicity. If the concentration of particles is the same both inside and outside of the cell, the environment is **isotonic** (*iso* means "same"). In this case, water still moves in and out of the cell but the net movement of water in and out is the same in both directions.

An environment in which there are more particles outside of the cell than inside is called **hypertonic** (*hyper* means "more"). In this case water will leave the cell, moving in the direction of the higher solute concentration, causing the cell to shrink, giving it a spiked appearance. The term that is used to describe a cell that has shrunk as a result of hypertonicity is **crenation**.

If there are fewer particles outside of the cell than inside, the environment is **hypotonic** (*hypo* means "less"). The net movement of water is into the cell, causing it to swell and possibly burst. The bursting of a cell due to the net influx of water is called **lysis**.

Tonicity is always measured as outside of the cell compared to inside. Thus, if we place a cell in a hypertonic solution, the concentration of particles is higher outside of the cell than inside. If we place a cell in a hypotonic solution, the concentration of particles is lower outside of the compared to inside of the cell.

EXERCISE 1 Demonstrating Brownian Motion

Macromolecules in biological systems do not settle out, but remain uniformly dispersed in the solvent. One reason for this is the constant agitation of the colloidal particles produced by the collision of water molecules with the colloidal particles. The resulting constant random movement of colloidal particles in solution is called Brownian motion which can be defined as the random motion of particles suspended in a fluid (a liquid or a gas) resulting from their collision with the atoms or molecules in the gas or liquid.

Purpose

To demonstrate Brownian Motion.

Materials

- Carmine indigo crystals
- Slides and cover slips
- Scoop
- Compound light microscope
- Hot plate

Methods

- Using a scoop, put a few carmine indigo crystals on a slide.
- Mix with a drop of water, cover, and examine under high power.
- Observe the oscillation of the carmine indigo crystals.
- Warm the slide over a hot plate and then immediately examine under high power.

EXERCISE 2 Demonstrating Diffusion

The external and internal environments of cells are aqueous solutions of dissolved organic and inorganic molecules. Within these environments, molecules move by diffusion which may be defined as the process by which molecules (or ions) move from a region in which they are highly concentrated to a region in which their concentration is lower.

Purpose

To demonstrate diffusion.

Materials

- Petri plate
- M&Ms

Methods

- Put enough water into a Petri plate to cover the entire surface and place 4 M&Ms equally distributed around the Petri plate.
- Observe the appearance of the M&Ms and the water in the Petri plate.

EXERCISE 3 Demonstrating Diffusion Across a Selectively Permeable Membrane

Dialysis is the separation of smaller molecules from larger molecules in solution by selective diffusion through a selectively permeable membrane. Dialysis tubing is made of selectively permeable cellulose tubing perforated with microscopic pores. The pores are small enough for the tubing to be used to model the behavior of the cell membrane with respect to the sizes of molecules that can or cannot diffuse through it.

If two solutions of different molecular weights are separated by a selectively permeable membrane, some substances may be able to pass through the membrane but others may be prevented from passing through.

Purpose

To demonstrate the selective permeability of dialysis tubing using glucose, starch, and iodine potassium iodide.

Materials

- Dialysis tubing
- Thread or string
- Glucose solution
- Starch solution
- Iodine potassium iodide solution
- Benedict's reagent
- Hot plate
- 500 mL beaker
- Smaller beakers
- Three test tubes
- Disposable pipettes

Methods

- Cut a piece of dialysis tubing approximately 25 cm in length and put it in a small beaker half filled with water.
- After the dialysis tubing has soaked in water for a few minutes, form it into a tube and tightly tie off one end with thread or string or a double knot. Ensure that the end of the tube is tightly tied to prevent any solutions from leaking through.
- Pipette put 10 mL of glucose solution and 10 ml of starch solution into the dialysis tubing and tightly tie off the other end of the tube with thread or string or a double knot.
- Fill the 500 mL beaker with approximately 300 mL of water.
- Add iodine potassium iodide to the water in the 500 mL until the water becomes very yellow.
- Observe the color of the bag and the color of the solution in the beaker. Record this data in the table below.
- Place the dialysis bag in the 500 mL beaker as shown below:

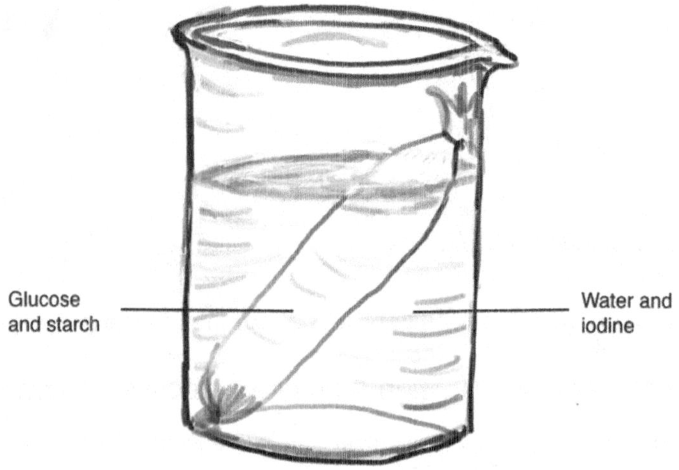

Glucose and starch

Water and iodine

Courtesy of Holly J. Morris

- Let the dialysis bag remain in the beaker for 30 minutes.
- After 30 minutes, observe the color of the bag and the color of the solution in the beaker. Record this data in the table below.
- Perform the Benedict's test on the contents of the dialysis bag, the contents of the beaker, and a control (water).
- Record your data in the **Permeability of Dialysis Tubing Experiment** table.

EXERCISE 4 Observing Osmotic Behavior in Animal Cells – Part 1

Animal cells are enclosed by a selectively permeable membrane. In this exercise, we will focus on red blood cells whose membrane is permeable to small molecules such as oxygen and carbon dioxide and impermeable to large molecules such as proteins. Red blood cells live in an isotonic solution called blood plasma. Within this environment, they maintain their normal biconcave shape. This enables red blood cells to perform their primary function, gas exchange. However, when red blood cells are put into a different environment, they react in various ways. For example, when red blood cells are put into a hypotonic environment, such as water, the water moves into through the cell's selectively permeable membrane causing the cell to expand. Eventually the cell will become so large, that it will become a "ghost" and burst (lyse). When red blood cells are placed in a hypertonic environment, water leaves the cells and as a result, the cells shrivel. In this condition, the cells are said to be crenate.

Purpose

To investigate the osmotic behavior of red blood cells placed in hypertonic, isotonic, and hypotonic solutions.

Materials

- Test tube containing bovine blood in an unknown type of solution (A)
- Test tube containing bovine blood in an unknown type of solution (B)
- Test tube containing bovine blood in an unknown type of solution ©
- Printed material

Methods

- Observe the 3 test tubes containing the 3 test tubes containing bovine blood and the unknowns (A, B, C).
- Record the appearance of the solution in each of the test tubes in the table below.
- Attempt to read the printed material through each of the test tubes.
- Record your observations in the **Appearance of Test Tubes Containing Unknowns A, B, and C** table.

EXERCISE 5 Observing Osmotic Behavior in Animal Cells – Part 2

Purpose

To investigate the osmotic behavior of red blood cells placed in hypertonic, isotonic, and hypotonic solutions.

Materials

- Bovine blood
- Unknown solution (A)
- Unknown solution (B)
- Unknown solution (C)
- Compound light microscope.
- Cover slides and coverslips.
- Disposable pipettes.

Methods

- Make a wet mount of bovine blood and observe the appearance of the red blood cells with a compound microscope.
- Record your observations in the table below and draw several of the red blood cells in the space provided.
- Make a wet mount of bovine blood using Unknown A and observe the appearance of the red blood cells with a compound light microscope.
- Record your observations in the table below and draw several of the red blood cells in the space provided.
- Make a wet mount of bovine blood using Unknown B and observe the appearance of the red blood cells with a compound microscope.
- Record your observations in the table below and draw several of the red blood cells in the space provided.

- Make a wet mount of bovine blood using Unknown C and observe the appearance of the red blood cells with a compound microscope.
- Record your observations in the **Appearance of Red Blood Cells in Test Solutions** table.
- Draw red blood cells as they appear in each of the unknown solutions in the space provided.

EXERCISE 6 Observing Osmotic Behavior in Plant Cells

Plant cells are enclosed within a selectively permeable membrane and an outer cell wall. If the cell is in a hypertonic solution, water leaves the cell. As a result, the protoplast (the cytoplasm enclosed by the plasma membrane) shrinks. Eventually, the protoplast may pull away from the cell wall resulting in a condition called plasmolysis. When a cell is placed in a hypotonic solution, water moves into the cell and the protoplast expands until it is restricted by the cell wall. At this point, the protoplast will actually begin to force water out of the cell to equalize the amount of water moving into the cell.

Purpose

To investigate the osmotic behavior of plant cells placed in hypertonic, isotonic, and hypotonic solutions.

Materials

- *Elodea*
- 0.9% saline solution
- 10% saline solution
- Compound light microscope.
- Cover slides and coverslips.
- Disposable pipettes.

Methods

- Make a wet mount of an *Elodea* leaf using distilled water and observe the appearance of the Elodea cells with a compound microscope.
- Record your observations in the in the space provided.
- Make a wet mount of an *Elodea* leaf using 0.9% saline solution and observe the appearance of the Elodea cells with a compound microscope.
- Record your observations in the space provided.
- Make a wet mount of an *Elodea* leaf using 10% saline solution and observe the appearance of the Elodea cells with a compound microscope.
- Record your observations in the space provided.

EXERCISE 7 Estimating the Osmolarity of Plant Cells

There is a solute concentration at which cells do not gain or lose water. This information is significant because cells need to maintain the ratio of water to osmotically active substances to be able to function effectively. In this exercise, you will use the change in the weight of potato pieces that have been submerged in various molar solutions to determine this point.

Purpose

To estimate the solute concentration of potato cells.

Materials

- 1 potato
- Cork borer
- Sucrose solutions (0.1 M, 0.2 M, 0.3 M, 0.4 M, 0.5 M, and 0.6 M)
- 7 plastic sups
- Electronic scale
- Petri plate
- Paper towels
- Forceps
- Scalpel

Methods

- Label the 7 plastic cups from 1 to 7.
- Fill cup # 1 with approximately 125 nL of water.
- Fill cup # 2 with approximately 125 mL of 0.1 M sucrose solution.
- Fill cup # 3 with approximately 125 mL of 0.2 M sucrose solution.
- Fill cup # 4 with approximately 125 mL of 0.3 M sucrose solution.
- Fill cup # 5 with approximately 125 mL of 0.4 M sucrose solution.
- Fill cup # 6 with approximately 125 mL of 0.5 M sucrose solution.
- Fill cup # 7 with approximately 125 mL of 0.6 M sucrose solution.
- With the cork borer, obtain 7 potato pieces. Immediately put the potato pieces in a Petri plate to prevent them from drying out.
- Use a scalpel to remove the peels before cutting the potato pieces to a length of approximately 5 cm.
- After weighing each of the potato pieces, place each in one of the labelled cups and record the weight in the "Initial Weight" row in the **Estimating Osmolarity by Change in Weight** table.
- After 60 minutes, remove each of the potato pieces from their cups.
- Before weighing the potato pieces, very gently remove any excess solution from the potato pieces by tamping them with a paper towel.
- Record the weight in the "Final Weight", "Weight Change", and "Weight Change %" in the **Estimating Osmolarity by Change in Weight** table.
- Calculate the weight change for each potato piece and record the data in the "Weight Change" row.
- Calculate the percentage change for each potato piece and record the data in the "Weight Change %" row.

Notes

Name:

Exercise 1 Results

1. Explain how the random motion of water molecules causes the Brownian movement of the carmine indigo crystals.

2. After heating the slide, was there any difference in the Brownian movement of the carmine indigo crystals? If so, what differences did you see? How would you expect the rate of Brownian motion to be affected by temperature?

Exercise 2 Results

1. Explain what the results indicate regarding diffusion.

Exercise 3 Results

1. Complete the table below with your results.

Permeability of Dialysis Tubing Experiment				
Solution	**Original Contents**	**Original Color**	**Final Color**	**Benedict's Test Results**
Bag				
Beaker				
Control				

2. Describe any color changes in the bag and the beaker.

3. Explain what the color changes in the bag and the beaker indicate.

4. After performing the Benedict's Test on the contents of the bag, beaker, and control, explain what the results indicate regarding the contents of the bag and beaker.

Exercise 4 Results

1. Complete the table below with your observations.

Appearance of Test Tubes Containing Unknowns A, B, and C		
	Appearance of Solution (such as opaque, translucent)	**Can You Read the Print?**
Unknown A		
Unknown B		
Unknown C		

2. Based on your observations, which unknown (A, B, and C) is in a hypertonic solution? Explain your answer.

3. Based on your observations, which of the unknowns (A, B, and C) is in an isotonic solution? Explain your answer.

4. Based on your observations, which of the unknowns (A, B, and C) is in a hypotonic solution? Explain your answer.

Exercise 5 Results

1. Complete the table below with your observations.

Appearance of Red Blood Cells in Test Solutions	
Solution	**Appearance, Size, and Condition of Red Blood Cells**
Blood	
Unknown A	
Unknown B	
Unknown C	

2. Make detailed drawings of the red blood cells as you observed them under high total magnification in unknown solutions A, B, and C.

Red blood cells in unknown A	Red blood cells in unknown B
Total magnification:	Total magnification:

Red blood cells in unknown C
Total magnification:

3. Explain the appearance of the red blood cells in the unknowns as they relate to the hypertonicity, isotonicity, and hypotonicity of the unknown solutions.

____

____

____

Exercise 6 Results

1. Make detailed drawings of the *Elodea* cells as you observed them under high total magnification in distilled water, 0.9% saline solution, and 10% saline solution.

***Elodea* cells in distilled water**	***Elodea* cells in 0.9% NaCl solution**
Total magnification:	Total magnification:

Elodea cells in 10% saline solution
Total magnification:

2. Explain the appearance of the *Elodea* cells as they relate to the hypertonicity, isotonicity, and hypotonicity of the solutions.

____

____

____

Exercise 7 Results

1. Complete the table summarizing your results.

Estimating Osmolarity by Change in Weight							
	Sucrose Molarity						
	0.0	0.1	0.2	0.3	0.4	0.5	0.6
Final Weight							
Initial Weight							

(continued)

Estimating Osmolarity by Change in Weight							
Sucrose Molarity							
	0.0	0.1	0.2	0.3	0.4	0.5	0.6
Weight Change (g)							
Weight Change %							

2. What is the osmolarity of the potato cells based on your results from this exercise?

____

____

3. Explain how you determined the osmolarity of potato cells.

____

____

CHAPTER 10

Cellular Respiration

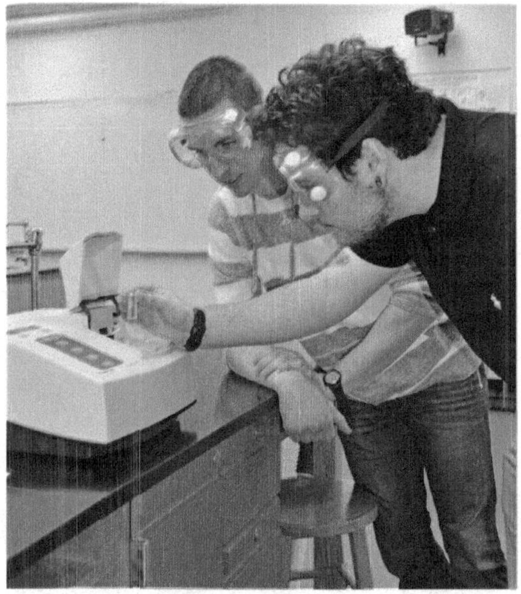

Photo courtesy of Holly J. Morris

OBJECTIVES

After completing these exercises, you will be able to:

- Understand the importance of cellular respiration to living organisms.
- Understand the process of alcoholic fermentation.
- Understand the process of aerobic cellular respiration.

INTRODUCTION

Cellular respiration is the process whereby cells break down organic molecules to release energy to be used for metabolic processes. The process can be aerobic (oxygen required) or anaerobic (oxygen not used). All cells carry out cellular respiration, whether prokaryotic or eukaryotic. In prokaryotic cells respiration occurs in the cytoplasm and the inner cell membrane. In eukaryotic cells respiration occurs in the cytoplasm and in the mitochondria. The final product of cellular respiration is adenosine triphosphate (ATP).

All organisms carry out anaerobic respiration. Most organisms then continue onto aerobic respiration, if sufficient oxygen is present, which is a process that is more efficient and produces far more ATP than anaerobic respiration.

The following equations compare fermentation with aerobic respiration, with glucose as the starting point.

Alcoholic Fermentation: $C_6H_{12}O_6 \rightarrow 2\ CH_3CH_2OH + 2\ CO_2 + 2\ ATP$

Aerobic Respiration: $C_6H_{12}O_6 + 6\ O_2 \rightarrow 6\ CO_2 + 6\ H_2O +$ approximately 38 ATP

EXERCISE 1 Investigating Alcoholic Fermentation

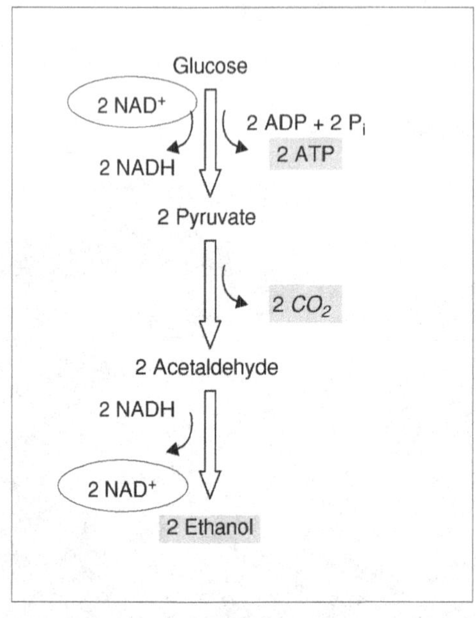

Overview of Alcoholic Fermentation (From Biological Investigations II: Lab Exercises for General Biology II by Edward E. Devine and Gretchen S. Bernard. Copyright © 2014 by Kendall Hunt Publishing Company. Reprinted by permission.)

Alcoholic fermentation is an anaerobic pathway carried out by yeasts in which simple sugars are converted to ethanol (C_2H_5OH) and carbon dioxide (CO_2). That being said, the main purpose of alcohol fermentation is to produce ATP, a usable source of energy for the cell, under anaerobic conditions. In actuality, the products of alcoholic fermentation, CO_2 and C_2H_5OH, are waste products.

Alcoholic fermentation can be divided into two parts. The first part of alcoholic fermentation is the glycolysis step. Glycolysis converts each molecule of glucose into two pyruvate molecules (a three carbon molecule). It is an anaerobic process that yields two ATP molecules per glucose molecule. For this step to occur, two molecules of nicotine adenine dinucleotide (NAD), a co-enzyme that carries the H^+ ions liberated as glucose is oxidized, are reduced to NADH.

The second part of alcoholic fermentation is the fermentation step. In this step, the two pyruvate molecules from the glycolysis step are converted into two CO_2 molecules and two C_2H_5OH molecules. Also in this step, the two NADH molecules are oxidized to NAD. This ensures that NAD will be available to the glycolysis step to keep the process operating.

Purpose

To measure the rate of alcoholic fermentation of yeast.

Materials

- Three large beakers
- Six large test tubes
- Three fermentation setups (plastic tubing inserted through rubber stoppers)
- Yeast
- Corn syrup

- Pipettes
- Metric ruler

Methods

- Prepare three fermentation set ups as shown in the picture below.
- Mix the fermentation solutions for the 3 test tubes according to the **Results of Alcoholic Fermentation** table.
- Place each of the reaction test tubes in one of the beakers.
- Push the fermentation set up (glass tubing inserted through a rubber stopper) in the reaction test tubes.
- Mark the water level of the collection test tubes for each of the 3 set ups.
- Every 5 minutes (for a total of 30 minutes), measure the distance (mm) from the base line to the current water line.
- Record the distance in the **Results of Alcoholic Fermentation** table.

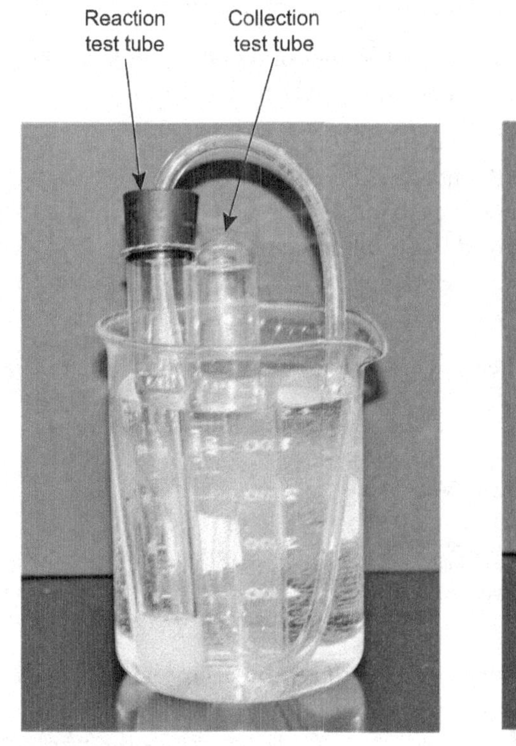

Start

Finish (Photos courtesy of Holly J. Morris and David T. Moat)

EXERCISE 2 Investigating Aerobic Cellular Respiration

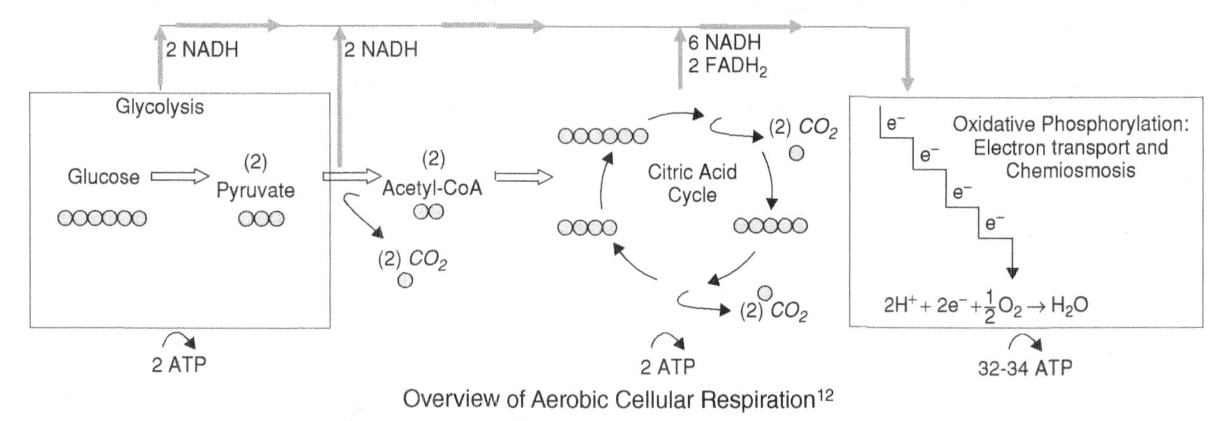

Overview of Aerobic Cellular Respiration[12]

Aerobic cellular respiration is a cellular process of fully oxidizing glucose to produce ATP. Aerobic cellular respiration is the opposite of photosynthesis. In photosynthesis, plants absorb the energy from sunlight, along with water and CO_2, and produce glucose and oxygen. In actuality, the oxygen is a waste product.

Aerobic cellular respiration is composed of four steps:

- Glycolysis
- Acetyl CoA formation
- Krebs Cycle
- Electron Transport System

Glycolysis, the first step in this process, occurs in the cytoplasm and converts each molecule of glucose into two pyruvate molecules (a three carbon molecule). It is an anaerobic process that yields a net two ATP molecules per glucose molecule. For this to occur, two nicotine adenine dinucleotide (NAD) molecules, a co-enzyme that carries the H^+ ions liberated as glucose is oxidized, are reduced to NADH.

Acetyl CoA formation, the next step in the process, occurs in the mitochondria and results in the formation of Acetyl CoA. This step requires the reduction of two NAD molecules per glucose molecule.

The Krebs Cycle also occurs in the mitochondria and yields a net of two ATP molecules per glucose molecule. To complete the oxidation of glucose, an additional 6 NAD molecules are required as well as two flavin adenine dinucleotide (FAD) molecules.

The Electron Transport System also occurs in the mitochondria. In this step, NADH and $FADH_2$ are oxidixed to NAD and FAD and carry their protons and electrons to a protein complex imbedded in the mitochondria's inner membrane. The protons are pumped to the inner membrane space and the electrons are then passed along the membrane proteins, giving up their energy to power a process that results in the generation of ATP. This process produces approximately 32 - 34 additional ATP molecules per glucose molecule.

The rate of aerobic cellular respiration can be measured by using the succinate to fumarate conversion step in the Krebs cycle as shown below.

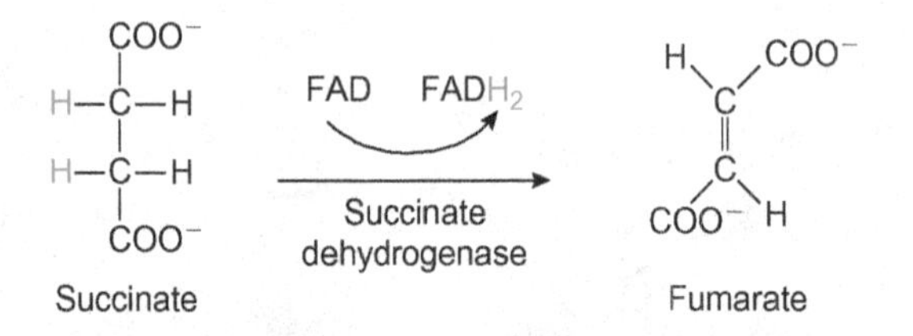

As this reaction occurs, succinate releases protons and electrons. These protons and electrons are transported to the Electron Transport System (ETS) to generate ATP. To investigate the rate of aerobic cellular respiration, we will use dichlorophenolindophenol (DCPIP). DCPIP intercepts the protons and electrons released by succinate before they can enter the ETS. When oxidized, DCPIP is blue with a maximal absorption at 600 nm and when reduced, DCPIP is colorless. When a DCPIP molecule picks up a proton and an electron, it loses its blue color and becomes colorless. This color change, which is a measure of the rate of aerobic cellular respiration, can be measured by using a spectrophotometer.

Purpose

To measure the rate of aerobic respiration using DCPIP and the succinate—fumarate reaction.

Materials

- Mitochondrial suspension
- DCPIP
- Succinate
- Phosphate buffer

- Four cuvettes
- Micropipettes
- Spectrophotometer

Methods

- Turn on the spectrophotometer.
  - Set the spectrophotometer to display "% transmittance."
  - Set the wavelength to 600 nm.
- Mix the solution for the "Blank" cuvette according to the **Results of the Mitochondrial Activity** table below.
- Mix the solution for the three numbered cuvettes according to the **Results of the Mitochondrial Activity** table below. Do not add the succinate to the cuvettes at this time.
- When all of the three solutions have been prepared, add the succinate to each of the cuvettes. Cover the cuvettes with parafilm and gently invert the cuvettes to mix the solutions.
- Insert the "Blank" cuvette into the spectrophotometer and zero the "% transmittance."
- Insert the three cuvettes into the spectrophotometer and record their "% transmittance" in the **Results of Mitochondrial Activity** table below in the 0 minute column.
- Every 5 minutes (for a total of 30 minutes), insert the "Blank" cuvette into the spectrophotometer and zero the "% transmittance." Then insert the three cuvettes into the spectrophotometer and record their "% transmittance" in the appropriate column in the **Results of Mitochondrial Activity** table.

Name:

Exercise 1 Results

1. Complete the following table.

Results of Alcoholic Fermentation							
	Minutes						
Test Tube	**0**	**5**	**10**	**15**	**20**	**25**	**30**
# 1 (3 mL corn syrup, 4 mL water)							
# 2 (3 mL corn syrup, 3 mL water, 1 mL yeast)							
# 3 (3 mL corn syrup, 1 mL water, 3 mL yeast)							

2. What was the hypothesis used in this experiment?

____

____

3. What was the independent variable for this experiment?

____

____

4. What was the dependent variable for this experiment?

____

____

5. Which fermentation tube was the control? Explain your answer.

____

____

6. Was your hypothesis proven false or supported by the experiment? Explain your answer.

____

____

7. Plot the results of the experiment on this graph.

Exercise 2 Results

1. Complete the following table.

Results of Mitochondrial Activity							
	Minutes						
Cuvette	0	5	10	15	20	25	30
Blank (0.3 mL mitochondrial suspension, 4.6 mL buffer, 0.1 mL succinate)							
# 1 (0.3 mL mitochondrial suspension, 0.3 mL DCPIP, 4.4 mL buffer, 0 mL succinate)							
# 2 (0.3 mL mitochondrial suspension, 0.3 mL DCPIP, 4.3 mL buffer, 0.1 mL succinate)							
# 3 (0.3 mL mitochondrial suspension, 0.3 mL DCPIP, 4.1 mL buffer, 0.3 mL succinate)							

2. What was the hypothesis used in this experiment?

3. What was the independent variable for this experiment?

4. What was the dependent variable for this experiment?

5. Which test tube was the control? Explain your answer.

6. What is the function of the DCPIP?

7. Why was the succinate added last?

8. Was your hypothesis proven false or supported by the experiment? Explain your answer.

9. Plot the results of the experiment on this graph.

CHAPTER 11

Mitosis

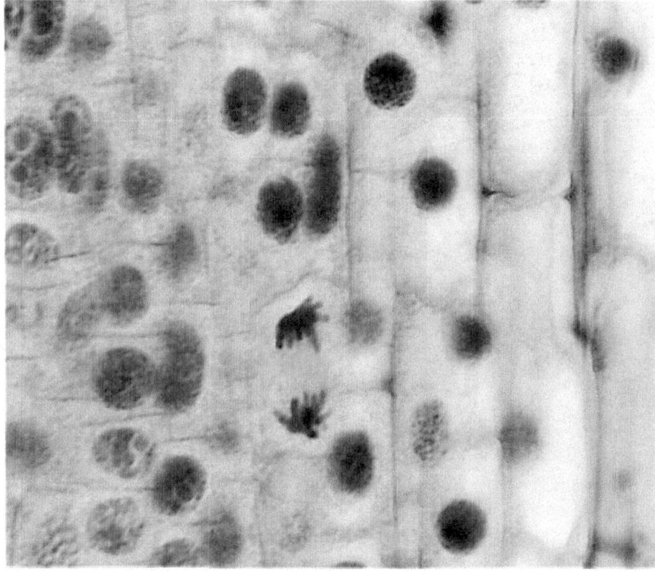

Photo courtesy of Holly J. Morris and David T. Moat

OBJECTIVES

After completing this exercise, you will:

- Understand the eukaryotic cell life cycle and its subdivisions.
- Understand the steps that occur in mitosis.
- Be able to describe the significant events that occur in each step in mitosis.
- Understand why mitotic divisions are necessary in living cells.
- Recognize the differences between plant and animal cytokinesis.
- Be able to recognize chromosomes in the various stages of mitosis.

INTRODUCTION

Cell division is central to the life of all organisms. All cells come from pre-existing cells as a result of cell division. Newly created cells are subject to wear and tear, as well as accidents and as a result, they are bound to die. If a multicellular organism is to continue to live, it must create new cells at a rate as fast as that at which its cells die. For example, many cells in adult humans must divide every second simply to maintain the status quo.[3]

All cells have a cell cycle which consists of two phases: *Interphase* (the phase during which the cell is not dividing) and *Mitotic* (the phase during which the cell is dividing).

A visual depiction of the phases of the eukaryotic cell life cycle follows.

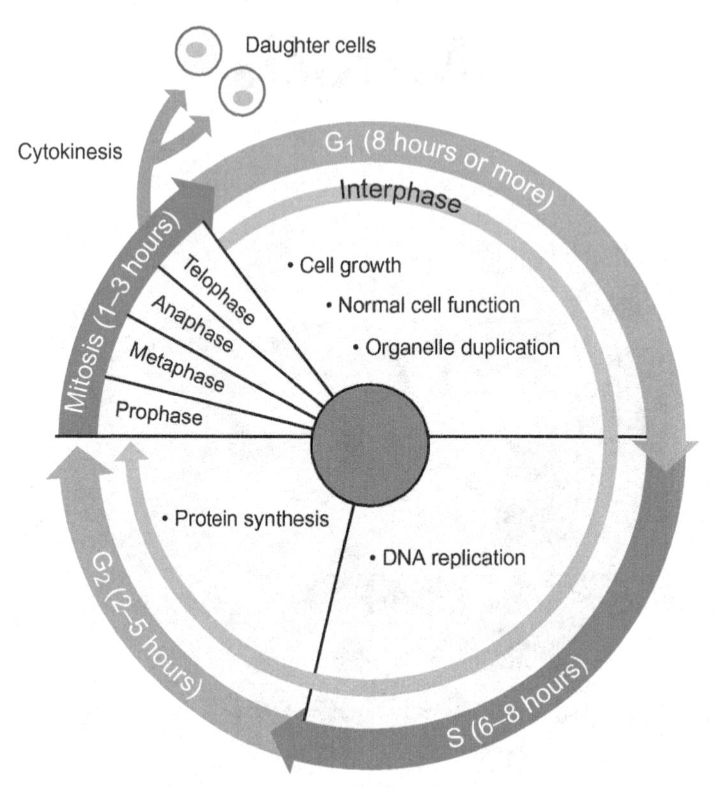

The main events that occur in each of the subphases of Interphase are:

G1 (First gap phase)	During this subphase, the newly divided cell is growing until it reaches its mature size.
Synthesis (S phase)	During this subphase, the cell's chromosomes are duplicated and upon completion of this phase, each chromosome consists of two identical chromatids.
G2 (Second gap phase)	During this subphase, the cell is preparing for the start of mitosis.
G0 (Temporary or permanent exit from the cell cycle)	Some cells pause during the G1 subphase and enter this phase where they may remain indefinitely. Some cells in this phase may never reenter the cell cycle.

A cell spends approximately 90% of its life in Interphase (G1, Synthesis, and G2) and only 10% of its life in the Mitotic phase (mitosis and cytokinesis). A cell may spend about 8 hours in the G1 phase because in this phase, the newly divided cell is growing to its mature size. The length of time a eukaryotic cell spends in the Synthesis phase depends upon the amount of DNA in the cell that needs to be copied and may require 6 to 8 hours. The cell may spend about 2 to 5 hours in G2 as it prepares for mitosis. The Mitotic phase (mitosis and cytokinesis) requires only about 1 to 3 hours.

Eukaryotic cells may experience two types of cell division: mitotic cell division and meiotic cell division. This exercise deals with mitotic cell division. Meiotic cell division will be considered in the next lab exercise. Prokaryotic cells do not use either mitotic cell division or meiotic cell division but instead, they divide by binary fission.

Mitotic cell division consists of two sequential processes: nuclear division (called mitosis) and cytoplasmic division (called cytokinesis). Mitotic cell division produces daughter cells that are genetically identical to each other and also genetically identical to the cell which divided to give rise to them.[3]

A visual depiction of the subphases of the mitotic phases follows.

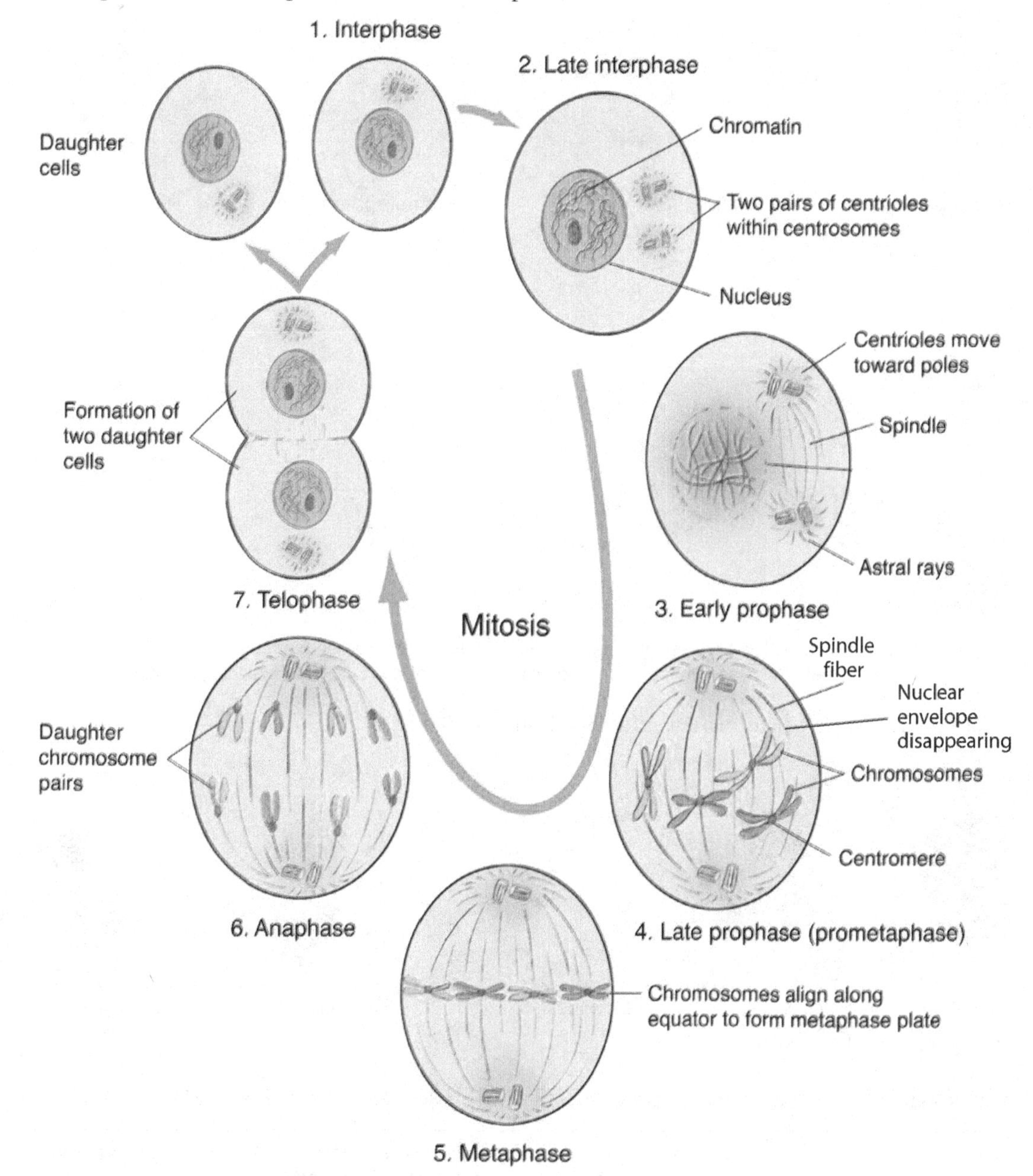

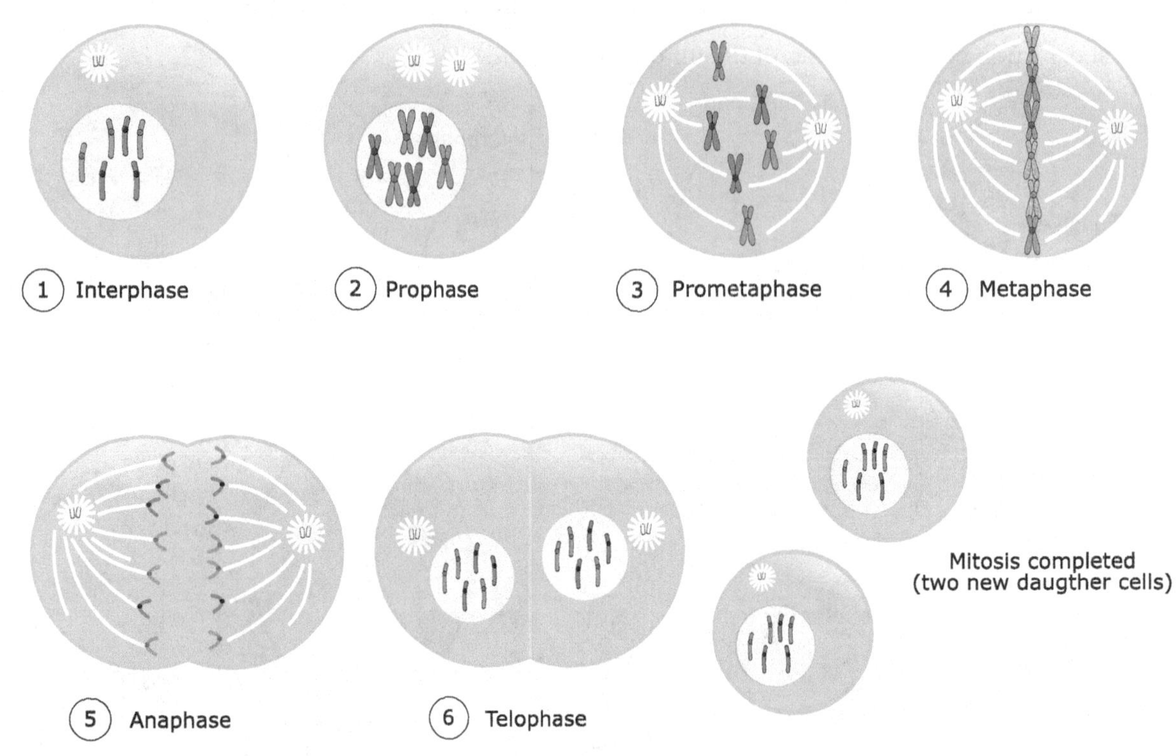

From Investigating Biology: The Diversity of Life Lab by Paul Florence and Annisa Florence. Copyright © 2013 by Paul Florence and Annisa Florence. Reprinted by permission of Kendall Hunt Publishing Company.

Photomicrograph images of the various mitotic subphases follow:

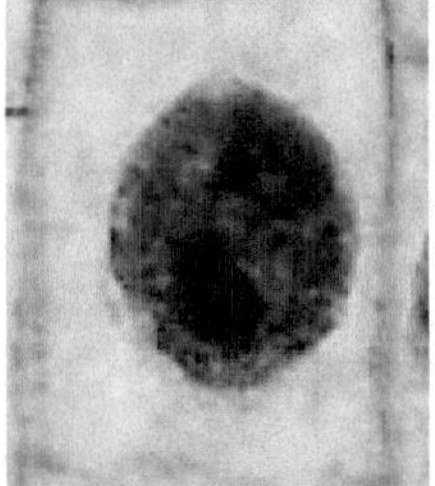
Interphase (400×)

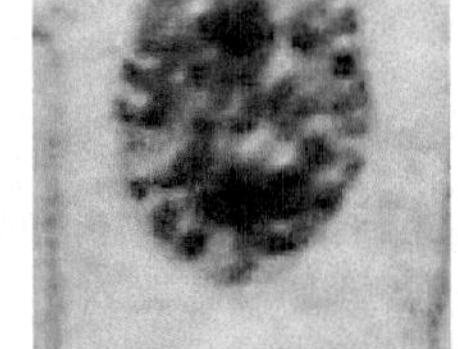
Prophase (400×)

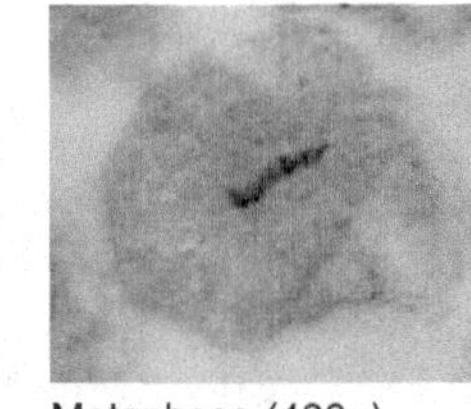
Metaphase (400×)

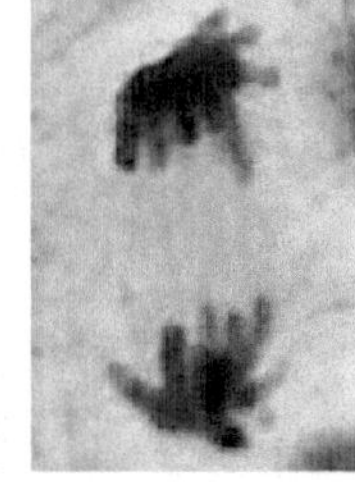
Anaphase (400×)

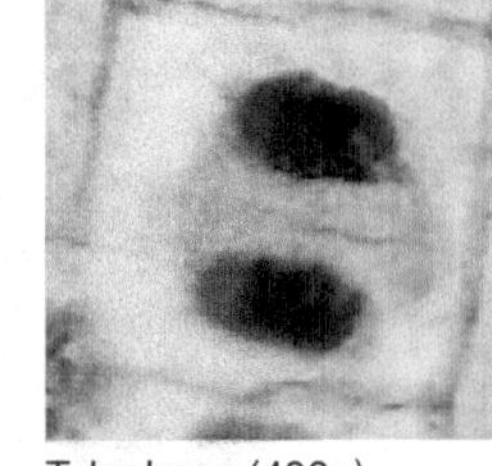
Telophase (400×)

(Photos courtesy of Holly J. Morris and David T. Moat)

The main activities that occur in each of the Mitotic sub-phases are described below[3]:

Prophase	• Chromatin slowly condenses into well-defined chromosomes. Each chromosome was duplicated during the S phase and consists of two sister chromatids that are joined together at a specific region known as the centromere. • Nucleolus disassembles and gradually disappears. • In animal cells, paired centrosomes (each containing paired centrioles) move to opposite poles. Centrosomes are microtubule organizing centers (MTOC). • Centrosome replication begins just prior to the S phase and continues through G2 producing two centrosomes each of which contains microtubules radiating from it called asters. Higher plant cells lack centrioles and asters. • The *mitotic spindle* begins to develop outside the nucleus. (The spindle consists of fibers constructed of microtubules and microtubule-associated proteins. Cytoskeletal microtubules disassemble, and their tubulin dimers start to reassemble to form the mitotic spindle. Each spindle fiber is, therefore, a cluster of microtubules. The first fibers to form are polar fibers, which extend from the two poles of the spindle toward the equator of the cell.)

Prometaphase	• The nuclear envelope breaks into membranous fragments. • Specialized structures called *kinetochores* (protein complexes) develop on either face of the centromeres and become attached to a special set of microtubules called kinetochore fibers. These fibers extend from poles on either end of the cell and attach to the kinetochore. • The chromosomes begin to become arranged in one plane (the *metaphase plate*) near the middle of the cell.
Metaphase	• The chromosomes become arranged so that their centromeres all lie in the metaphase plate. The kinetochore fibers seem to be responsible for aligning the chromosomes halfway between the spindle poles and for orienting them with their long axes at right angles to the spindle axis.
Anaphase	• Motor molecules associated with each kinetochore separate the sister chromatids of each chromosome. • Each chromatid (now called a chromosome) is moved slowly toward a spindle pole by the motor molecules of the kinetochore as they "walk" along the kinetochore microtubules. • Kinetochore fibers (microtubules) progressively shorten (by depolymerizing at their kinetochore ends) as the chromosomes approach the poles. • Polar fibers (microtubules) elongate and move the two poles of the spindle further apart elongating the whole cell along the polar axis.
Telophase	• As the separated chromosomes (formerly called sister chromatids) arrive at the poles, the kinetochore fibers disappear. • A new nuclear envelope reforms around each group of chromosomes. • The condensed chromatin disperses. • Nucleoli reappear.
Cytokinesis	• In animal cells, cytokinesis is accomplished by the development of a *cleavage furrow*. The cleavage furrow forms as a result of the interaction of actin and myosin filaments, which form a contractile ring just below the plasma membrane. • In plant cells, cytokinesis is accomplished by the formation of a *cell plate*, rather than a cleavage furrow. The cell plate is assembled as vesicles containing cell wall material are directed to the center of the dividing cell by microtubules. The vesicles then fuse with each other to create the cell plate.

[3]

EXERCISE 1 Observing Mitosis in Animal and Plant Cells

Cells come from pre-existing cells as a result of mitosis. Cells undergo mitosis for:

- Reproduction
- Growth and development
- Tissue renewal

It is, therefore, important to understand this process.

Purpose

To observe plant and animal cells in the various phases of mitosis to help you understand the process of cell division.

Materials

- Compound light microscope
- Prepared onion tip slides
- Prepared fish blastula slides

Methods

- Observe a prepared onion root tip slide and make a **detailed drawing** of 1 cell in each of the following stages: Interphase, Prophase/Prometaphase, Metaphase, Anaphase, and Telophase.
- Observe a prepared fish blastula slide and make a **detailed drawing** of 1 cell in each of the following stages: Interphase, Prophase/Prometaphase, Metaphase, Anaphase, and Telophase.

Name:

Exercise 1 Results

1. Observe a prepared onion root-tip slide and look for cells in interphase and the various stages of mitosis: interphase, prophase/prometaphase, metaphase, anaphase and telophase. Make **detailed drawings** of each mitotic stage as you observed them. Label the nucleus (if present), chromosomes (if visible), cell plate (if visible), cell membrane, and cell wall (if present).

Interphase	Prophase/Prometaphase
Total magnification:	Total magnification:

Metaphase	Anaphase
Total magnification:	Total magnification:

Telophase
Total magnification:

2. Observe a prepared fish blastula slide and look for cells in interphase and the various stages of mitosis: interphase, prophase/prometaphase, metaphase, anaphase, and telophase. Make **detailed drawings** of each mitotic stage as you observed them. Label the nucleus (if present), chromosomes (if visible), and cell membrane.

Interphase	Prophase/Prometaphase
Total magnification:	Total magnification:

Metaphase	Anaphase
Total magnification:	Total magnification:

Telophase
Total magnification:

3. List two reasons why a cell may have to divide from time to time.

____

____

4. Indicate the appropriate cell cycle stage (interphase, prophase, prometaphase, metaphase, anaphase, and telophase) in which the following **mitotic** events occur.

	Chromosomes align on the equator
	Cell plate starts to become visible in onion cells
	Distinct chromosomes are not visible in the cell at 40×
	Nuclear envelope reforms
	Chromatin coils into chromosomes
	Chromosome copies move to poles
	Nuclear envelope breaks up

CHAPTER 12

Meiosis

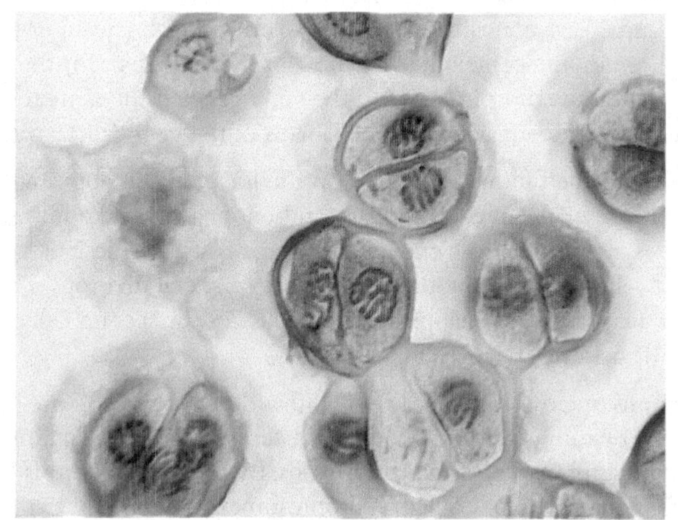
Photo courtesy of Holly J. Morris and David T. Moat

OBJECTIVES

After completing this exercise, you will:

- Understand the steps that occur in meiosis.
- Be able to describe the significant events that occur in each step of meiosis.
- Be able to explain the necessity for meiotic division in sexually reproducing organisms.
- Be able to define and use the terms haploid and diploid.
- Be able to distinguish between oogenesis and spermatogenesis as to the number and the types of cells produced.
- Understand how the meiotic process produces genetic variation by means of random assortment of chromosomes and crossing over.
- Be able to compare and contrast the events of mitosis and meiosis.

INTRODUCTION

Meiosis, which occurs only in eukaryotic organisms, is a process that forms *4 haploid cells* (cells with a single set of chromosomes) from a *single diploid cell* (cells with two sets of chromosomes) by two successive nuclear divisions called Meiosis I and Meiosis II. The number of chromosomes per nucleus is constant for all the individuals of a species but varies from one species to another. For example, humans have

46 chromosomes in each of their diploid cells while frogs have 26 chromosomes, onions have 16 chromosomes, and watermelon has 22 chromosomes.

To maintain the number of chromosomes from generation after generation, sex cells (gametes) contain only half the number of chromosomes found in the somatic (diploid) cells. In humans, each mature ovum and each mature sperm cell are haploid and have 23 chromosomes. The uniting of a sperm cell and an egg cell during fertilization forms a fertilized egg (zygote) with 46 chromosomes. As such, the zygote is, therefore, a diploid cell with two sets of 23 chromosomes. As a result, the original chromosome number is then restored.

There are two types of gametogenesis which is the process producing gametes in animals - spermatogenesis and oogenesis. Spermatogenesis produces sperm and oogenesis produces ova (eggs).

A zygote contains 46 chromosomes consisting of 23 pairs of homologous chromosomes of which 23 are contributed by the egg and 23 are contributed by the sperm. For each chromosome that the egg contributes, there is a similar chromosome contributed by the sperm cell. Except for the sex cells, these homologous chromosomes have the same morphology and the same genes at the same relative position (locus) on both chromosomes. However, there may be different forms of the gene (alleles) at the locus.

As with mitosis, meiosis is divided into stages based on the location and appearance of the chromosomes. The stages are named prophase, metaphase, anaphase, and telophase. Unlike mitosis, there are two rounds of cell division in meiosis, rather than just one, so each stage is further designated with a I or II, for first cell division and second cell division. Thus, the stages of the first cell division in meiosis are designated as prophase I, metaphase I, anaphase I, and telophase I, the stages of the second cell division are designated as prophase II, metaphase II, anaphase II, and telophase II.

At the beginning of meiosis (meiosis I), the cell is diploid and each chromosome contains two chromatids. By the end of the first cell division, there are two haploid cells, each containing one copy of each chromosome, but the chromosomes still consist of two chromatids. By the end of meiosis II, there are four haploid cells containing a single chromatid (now called a chromosome).

A depiction of the major subphases of meiosis follows.

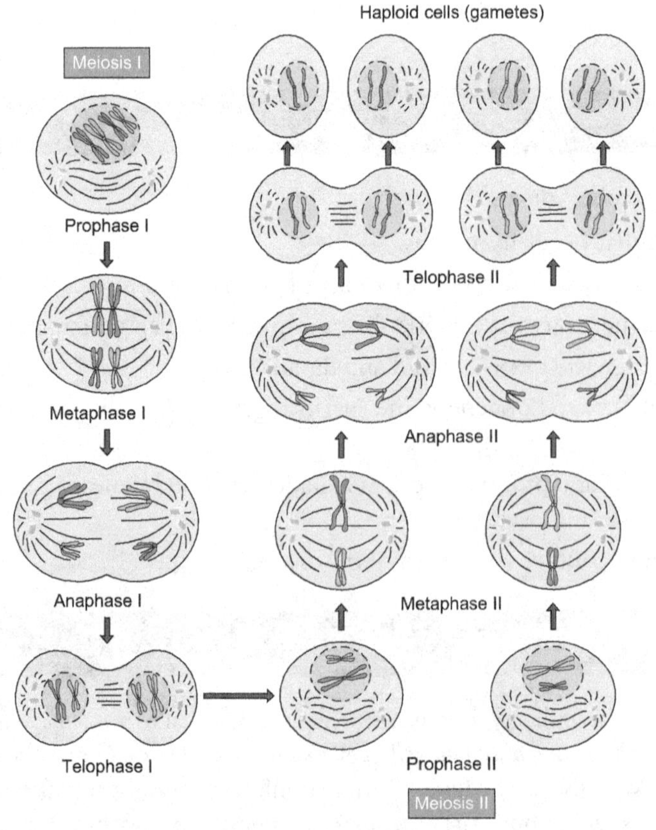

Photomicrograph images of the various subphases of meiosis follow.

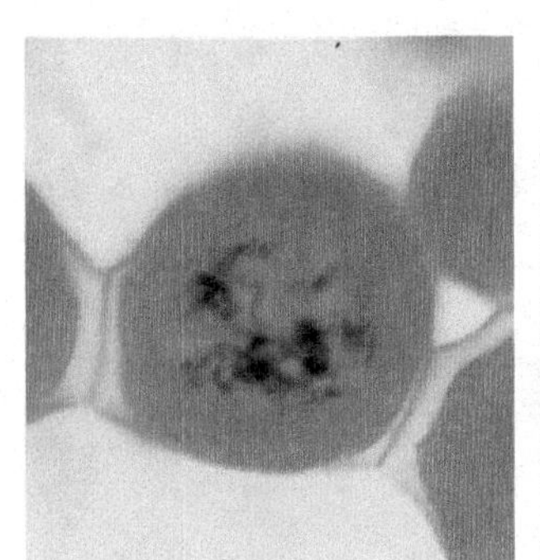
Prophase I (400×)

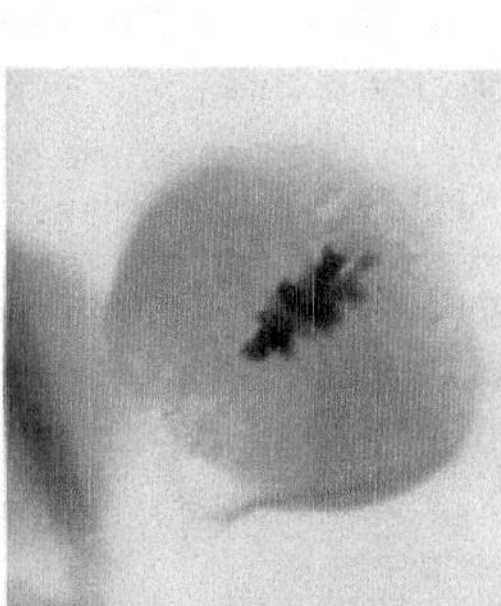
Metaphase I (400×)

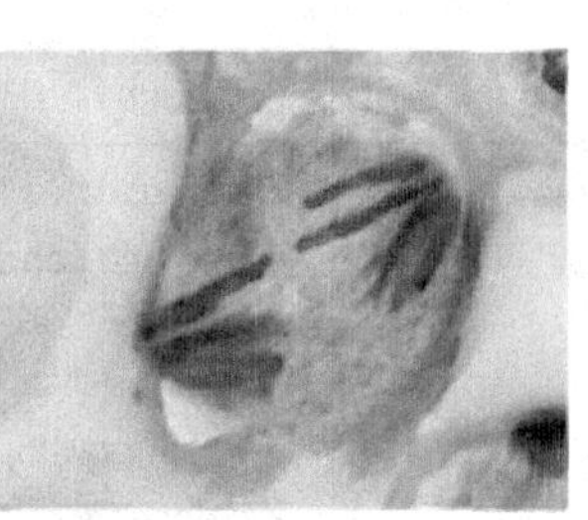
Anaphase I (400×)

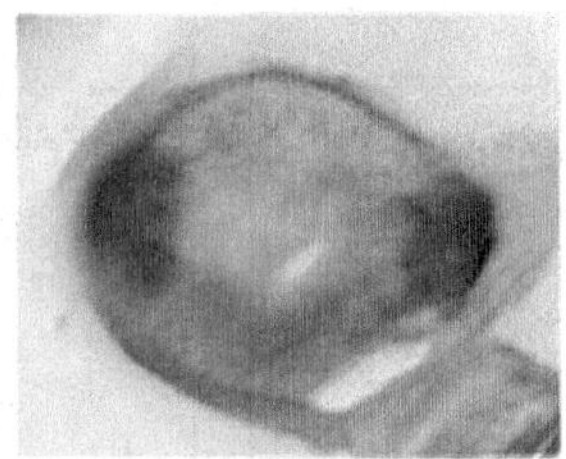
Telophase I (400×) (Photos courtesy of Holly J. Morris and David T. Moat)

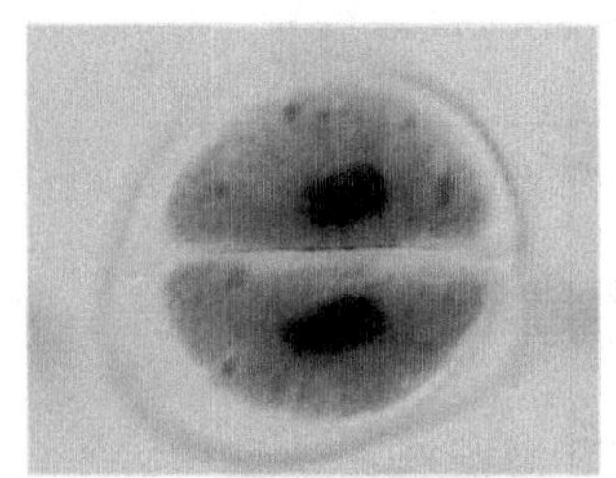
Prophase II (400×)

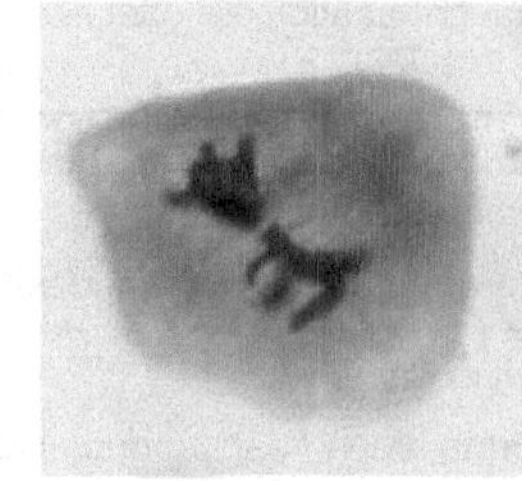
Metaphase II (400×)

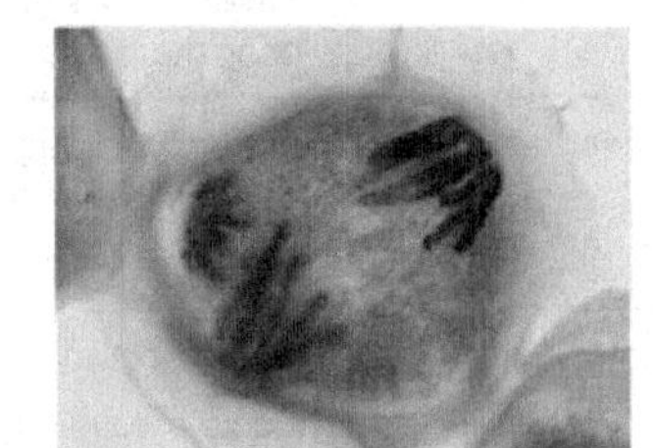
Anaphase II (400×)

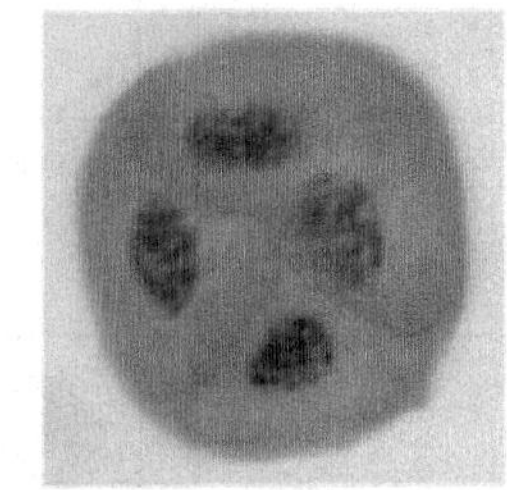
Telophase II (400×) (Photos courtesy of Holly J. Morris and David T. Moat)

The main events that occur in each of the meiotic subphases are described below:

Meiosis I	Prophase I	• Chromosomes condense and become visible under the microscope. • *Synapsis*, the pairing of homologous chromosomes, occurs. This results in a tetrad of chromatids that appear as two X's sitting next to each other. This arrangement allows for crossing over of nonsister chromatids. • Sister chromatids belong to the same chromosome. Crossing over can occur between adjacent nonsister chromatids. The point where crossing over occurs is called a *chiasmata*. • Nucleolus disappears. • Nuclear envelope breaks down. • Late in prophase I, microtubules from one pole or the other become attached to the kinetochores of each chromosome.
	Metaphase I	• Homologous pairs of chromosomes line up somewhere along the metaphase plate. • The arrangement of tetrads at the metaphase plate is independent of the orientation of the other tetrads, which results in a random assortment of chromosomes from mother or father to the daughter cells. • Both chromatids of one of the homologous pair are attached to kinetochore proteins from one pole. The other set of chromatids are attached to kinetochore proteins from the opposite pole. The homologous chromosomes still look like two side-by-side X's
	Anaphase I	• Proteins holding the homologous pairs of chromosomes together break down, which allows the homologs to separate. • Microtubules that are attached to the respective centromeres, pull the homologs toward opposite poles, that is, the X's separate.

	Telophase I and Cytokinesis I	• Chromosomes cluster at the poles as division of the cytoplasm occurs. In some species, chromosomes decondense and nuclear envelopes form around the complete haploid set of chromosomes. • No chromosome duplication occurs. • In some species there may be an interphase between Meiosis I and Meiosis II, though no DNA synthesis (S phase) occurs.
Meiosis II	Prophase II	• Chromosomes become visible as long filaments. • Chromatin condenses into chromosomes that consist of two chromatids that are joined at the centromere. • Nucleolus disappears. • Nuclear envelope disappears. • Kinetochore microtubules become attached to the kinetochores of sister chromatids. • Chromosomes begin to assemble at the metaphase plate.
	Metaphase II	• Chromosomes are positioned at the metaphase plate in single file, random order. • Kinetochores of sister chromatids are attached to microtubules from opposite poles.
	Anaphase II	• Proteins holding the chromatids together at the centromere break down, which allows the chromatids to separate. • Microtubules pull the chromatids toward opposite poles.
	Telophase II and Cytokinesis II	• Chromosomes arrive at the poles. • Nuclear envelope forms around each group of chromosomes. • The chromatin decondenses. • Nucleolus reappears. • Division of cytoplasm occurs. • Four haploid daughter cells that are genetically different from the parent cell have been produced.

14

VERTEBRATE GAMETOGENESIS

Gamatogenesis means "creation of gametes," that is, production of eggs and sperm. In the testes, diploid cells called primary spermatocytes undergo meiosis, as described above, resulting in four haploid sperm.

The production of eggs is more complicated. In the ovary, the corollary to primary spermatocytes are cells called primary oocytes. Primary oocytes begin meiosis I embryologically, but stop part way through prophase I and remain dormant until sexual maturity (puberty, in the case of human females). Primary oocytes resume meiosis I just prior to being released from the ovary. The two resulting cells are haploid, as described above, but cytokinesis is uneven, resulting in one large cell, called the secondary oocyte, and one very small cell, called a polar body. The polar body may divide further or may simply degenerate.

The secondary oocyte continues onto meiosis II but stops at metaphase II. The secondary oocyte is released during ovulation but only continues meiosis II if it is penetrated (fertilized) by a sperm. As meiosis II resumes, telophase II again results in a large cell, which is the fertilized egg, and a second polar body, which degenerates. Thus, meiosis in the female starts with one diploid cell and ends with one haploid cell that is fertilized, resulting in it becoming diploid.

These events are depicted below:

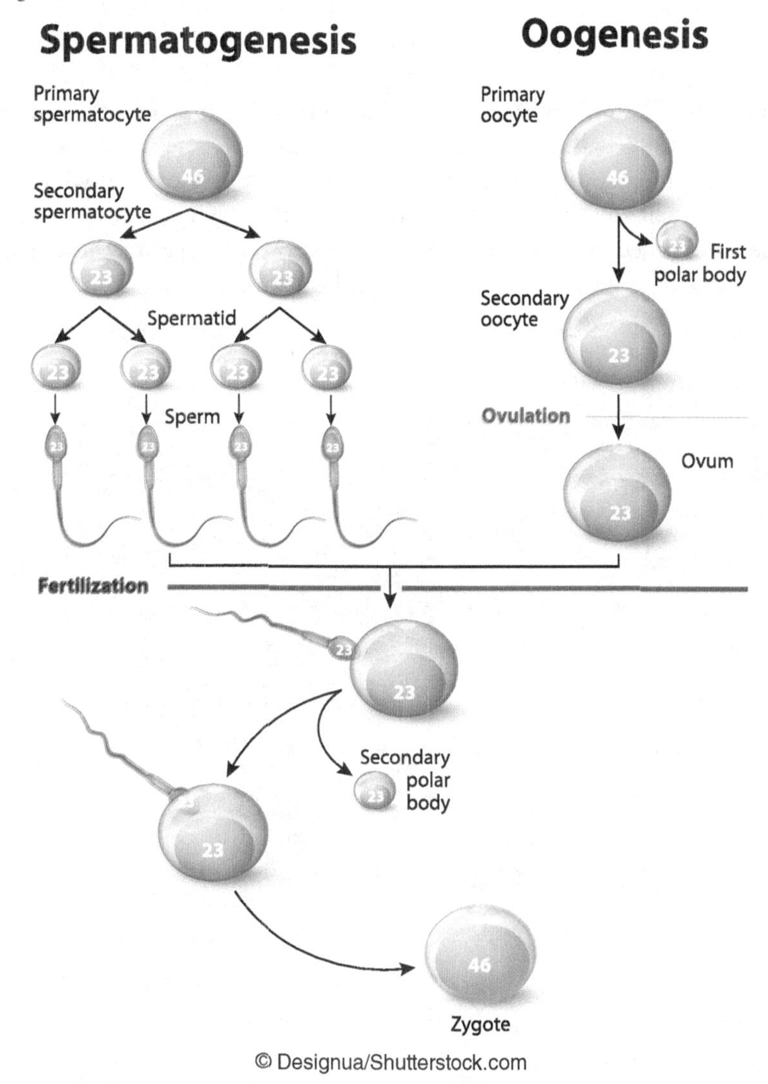

EXERCISE 1 Observing Meiosis in Plant Cells

The purpose of meiosis is to provide for genetic variation. As a result of meiosis, four genetically different haploid cells are produced. This provides the mechanism for genetic diversity to occur.

Purpose

To observe plant cells in the various phases of meiosis to help you understand the process of meiosis.

Materials

- Compound light microscope
- Prepared *Lilium* anther meiosis slides

Methods

- Observe prepared *Lilium* anther meiosis slides and make **detailed drawings** of four cells in any of the meiotic subphases (Prophase I, Metaphase I, Anaphase I, Telophase I, Prophase II, Metaphase II, Anaphase II, and Telophase II).

EXERCISE 2 Observing Crossing Over During Meiosis in *Sordaria fimicola*

Sardoria fimicola is a microscopic fungus in the phylum Ascomycota, and serves as a model to study the process of crossing over during meiosis. *Sordaria* is haploid for most of its life cycle, but under favorable environmental conditions will undergo sexual reproduction in which filaments from two different mating types fuse, forming a structures with two haploid nuclei. Ultimately the nuclei will fuse, forming diploid zygotes. The zygotes each undergo meiosis, resulting in four ascospores, each of which then undergo mitosis, resulting in a total of eight ascospores that are lined up within an ascus. Several asci are protected with a fruiting body called a perithecium (ascocarp). When the ascospores mature, the asci and the perithecium rupture, releasing ascospores to develop into new haploid organisms.

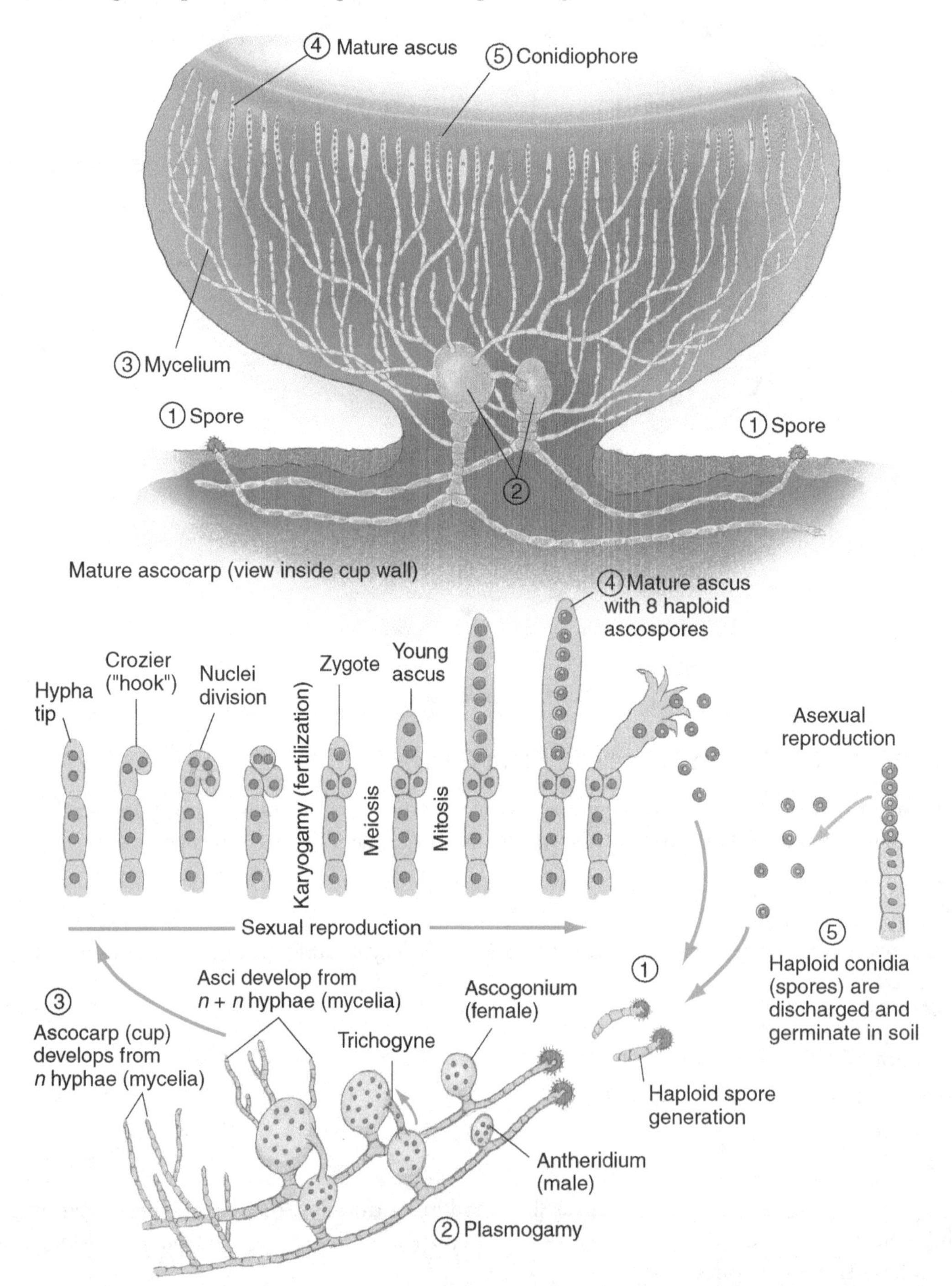

The normal (wild type) color for *Sordaria* spores is black. The black color is due to the pigment melanin that is deposited in the cell walls. One mutant form of the gene produces tan spores. When spores from the black strain mate with spores from the tan strain, the resulting asci contain ascospores arranged in such a way that demonstrates whether or not crossing over occurred during prophase of meiosis I.

The eight ascospores produced inside an ascus provides information regarding the amount of crossing over that occurs during meiosis. For example, if no crossing over occurs, then the 4 black spores line up together and 4 tan spores line up together (a 4:4 pattern) as shown below.

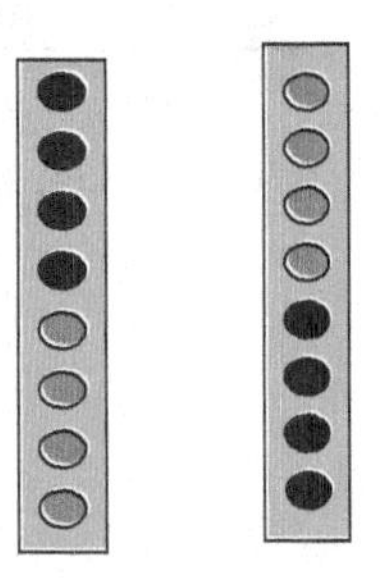

If crossing over does occur there is a 2:2:2:2 pattern or a 2:4:2 pattern of the ascospores as shown below.

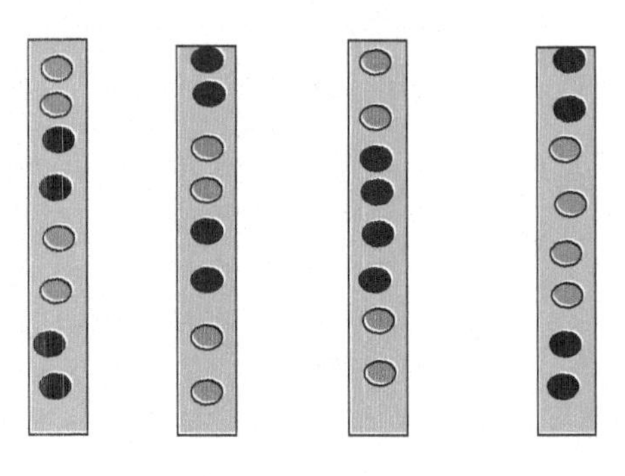

Purpose

To determine the frequency of crossing over that occurs during *Sordaria fimicola* meiosis.

Materials

- Petri plates containing mycelia resulting from a cross between *Sordaria* black and tan spores.
- Compound light microscope
- Slides and coverslips
- Toothpick
- Pencil

Methods

- Place a drop of water on a clean slide.
- Using a toothpick, remove several of the *Sordaria* perithecia from the area where the black and tan strains have grown together.
- Place the perithecia in the drop of water on the slide and cover with the coverslip.
- Using the eraser of a pencil, press gently on the coverslip to flatten the perithecia.
- Place the slide on the microscope and locate the clusters of asci.

- Switch to high power to observe 100 mixed color asci and determine whether recombination has occurred.
    - Black and tan spores with a 4:4 pattern of the ascospores (indicates recombination has not occurred – Nonrecombinant Asci).
    - Black and tan spores with a 2:2:2:2 or 2:4:2 pattern of the ascospores (indicates recombination has occurred – Recombinant Asci).
- Complete the ***Sordaria* Asci Analysis** table summarizing your results.
- Calculate the recombination frequency.

$$\text{Recombination Frequency} = \frac{\text{Recombinant Asci}}{\text{Total Asci}} \times 100$$

Name:

Exercise 1 Results

1. Observe prepared Lilium anther meiosis slides and make **detailed drawings** of four cells in any of the phases of meiosis (Prophase I, Metaphase I, Anaphase I, Telophase I, Prophase II, Metaphase II, Anaphase II, and Telophase II). Label the nucleus (if present), chromosomes (if visible), cell plate (if visible), cell membrane, and cell wall (if present).

Phase: ___________	**Phase:** ___________
Total magnification:	Total magnification:

Phase: ___________	**Phase:** ___________
Total magnification:	Total magnification:

2. Why must the cells that result from meiosis be haploid?

____

____

3. How are mitosis and meiosis involved in the human life cycle?

____

____

4 Complete the table comparing mitosis and meiosis.

Comparison of Nuclear & Chromosomal Activities in Mitosis and Meiosis		
	Mitosis	**Meiosis**
Synapses Occurs? (Yes / No)		
Crossing Over Occurs (Yes / No)		
Phase when Centromeres Split		
Chromosome Structure and Movement during Anaphase	Anaphase:	Anaphase I: Anaphase II:
Number of Nuclear Divisions		
Number of Cells Resulting		
Number of Chromosomes in Daughter Cells		
Genetic Similarity of Daughter Cells to Parents		

Exercise 2 Results

1. Complete the table summarizing the results of your asci observations:

Sordaria Asci Analysis	
	Number of Asci
Black and tan spores with a 4:4 pattern (Nonrecombinant Asci)	
Black and tan spores with a 2:2:2:2 or 2:4:2 pattern (Recombinant Asci)	
Total	**100**

2. Calculate the recombination frequency of *Sordaria*:

$$\text{Recombination Frequency} = \frac{(\quad)\ \text{Recombinant Asci}}{(\quad)\ \text{Total Asci}} \times 100 = \underline{\qquad\qquad}$$

CHAPTER 13

Genetics

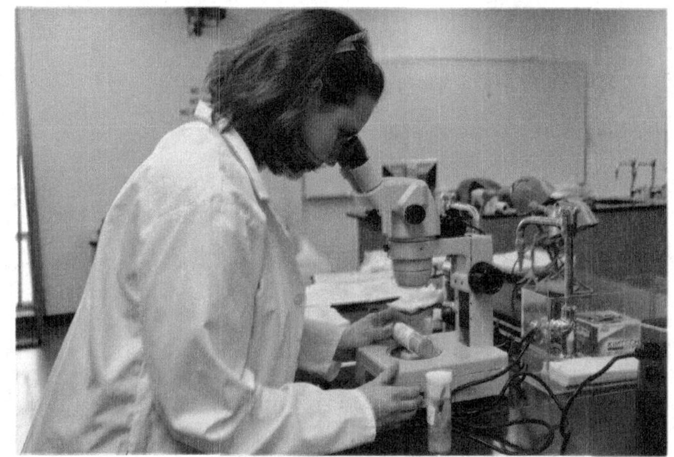
Photo courtesy of Holly J. Morris

OBJECTIVES

After completing this exercise, you will:

- Be able to correctly analyze genetics problems including monohybrid and dihybrid crosses for:
  - Complete dominance.
  - Incomplete dominance.
  - Codominance.
  - X linked recessive.
- Understand how the gametes that result from a diploid cell undergoing meiosis are formed.
- Understand the use of pedigrees as a genetic tool.

INTRODUCTION

Genetics is the branch of biology that studies how genes control the characteristics of living organisms. However, before we can discuss genetics, we need to become familiar with genetics related terms.

Character: An observable heritable feature; e.g., flower color.

Trait: Variations of a specific character; e.g., purple versus white flower color.

Alleles: Alternative versions of genes; e.g., the gene that codes for purple pigment in a flower versus the gene that codes for white pigment.

Phenotype: The outward appearance of a given trait.

Genotype: The genetic makeup of the specific genes for a trait.

Dominant: If one allele is expressed, masking the expression of the other allele, it is labeled as dominant. It is often represented with an upper case letter.

Recessive: If expression of one allele is masked by another allele, it is labeled as recessive. It is often represented by a lower case letter.

Homozygous: Both copies of a gene are the same, and can be homozygous dominant or homozygous recessive.

Heterozygous: The two alleles for a specific gene are different.

True-breeding: The alleles are homozygous.

P generation: "P" stands for parental, and is the starting point from which a cross is made. The individuals being crossed are usually true-breeding for the characteristics being examined.

F_1 generation: "F" comes from filial. These are the offspring produced by crossing the two parents of the P generation.

F_2 generation: These are the progeny that result from crossing two members of the F_1 generation.

One additional item that needs to be discussed is the Punnett Square. A Punnett Square is a visual representation that shows all of the possible genetic combinations resulting from a test cross. This information can be used to examine the phenotypic and genotypic outcome probabilities of the offspring of a cross involving a single trait or multiple traits.

SETTING UP A PUNNETT SQUARE

Monohybrid Cross

Organisms inherit 2 alleles for each trait – one from each parent. Suppose that round seed trait (**R**) in pea plants is dominant to wrinkled seeds (**r**). Therefore, we know that the genotype of heterozygous pea plants is **Rr**. We also know that the presence of the dominant allele (**R**) in homozygous (**RR**) or heterozygous (**Rr**) plants results in round seeds. Only homozygous recessive (**rr**) plants have the wrinkled seed trait. If we wanted to cross two of these heterozygous pea plants, we would set up a *monohybrid cross*, a mating between two individuals with different alleles at a genetic locus of interest. To determine the fraction of the offspring that would have round seeds and wrinkled seeds, we would set up a Punnett square to examine the following cross: **Rr** × **Rr**.

To set up a Punnett Square for this cross…

Since each parent has two allele combinations, the Punnett square needs to be a 2 × 2 Punnett square:

Determine the allele combinations for the seed type characteristic. Since each gamete has one copy of the characteristic, there are two allele combinations. For this example, the two allele combinations are: **R** and **r**.

Write the allele combinations for parent 1 on the left side of the Punnett square, and the allele combinations for parent 2 above the Punnett square. To keep track of the alleles, we will use the colors red for parent 1 and green for parent 2. By convention, letters from the left side are written first, but upper case always comes before lower case.

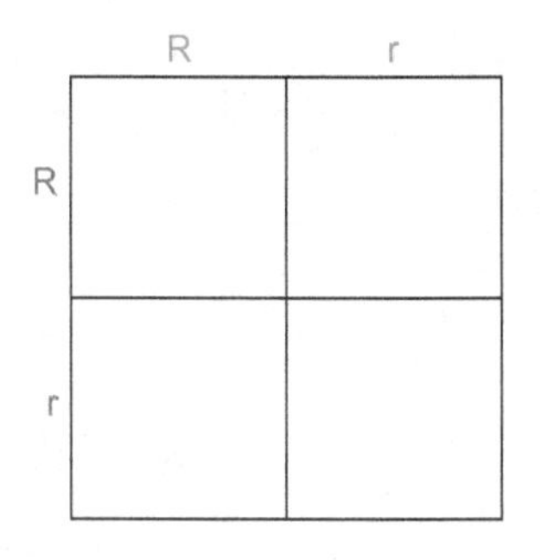

The results of the cross are:

	R	r
R	RR	Rr
r	Rr	rr

In evaluating the phenotype, or the outward appearance of the seeds in this cross, since **R** is dominant, as long as there is at least one **R**, the seeds will have the dominant trait, which is round. In this case 75% of the seeds are expected to be round and 25% are expected to be wrinkled.

The ratio of the phenotype is 3 round: 1 wrinkled.

The genotype looks at the actual genes. In this case 25% of the offspring are expected to be homozygous dominant, **RR**, 50% are expected to be heterozygous, **Rr**, and 25% are expected to be homozygous recessive, **rr**.

The ratio of the genotypes is 1:2:1

Dihybrid Cross

If we wanted to cross 2 pea plants heterozygous for 2 traits (such as seed type and flower color), we would perform a *dihybrid cross*. Assuming that **R** is a dominant trait for round seeds while **r** is a recessive trait for wrinkled seeds, and **Y** is a dominant trait for yellow flower color while **y** is a recessive trait for white flower color. we would set up a Punnett square to examine the following cross: **RrYy** × **RrYy**

To set up a Punnett Square for this cross…

Since each gamete has one copy of each characteristic, there are four allele combinations when looking at two characteristics. For this example, the four allele combinations are:

Parental Genotype:	Allele combinations:
Rr Yy	RY Ry rY ry

Until you become comfortable determining the gametes, write them out as shown above, then place them on the sides of the Punnett square.

Since each parent has four allele combinations of the two characteristics, the Punnett square needs to be 4 × 4. Write the alleles for parent 1 on the left side of the Punnett square, and the alleles for parent 2 above the Punnett square. To keep track of the alleles, we will use the colors red for parent 1 and green for parent 2.

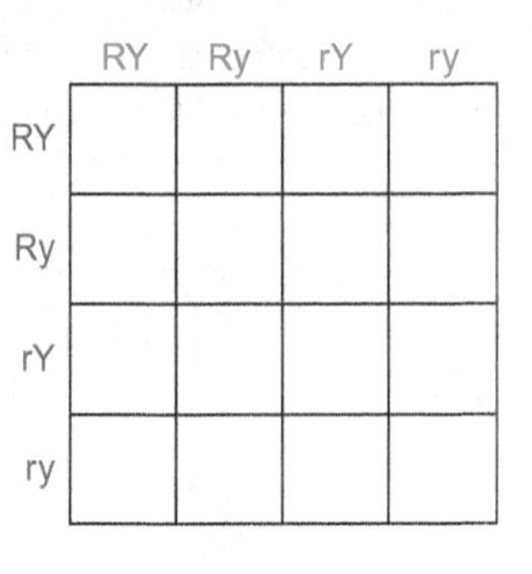

The results of the cross are:

	RY	Ry	rY	ry
RY	RRYY	RRYy	RrYY	RrYy
Ry	RRYy	RRyy	RrYy	Rryy
rY	RrYY	RrYy	rrYY	rrYy
ry	RrYy	Rryy	rrYy	rryy

The expected phenotypes are:

R_Y_: Round seed, Yellow flower: 9

R_yy: Round seed, White flower: 3

rrY_: Wrinkled seed, Yellow flower: 3

rryy: Wrinkled seed, White flower: 1

The expected genotypes are:

RRYY: 1/16

RRYy: 2/16

RRyy: 1/16

RrYY: 2/16

RrYy: 4/16

Rryy: 2/16

rrYY: 1/16

rrYy: 2/16

rryy: 1/16

Using these expected genotypes, determine the phenotypic appearance for each of the genotypes above.

There are four types of inheritance situations:

- Complete dominance
- Incomplete dominance
- Codominance
- X linked recessive

COMPLETE DOMINANCE

The examples discussed above (with the pea seed shape and flower color) are examples of complete dominance. Complete dominance can be defined as a kind of dominance in which the dominant allele completely masks the effect of the recessive allele in heterozygous condition. Specifically, if an individual has two dominant alleles (for example, round pea shape (**RR**)), the trait will be expressed. However if the individual has one dominant allele (**R**) and one recessive allele (**r**), the dominant allele will be expressed while the recessive allele will appear to be suppressed. As a result, the heterozygote (**Rr**) will have the same phenotype as that of the homozygous dominant (**RR**). Only if both alleles are recessive (**rr**) will the recessive trait be expressed.

INCOMPLETE DOMINANCE

Incomplete dominance (also referred to as partial dominance) is a type of dominance in which the heterozygous phenotype is distinct from and often an intermediate to the homozygous phenotypes.

An example of incomplete dominance is carnation flower color. Carnations have only two alleles for color - a dominant allele (**R**) for red pigment and a recessive allele (**r**) for no pigment. However, carnations have 3 flower colors - red, pink, and white. In a heterozygous plant carrying both alleles (**Rr**), the dominant allele is only partially expressed and as a result, the dominant allele (**R**) is not be able to produce enough red pigment to produce a red flower. A result, the flowers will be pink. For example, when a red carnation (homozygous dominant – **RR**) is crossed with a white carnation (homozygous recessive – **rr**), all of the offspring will be pink carnations (heterozygotes – **Rr**).

CODOMINANCE

Codominance is a type of inheritance in which both alleles are fully expressed in the heterozygote. An example of codominance is the ABO blood group system. If an individual has both the A allele and the B allele, they will have blood type AB because blood type A and blood type B are both expressed morphologically, neither allele is suppressed.

X LINKED RECESSIVE

X linked recessive is a type of inheritance in which an abnormal recessive gene on the X chromosome is the cause of a disease such as hemophilia or color blindness.

X linked recessive diseases usually occur in males although they can occur in females. Males have one X chromosome and one Y chromosome, however the Y chromosome does not contain most of the genes of the X chromosome. As a result, an abnormal recessive gene on the portion of the X chromosome for which there is no corresponding Y chromosome will cause the disease.

Females can have an X linked recessive disorder but it is a very rare occurrence. For this to occur, the offspring would have to inherit an abnormal gene on the X chromosome from each parent. This could occur if the mother is a carrier and the father has the disease, or if both the mother and the father have the disease.

ANALYZING PEDIGREES

In many situations, such as those involving humans, controlled crosses cannot be performed to provide data for analysis. In these situations, pedigree analysis can be used to study the inheritance of genes. Pedigree analysis is also useful when studying populations in which population data is limited or the populations have a long generation time.

A pedigree uses a series of symbols to represent different features of a population. See the handout for some of the main symbols.

When the appropriate phenotypic data is collected, preferably from several generations, a pedigree can be drawn and analyzed.

If a trait is dominant:

- Individuals with the trait will have at least one parent with the trait
- The phenotype appears in each generation
- If both parents are unaffected, their offspring will be unaffected

If the trait is recessive:

- Parents who do not express the trait can have affected offspring
- Affected offspring can be male or female

EXERCISE 1 Analyzing Genetics Problems

Genetics is the study of genes and their role in passing traits from one generation to the next. These principles of genetics are based on Gregor Mendel's work on the transmission of traits with pea plants. The field of genetics studies the role of genes, heredity, and variation within populations of living organisms.

Purpose

To apply your knowledge of genetics to correctly analyze different types of genetics problems.

Materials

- Maize

Methods

- Apply your knowledge of genetics to analyze questions 1 → 6.
- Obtain one of the maize cobs.
- Count the number of different color kernels and record your counts in the **Maize Monohybrid Cross** table (question 7a).
- Based on your results to question 7a, apply your knowledge of genetics to analyze the questions 7b, 7c, and 7d.
- Apply your knowledge of pedigrees to analyze pedigrees. (Exercises will be distributed in lab.)

Name:

Exercise 1 Results

1. The ability to taste the chemical PTC results from the presence of the dominant allele T and the inability to taste this chemical results from the presence of the recessive allele t. If two heterozygous tasters have a large family:

   (a) What is the proportion of their children who will be tasters and nontasters.

   (b) What is the probability that their first child will be a nontaster?

   (c) What is the probability that their fourth child will be a nontaster?

   (d) What is the probability that the first three children will be tasters?

2. A male rabbit with medium length fur is mated to a female rabbit with medium length fur. Among their offspring are 30 rabbits with medium length fur, 16 with long fur, and 12 with short fur.

   (a) What is the simplest explanation for the inheritance pattern of fur length in rabbits?

   (b) What offspring would you expect from the mating of a male rabbit with medium length fur and a female rabbit with long fur?

3. Blue (B) flower color is dominant to white (b). If you have a blue flowered plant and a white flowered plant:

   (a) What is the genotype and phenotype of the blue flowered plant and the white flowered plant?

   (b) If all the offspring of a cross between these plants are blue, what does this tell you about the genotypes of the original plants?

   (c) What gametes will be produced by the offspring of a cross between two of the blue flowered plants from 3(b)?

(d) What phenotypes will be produced by a cross between two of the blue flowered plants from 3(b)?

(e) What genotypic proportions will be produced by a cross between two of the blue flowered plants from 3(b)?

4. Hemophilia is an X linked trait. X^H is a dominant allele that causes normal blood clotting while X^h is a recessive allele that causes hemophilia.

(a) What are the genotypes of:

(1) A woman with normal blood clotting whose father had hemophilia

(2) A normal man whose father had hemophilia.

(b) What is the probability that a child produced by these two individuals will have hemophilia?

(c) What is the probability that a daughter produced by these two individuals will be a carrier of the hemophilia trait?

(d) What is the probability that a daughter produced by these two individuals will have hemophilia?

(e) What is the probability that a son produced by these two individuals will have hemophilia?

5. A human female with a blood type of A has a child with a blood type of B. Indicate whether each statement is **TRUE** or **FALSE**.

(a) The genotype of the father is AB.

(b) The genotype of the mother is AA.

(c) The father of the child has type O blood.

(d) The father of the child cannot have type BB blood.

(e) The father of the child can have type A blood.

6. In a certain plant, tall plants (T) are dominant to dwarf plants (t) and purple flowers (P) are dominant to white flowers (p). Cross a plant heterozygous for both traits with a plant homozygous recessive for both traits.

   (a) What gametes are produced for the **heterozygous** plant?

   (b) Set up a Punnett Square for this cross.

   (c) What is the probability of producing dwarf plants with purple flowers?

   (d) What is the probability of producing plants with white flowers?

7. Assume that purple color in corn kernels is dominant to yellow.

   (a) Count the number of different color kernels in your maize cob and complete the table below.

Maize Monohybrid Cross	
F_2 Phenotypes	**Actual**
Total	

(b) Using the actual counts from 7 (a) above, what are the most likely genotypes of the F_1 cross?

(c) What are the most likely phenotypes of the F_1 cross?

(d) What are the most likely genotypes of the parental cross?

CHAPTER 14

Extracting DNA from Cheek Cells

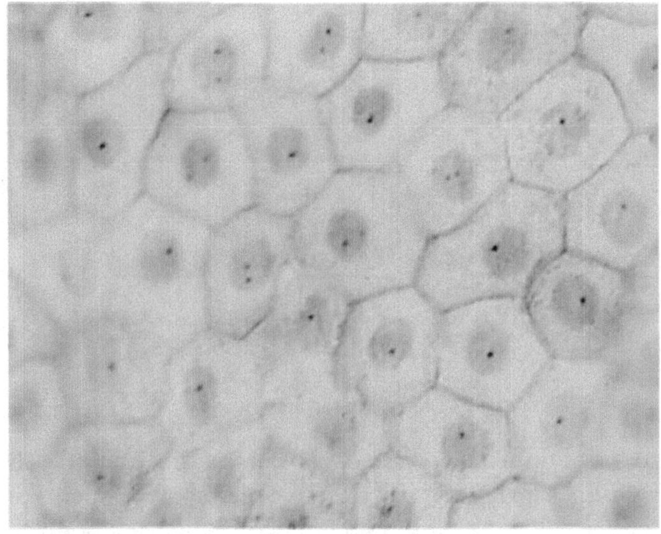

Photo courtesy of Holly J. Morris and David T. Moat

OBJECTIVES

After completing this exercise, you will be able to:

- Understand the process to extract deoxyribonucleic acid (DNA) from cells.
- Describe the steps to extract DNA from cells.
- Observe the physical appearance of DNA.

INTRODUCTION

All cells contain a set of instructions that determines their characteristics. Cells use this genetic information to synthesize the molecules that direct their development, functioning, and reproduction.

The storage medium for this genetic information is DNA. DNA consists of two strands that run in opposite directions and are coiled around each other forming a double helix which is held together by hydrogen bonds. The DNA strands are polymers composed of monomers called nucleotides. Each nucleotide is composed of a monosaccharide sugar (deoxyribose), a phosphate group (which give DNA its slightly negative charge), and a nitrogen base (cytosine [C], guanine [G], adenine [A], or thymine [T]). According to base pairing rules, adenine of one strand binds with thymine of the other and similarly, cytosine of one strand binds with guanine of the other. The nucleotides in a strand are joined together by covalent bonds between the sugar of one nucleotide and the phosphate of the next nucleotide creating a sugar phosphate backbone. It is the nucleotide sequence of the DNA that encodes the cell's genetic information.

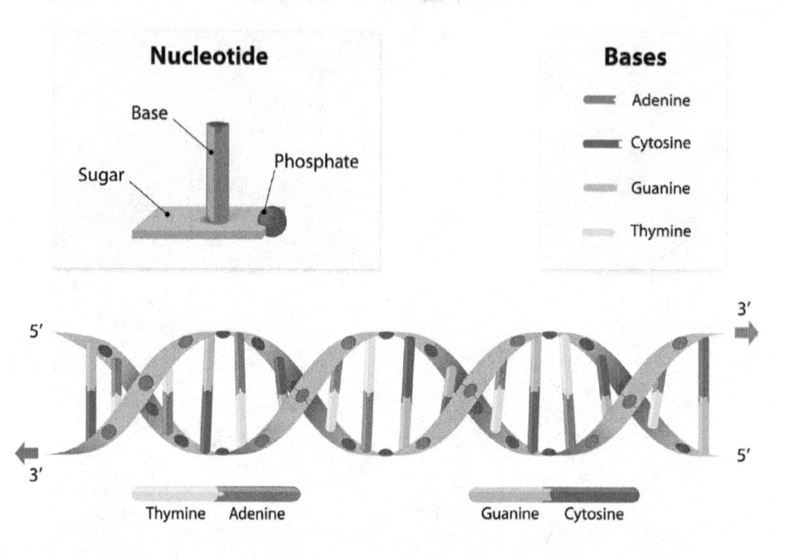

© Designua/Shutterstock.com

DNA is packaged into macromolecules called chromatin that includes both the DNA and its associated proteins. Chromatin is further organized into genes, short segments of DNA that code for the production of specific proteins. Prior to mitosis, chromatin replicates, doubling the amount of DNA, then early in mitosis, chromatin condenses into chromosomes, the familiar structures that can be seen under a microscope. During mitosis a complete set of chromosomes is passed onto each daughter cell.

In eukaryotic cells, the DNA is stored in the cell nucleus while in prokaryotic cells, the DNA is found in the cytoplasm. Organelles, such as mitochondria and chloroplasts, also contain their own DNA.

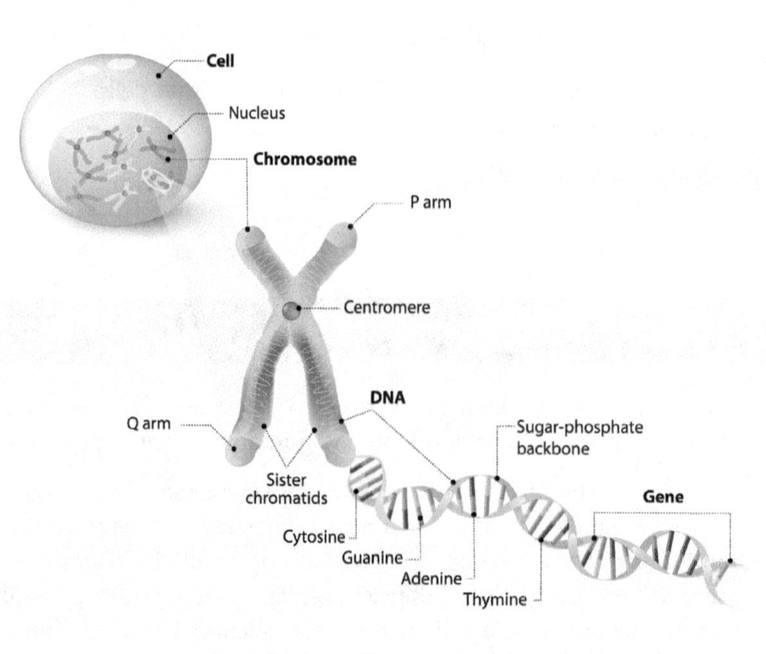

© Designua/Shutterstock.com

EXERCISE 1 Extracting DNA from Cheek Cells

To extract DNA, it must first be removed from various structures and molecules. The process we will utilize performs the following functions:

- Release the DNA from the cell and the cell nucleus.
- Separate the DNA from the histones.
- Inactivate DNAses which are enzymes that cut the long strands of DNA apart thereby making aggregation less likely to occur.
- Make the DNA electrically neutral so that the DNA strands will be able to aggregate.

Purpose

To extract DNA from human cheek cells.

Materials

- Cheek cells
- 15 mL centrifuge test tube
- Paper cups
- Bottled water
- Lysis buffer
- Protease
- Ice cold ethanol
- Disposable pipettes
- 55°C water bath

Methods

- Put 3 mL of bottled water in a paper cup and rinse your mouth for 1 minute. Gently chew the insides of your cheeks to help extract your cheek cells.
- After 1 minute, spit the water back into your paper cup and then pour it into a 15 mL centrifuge test tube.
- Add 2 mL of lysis buffer to your extract solution in the test tube. Tighten the cap and slowly invert the test tube five times.
- Add 3 drops of protease to your extract solution in the test tube and slowly invert the test tube five times.
- Put your extract solution in the test tube in a 55°C water bath for 10 minutes.
- Remove your test tube from the water bath. Hold the test tube at a 45° angle and slowly add 10 mL of ice cold ethanol to your test tube. Do not shake or mix the alcohol with the test tube's contents. The ethanol will form a layer on top of the extract solution. Since DNA is insoluble in ethanol, many individual DNA strands will aggregate and precipitate together forming whitish clumps.
- Observe the whitish material (your DNA) which appears in your test tube.

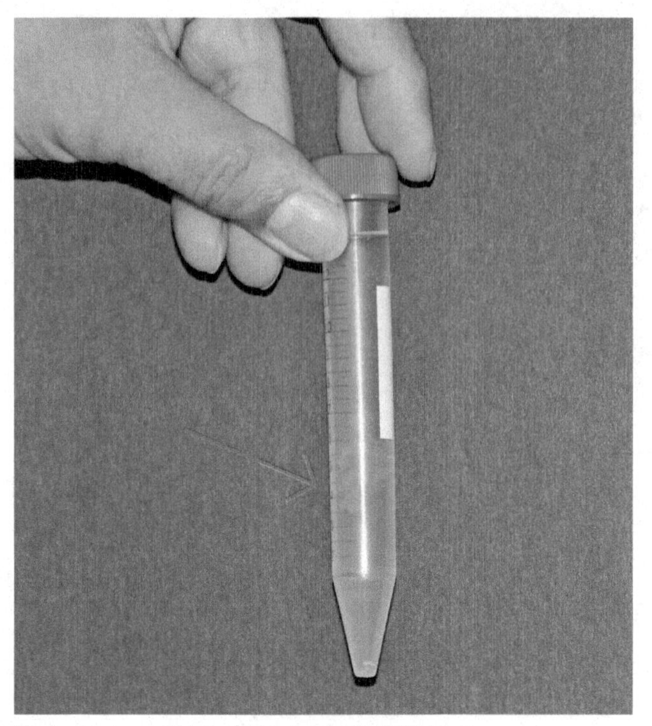

Photo courtesy of Holly J. Morris and David T. Moat

EXERCISE 2 Viewing DNA

DNA is a long stringy molecule. The whitish DNA that you see in your test tube is not a single strand of DNA but is a clump of many individual DNA strands that are tangled together. As DNA begins to precipitate, many DNA strands are pulled together forming the tangled whitish clumps.

Purpose

To observe DNA extracted from your cheek cells.

Materials

- Extracted DNA
- Methylene blue
- Slides and cover slips
- Disposable pipettes
- Compound light microscope

Methods

- Using a pipette, remove a small amount of your DNA from your test tube and place it on a slide with a drop of methylene blue and a cover slip.
- Observe your DNA using a compound light microscope at 400× total magnification.
- Dispose of your slides by placing them in the designated waste container.

Name:

Exercise 1 Results

1. What cell structures must be broken to release the DNA?

2. What was the purpose of adding the lysis buffer during the extraction process?

3. Why is salt added during the extraction process?

4. What was the purpose of putting the extract solution in the water bath?

5. What was the purpose of adding the protease during the extraction process?

6. What was the purpose of adding the ethanol during the DNA extraction process?

Exercise 2 Results

1. Describe the appearance of your DNA.

2. Make a **detailed drawing** of your DNA.

Extracted DNA
Total magnification:

CHAPTER 15

Catechol Oxidase: An Independent Investigation

Photo courtesy of Holly J. Morris

OBJECTIVES

After completing this exercise, you will be able to:

- Apply the scientific method using procedures and techniques performed in previous labs to determine the type of inhibition exhibited by catechol oxidase.

INTRODUCTION

Enzymes are biological catalysts that speed up the rate of specific biochemical reactions in living cells without being used up in the reaction. They accomplish this by lowering the energy of activation needed for the reaction, therefore making the reaction more likely to occur. Because enzymes are reusable, a small number of enzymes can catalyze an enormous number of reactions.

The active site of the enzyme is shaped in such a way that allows the substrate to bind with the active site, temporarily forming an enzyme-substrate complex. The enzyme strains the substrate, forcing a change to occur, either forcing two smaller molecules together, or breaking a larger molecule into two smaller molecules. When the reaction is complete, the complex dissociates into the enzyme and the product or products.

Catechol oxidase (or catecholase) is an enzyme that is common in plants. In undamaged cells, catechol oxidase is stored in vesicles and does not interact with catechol. However, in damaged cells, catechol oxidase reacts with catechol and oxygen to produce benzoquinone and water. Benzoquinone, the main product of this reaction, produces a brown color that indicates that this reaction (called enzymatic browning) has occurred.

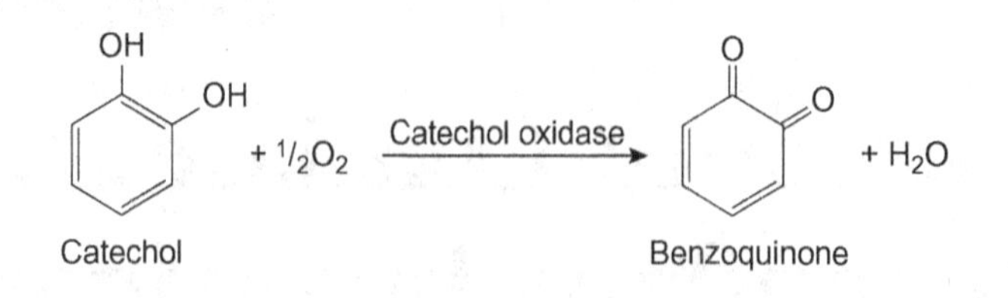

The action of enzymes can be inhibited by specific chemicals called inhibitors. An inhibitor molecule can affect an enzyme in two ways:

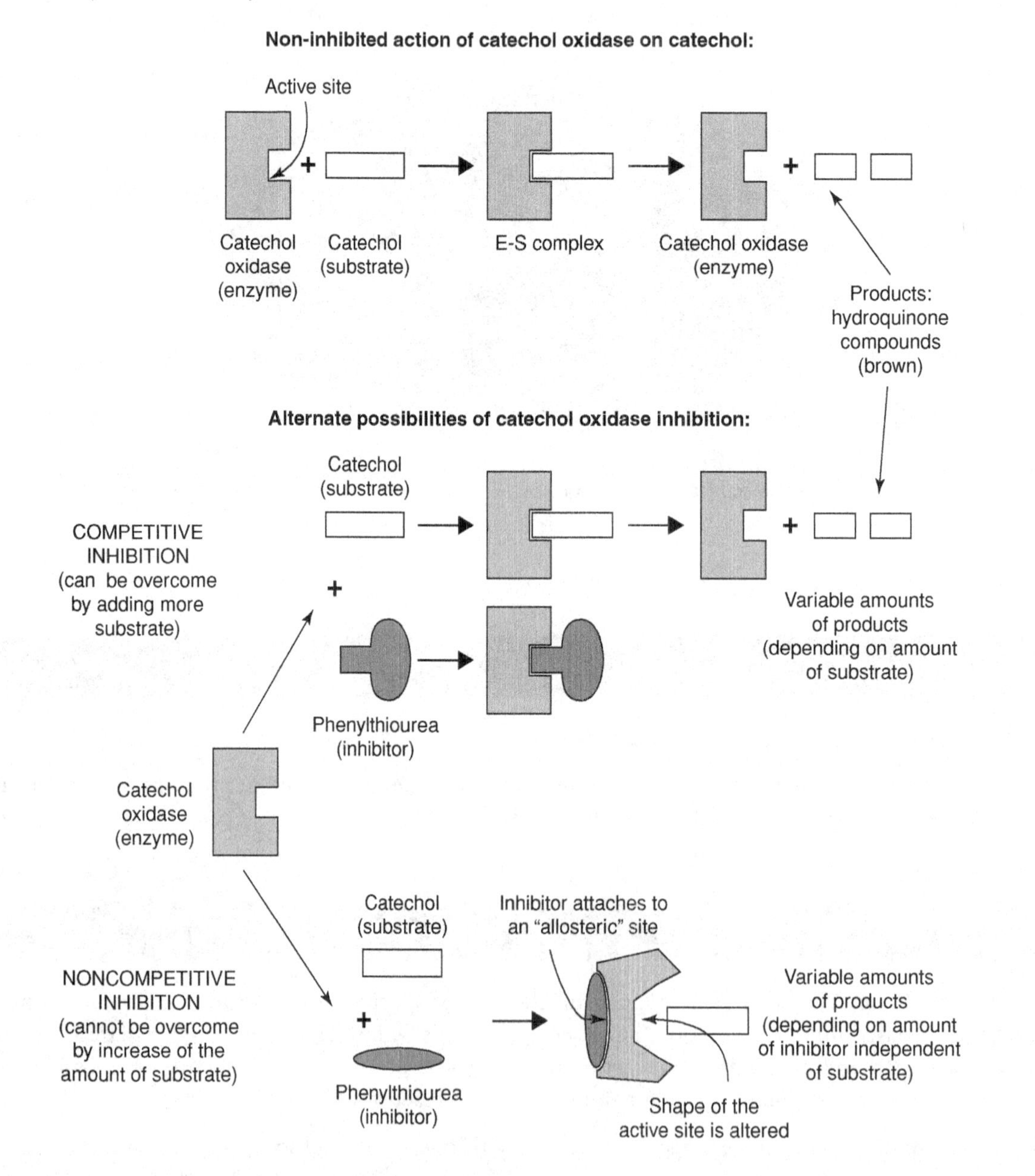

- **Competitive inhibition**: A competitive inhibitor is a molecule that has a similar shape to the actual substrate and competes for the active site. The competing molecule binds to the active site, preventing the substrate from binding and, therefore no reaction takes place. Competitive inhibition is reversible. If the concentration of actual substrate is significantly increased and if the concentration of the inhibitor molecule is held constant, competitive inhibition can be reversed.

- **Noncompetitive inhibition:** This type of inhibition occurs when the inhibitor molecule binds to the enzyme at a location other than the active site. This binding changes the overall shape of the enzyme, thereby changing the shape of the active site. The result is that the active site is no longer conducive to catalyzing the reaction. Noncompetitive inhibition can be reversed, but because the substrate is not competing with the inhibitor for a place on the active site, simply increasing the concentration of substrate will not reverse noncompetitive inhibition.

Catechol oxidase requires copper as a cofactor, a nonprotein component of an enzyme. Phenylthiourea (PTU) is a chemical that combines with the copper in catechol oxidase and inhibits its enzymatic activity.

EXERCISE 1 Determining the Presence of Catechol Oxidase in Potatoes

Catechol oxidase is a common plant enzyme. The overall scope of this independent investigation is to determine whether catechol oxidase is competitively or noncompetitively inhibited. As such, the first step is to ensure that catechol oxidase is present in potatoes, the plant being used in this investigation.

Purpose

To utilize the scientific method to design and perform an experiment to determine whether potatoes contain catechol oxidase.

Materials

- 1% solution potato extract solution
- 1% catechol solution
- 1% phenylthiourea (PTU) solution
- Spectrophotometer (to quantify the results)

Methods

- Members of the lab are separated into groups of three or four.
- Each group independently develops a hypothesis that they will test and then develops the experimental procedures they will use to test their hypothesis.
- Each group performs their investigation to determine whether potatoes contain catechol oxidase.
- Each group uses a spectrophotometer to quantify the results of their investigation.
- Each group member records the results in the table below.

EXERCISE 2 Determining How Catechol Oxidase Is Inhibited

The action of enzymes can be inhibited by specific chemicals called inhibitors. An inhibitor molecule can affect an enzyme in two ways:

- Competitive inhibition
- Noncompetitive inhibition

PTU combines with the copper in catechol oxidase to inhibit its enzymatic activity. You will investigate how PTU inhibits the activity of catechol oxidase.

Purpose

To utilize the scientific method to design and perform an experiment to determine how catechol oxidase is inhibited.

Materials

- 1% solution potato extract solution
- 1% catechol solution
- 1% phenylthiourea (PTU) solution
- Spectrophotometer (to quantify the results)

Methods

- Members of the lab are separated into groups of three or four.
- Each group independently develops a hypothesis that they will test and then develops the experimental procedures they will use to test their hypothesis.
- Each group performs their investigation to determine whether how catechol oxidase is inhibited.
- Each group uses a spectrophotometer to quantify the results of their investigation.
- Each group member records the results in the table below.

Name:

Exercise 1 Results

1. For the experiment to determine the presence of catechol oxidase in potatoes, list:
   - Hypothesis:
   - Independent variable:
   - Dependent variable:
   - Control Treatment:

2. List in numerical order each step in your procedure:

3. Record the results of your experiment.

4. Prepare a graph of your results.

Exercise 2 Results

1. For the experiment to determine how catechol oxidase is inhibited, list:

- Hypothesis:

  ___

- Independent variable:

  ___

- Dependent variable:

  ___

- Control Treatment:

  ___

- Prediction (predict the results of your experiment based on your hypothesis):

  ___

2. List in numerical order each step in your procedure:

___

___

___

___

___

___

3. Record the results of your experiment.

4. Prepare a graph of your results.

CHAPTER 16

Evolution

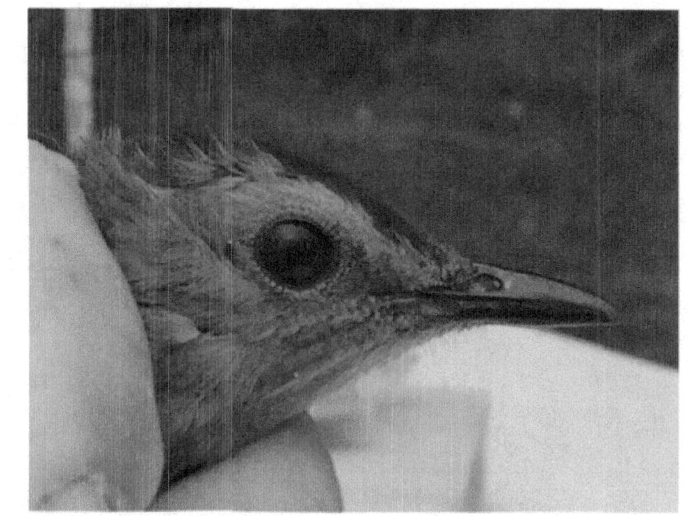

Photo courtesy of Holly J. Morris

OBJECTIVES

After completing these exercises, you will be able to:

- Define natural selection and evolution.
- Identify the conditions under which natural selection can occur.
- Explain how adaptation results from natural selection.
- Explain evolution in terms of alleles within a population.

INTRODUCTION

Evolution occurs through natural selection and is the process whereby populations adapt to changes in their environment. Evolution is "descent with modification," which means that organisms descended from ancestral species that were different than present-day organisms.

One important component of evolution is that individuals do not evolve. Only populations evolve. If the environment is stable, the population within that environment will most likely remain stable. However, if a change occurs in an environment, those organisms within the population that are naturally better adapted to that change are more likely to survive and reproduce, thus passing those advantageous genes onto their offspring. In order for evolution to occur, the genes (alleles) within a population must be diverse.

Natural selection can occur in multiple ways:

- Directional selection
- Stabilizing selection
- Disruptive selection

Directional selection favors one end of the phenotypic distribution. As a result, over time, the favored extreme will become more common but the intermediate and the other extreme will become less common or will be lost. For example, a population of mice migrates to a wooded area with dark soil and decomposing logs. Those members of the population that are better camouflaged are less likely to be prey, passing their genes onto offspring, and the appearance of the population will change over time.

Drawings courtesy of Holly J. Morris

Picture, instead, a population in a habitat with patchy colored rocks. Mice with an intermediate color will have an advantage. This is stabilizing selection. Stabilizing selection favors the middle of a phenotypic distribution. As a result, over time, the intermediate characteristic becomes more common and the extremes will become less common or will be lost.

Drawings courtesy of Holly J. Morris

A population of mice living in an environment where light colored sand abuts darker rocks may experience disruptive selection. Mice living in the sand will have an advantage if they are light, like the sand, while mice living in the darker rocks will have an advantage if they are darker. Disruptive selection favors both of the extremes of a phenotypic distribution. Over time, the two extremes will become more common and the intermediates will be become less common or will be lost.

Drawings courtesy of Holly J. Morris

Life has existed for about 3.5 billion years and has changed over time. The changes that have occurred are documented in many ways. As you learned in Chapter 13, organisms inherit two alleles for each trait – one from each parent, though there can be several alleles for any given characteristic within a population. The frequency of specific alleles can change over time, depending on factors, or stresses, within the environment.

The exercises that we will do examine the lines of evidence that support evolution. These lines of evidence include:

- Paleontological evidence.
- Comparative anatomical evidence.
- Embryological evidence.
- Comparative biochemical and genetic evidence.

EXERCISE 1 Evidence of Adaptation

There are many different types of birds in the world. Birds can be identified by their characteristics such as where they live, how they look, and how they behave. Birds can also be classified by their beak types since beak type is related to the bird's diet and their ecological role. The diets of birds vary depending on their environment. Bird beaks have adapted over the years to help them survive in their environment. Some birds have beaks that are adapted for a specific environment. Other birds have beaks that are not adapted for a specific environment but enable them to live in many types of environments. Some examples of bird beaks are shown below.

© Tomacco/Shutterstock.com

Here are some common beak shapes and the food that they are especially adapted to eat.

Type	Examples	Adaptations
Cracker	Seed eaters such as sparrows and cardinals	Short thick conical beaks for cracking seeds
Shredder	Birds of prey such as owls and hawks	Sharp curved beaks for tearing meat
Chisel	Woodpeckers	Long chisel-like bills for boring into wood to find insects
Probe	Hummingbirds	Long slender bills for probing flowers to find nectar
Strainer	Some types of ducks	Long flat bills to strain small plants and animals from the water
Spear	Herons and kingfishers	Spear like bills for "fishing"
Tweezers	Insect eaters like warblers	Thin pointer bills for finding insects
Swiss Army Knife	Crows	Multi-purpose bill for eating fruit, seeds, insects, fish, and other animals.

Purpose

To illustrate the observation that the shape of a bird's beak is evidence of natural selection.

Materials

- Toothpicks (simulating insets)
- Paper clips (simulating insects)
- Rice (simulating ants)
- Unpopped popcorn (simulating seeds)
- Rubber bands (simulating worms)
- Dissecting needles
- Tweezers
- Scissors
- Straws
- Plastic cups

Methods

- Work in groups of five or six.
- Place the five different "foods" (toothpicks, paper clips, rice, unpopped popcorn, and rubber bands) on your desk and mix them together.
- Have each bird type pick up their "beak" and a plastic cup.
- For 30 seconds, each bird type picks up as many of all "foods" as possible and places them in their plastic cup. Note that one bird type cannot help another bird type collect "foods."
- Each bird type counts the number of each type of "food" captured and records results in the **Foods Captured by Each Bird Type** table.

EXERCISE 2 Paleontological Evidence of Evolution

Fossils record the changes in organisms that have occurred over time. The fossil record provides another evidentiary source for evolution. A fossil can be a complete organism or simply a piece of an organism such as a bone, tooth, shell, or leaf. Fossils also provide information about the environment in the past. This is important because environmental factors are considered to be major evolutionary drivers.

Fossils can show how an organism lived. For example, the fact that the fossil record at a single site contained the remains of more than 10,000 duck-billed dinosaurs led scientists to hypothesize that these dinosaurs lived in large herds.

Amber is hardened fossilized tree resin that can drip from a tree trunk and as it drips, it can trap many different kinds of organisms such as small insects. The fossils of ants, trapped in amber while they were eating leaves, provides information enabling scientists to know exactly what they ate and how they ate it.

The most common way in which fossils are formed occurs when an organism dies and after its soft parts are eaten and broken down by scavengers and decomposers, the remains are buried rapidly. If this occurs in a river or other moving water body, sediment carried by the water is deposited on top of the remains preserving them as a fossil.

Determining the age of fossils is important evidence for evolution. Fossils themselves are usually not able to be dated. Rather, fossils are dated by the age of the rock in which they are found. There are two methods of dating rocks:

- Law of superposition
- Radioactive dating

The law of superposition says that in undisturbed rock layers, the youngest layer is on top and the oldest layer is on the bottom. Each layer is younger than the layer beneath it and older than the layer above it. As such, the relative age of a fossil can be determined. For example, the lower, or older, rock layers contain fossils of primitive forms of life while the fossils found in the younger rock layers above contain progressively more complex life forms.

It is also possible to determine the age of fossils by dating the rocks in which the fossils are found. Dating the rocks is done by measuring the radioactive traces of elements, such as carbon (C) and potassium (K) present in the rock. Carbon and potassium have radioactive isotopes (C^{14} and K^{40}), respectively which are unstable and decay at a known rate (their half-life). As such, dating rocks also provides a time scale of evolution.

Many times fossils are unable to be classified into any one group because the fossils are intermediate between groups. These transitional fossils provide an important link between groups and suggest that one group gave rise to another group by evolutionary mechanisms.

An example of transitional fossils is *Archaeopteryx,* the earliest known bird. Fossil skeletons of date back 150 million years and show a mixture of reptilian and bird features such as the impressions of feathers attached to the forelimb suggesting that birds evolved from a group of reptiles.

Purpose

To illustrate how the fossil record provides support for evidence.

Materials

- Fossils (actual or models)

Methods

- Select one of the fossils and answer the questions relating to your fossil.

EXERCISE 3 Comparative Anatomical Evidence of Evolution

Comparative anatomy, comparing the structural features of different groups of organisms, is another method providing evidence for evolution. Structural comparisons are useful when trying to find a common ancestor because organisms having a common ancestor are likely to have certain basic structural features in common. The more similar the structural features of different groups of organisms are, the more closely related the groups are. For example, although the wings of a bee and the wings of a bird have the same function, they differ greatly in structure, indicating that the two groups are evolutionarily distant.

Structures that have developed from the same part are called homologous structures. For example, insect mouthparts vary in function but are based on similar basic components. Early insects, such as the giant dragonfly, had the following basic mouthparts:

- Labrum and labium (upper and lower lips) to hold food for chewing
- Mandibles (jaws) for chewing
- Maxillae (secondary jaws) for cutting.

Over time, these mouthparts have been modified in different insects for specific functions such as siphoning, piercing, and sucking. For example, butterflies have a long coiled proboscis which it uncoils to suck the nectar located deep in flowers. The development of this proboscis resulted in the reduction, but not elimination, of most of these mouthparts.

© suns07butterfly/Shutterstock.com

Another example of the use of comparative anatomy as a method providing evidence of evolution is the vertebrate 5 digit limb. Most vertebrates have a 5 digit limb that has been modified for specific functional requirements. For example, the forelimbs of a human, a bird and a lion, have basic similarities in bone arrangement. The forelimbs have homologous structures because each forelimb has a humerus, a radius, an ulna, carpals (wrist bones) and a set of phalanges (digits).

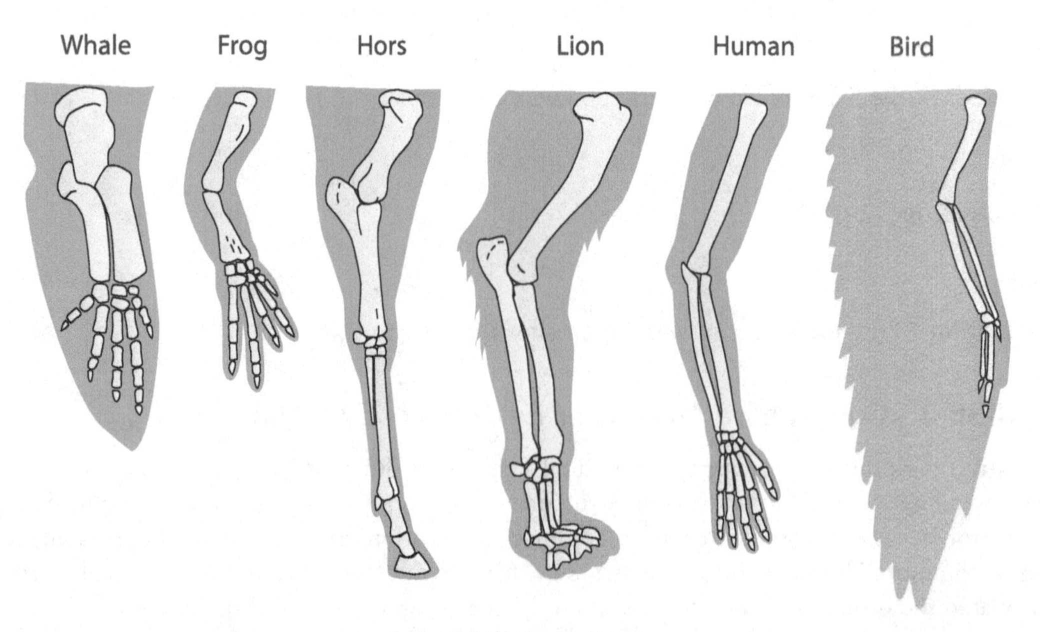

© Usagi-P/Shutterstock.com

Purpose

To illustrate how comparative anatomy supports evolution.

Materials

- Models of various vertebrate forearms.

Methods

- Examine the forelimbs of each of the displayed vertebrates.
- Make a detailed drawing of each of the forelimbs in the **Vertebrate Forelimb Adaptations** table below.
- Describe the functions of each of the forelimbs in the **Vertebrate Forelimb Adaptations** table.

EXERCISE 4 Embryological Evidence of Evolution

Embryology is a branch of biology that studies the development of gametes, zygotes, and embryos. As late as the 18th century, embryonic organisms were thought to be preformed miniature versions of the adult organism. The picture (below) shows an artist's rendition of a miniature human preformed within a sperm cell.

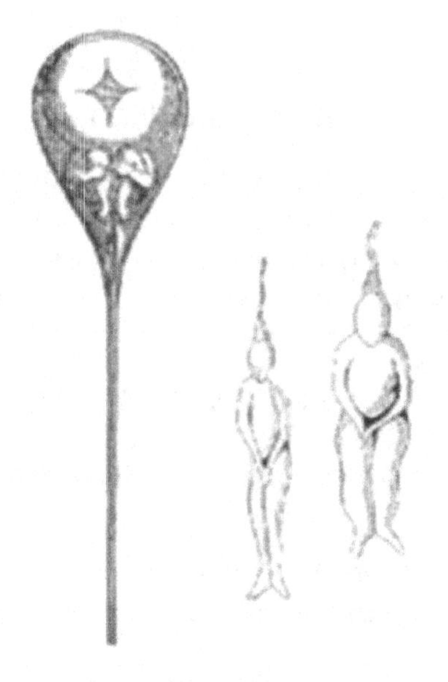

Public Domain

Embryological development is another discipline providing evidence for evolution. Charles Darwin recognized that evolutionary changes could be inferred by observing changes in embryo development. He suggested that primitive features in organisms tend to be generalized while derived features tend to be specialized.

In actuality, in the early embryonic development of many different types of organisms, the characteristics are the same while more specialized characteristics, which distinguish the groups, appear later in embryonic development. For example, a human embryo in the early stage of development has the same characteristics of all vertebrates. The human embryo then develops mammalian characteristics followed by characteristics relating to the primate order, the human species and then finally the traits of the individual.

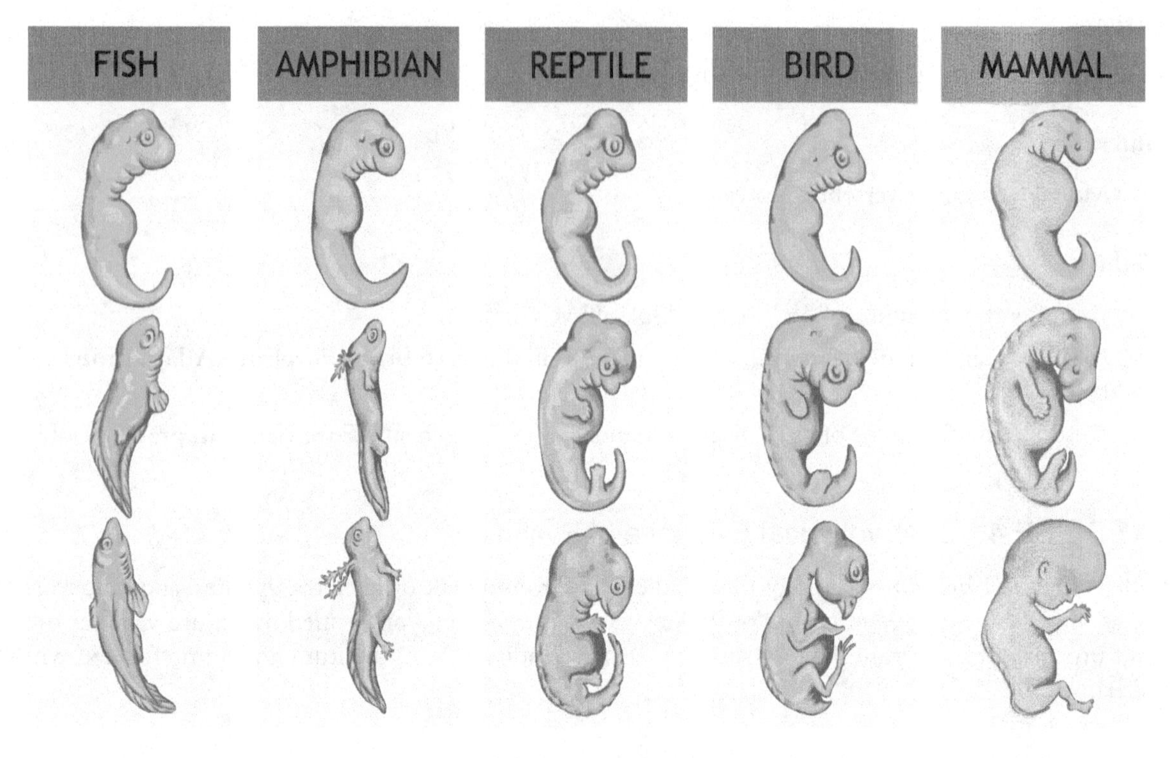

© stihii/Shutterstock.com

EXERCISE 5 Comparative Biochemistry and Genetic Evidence of Evolution

Comparative biochemistry and genetics also provide evidence for evolution. Biologists use structural and physiological characteristics to determine levels of similarity and as a result, establish evolutionary links between groups of organisms. However, with the improved knowledge of DNA, this biochemical information is being used to determine evolutionary relationships between groups of organisms.

Proteins are long chains of amino acids. Every protein contains a different number of amino acids, different types of amino acids, and a different sequence of amino acids. A small difference in a protein between 2 different groups of organisms indicates a recent divergence from a common ancestor while a large difference indicates a more distant, evolutionary relationship.

The blood protein hemoglobin is made up of 146 amino acids. Comparing the differences in this protein is an indicator of the relationships between humans and other primates.

Primate	Number of differences in amino acid sequence compared to human hemoglobin
Chimpanzee	0
Gorilla	1
Gibbon	3
Orangutan	4
Rhesus monkey	8
Lemur	5

The very close genetic relationship between humans and chimpanzees is shown by the fact that the sequence of amino acids in hemoglobin is identical. The 67 differences in the sequence of amino acids between humans and frogs indicate a distant evolutionary relationship.

These biochemical comparisons lead to the concept of a "molecular clock." The "molecular clock" concept is based on the assumption that genetic mutations occur at a standard rate and as a result, the time at which two groups of organisms diverged, can be determined.

The comparison of amino acid sequences in a protein as a method to determine the evolutionary relationships between groups of organisms is being replaced by the comparison of the DNA sequences that makes up genes.

Purpose

To illustrate how comparative biochemical and genetic evidence supports evolution.

Materials

- Fictionalized DNA sequence of a region on a chromosome of *Au afarensis* and five other hominid species.

Au afarensis	*H. sapiens*	*H. erectus*	*H. habilis*	*H. neanderthalensis*	*Au robustus*
A	A	A	A	T	A
A	T	T	T	T	A
T	C	G	G	G	T
G	T	T	T	T	G
C	C	C	C	C	G
C	C	C	C	C	G
C	C	C	C	C	G
C	C	C	C	C	G
A	A	A	A	A	G
A	T	T	A	T	A
T	A	A	T	A	T
G	G	G	G	G	G
G	C	C	C	C	G
C	G	G	G	G	C
T	T	T	T	T	T
G	C	C	C	C	G
T	T	T	T	T	T
A	C	C	T	C	A
G	T	G	G	G	G
G	C	C	G	C	G

Methods

- Determine the number of differences in the DNA sequences between the ancestor (*Au afarensis*) and each of the other hominid species.
- Record your results in the **Differences between *Au afarensis* and Other Hominid Species** table.

Name:

Exercise 1 Results

1. Complete the table with your results.

Foods Captured by Each Bird Type					
Bird Type	**Paper Clips**	**Unpopped Popcorn**	**Rubber Bands**	**Toothpicks**	**Rice**
Insect Eater (Tweezer)					
Strainer (Scissors)					
Fisher (Dissecting Needle)					
Prober (Straws)					

2. Based on beak shape and size, identify which food type is best for each type of beak. Explain your answer.

___

___

___

3. Based on beak shape and size, identify which food shape is least likely to be captured by each type of beak. Explain your answer.

4. If an event occurs that destroys all food sources except toothpicks, what will happen? As a result, what kind of bird beaks will comprise the bird population in several years?

5. In nature, why is being a generalist sometimes an advantage over being a specialist?

Exercise 2 Results

1. What kind of fossil do you have?

2. Is your fossil a complete organism, a partial organism, or a piece of an organism?

3. In what kind of environment might this fossil have lived?

4. How might this organism become fossilized?

5. How do fossils provide valuable information to scientists about evolution?

Exercise 3 Results

1. Complete the table with your observations.

Vertebrate Forelimb Adaptations		
Animal	**Forelimb Structure**	**Forelimb Function**
Human		
Frog		
Chicken		
Bat		
Mole		
Rat		

Exercise 4 Results

1. How does embryological development support evolution?

____

____

Exercise 5 Results

1. Complete the table with your results.

Differences between *Au afarensis* and Other Hominid Species					
***Au afarensis* DNA compared to:**	***H sapiens***	***H erectus***	***H habilis***	***H neanderthalensis***	***Au robustus***
Differences					

2. Which species is the least closely related to the ancestor species (*Au afarensis*)? Explain your answer.

3. Which species is the most closely related to *H sapiens*? Explain your answer.

4. What is the "molecular clock" theory and why is it important in establishing ancestry?

Summary Results

1. How can natural selection cause 2 populations to become different over time?

2. Describe how the effects of directional selection may be offset by immigration

3. In humans, birth weight is an example of a characteristic affected by stabilizing selection. What does this mean to the long term average birth weight of humans? How might the increasing number of caesarean sections be affecting this characteristic?

4. What is the role of natural selection in the process of evolution?

CHAPTER 17

Population Genetics

Photo courtesy of Holly J. Morris

OBJECTIVES

After completing this exercise, you will be able to:

- Explain the significance of the Hardy—Weinberg Equation.
- State the conditions that must be present for the allele frequencies to remain constant from generation to generation.
- Calculate the percent of dominant and recessive genes in a population from the phenotypic ratios.
- Predict the phenotypic and genotypic ratios in successive generations when the starting generation is known.

INTRODUCTION

Evolution is defined as the change in the allele (gene) frequencies of a population over a period of time. In the early 1900s, G. H. Hardy, an English mathematician, and Wilhelm Weinberg, a German physician, independently published papers explaining why certain genetic traits continue in a population generation after generation. According to their research, the percentages of dominant and recessive genes in a population will remain the same generation after generation if:

1. There is no genetic drift (the population is large enough that allele frequency is not subject to random events such as floods, volcano eruptions, etc.).

2. There is no gene flow (the population does not have any emigration or immigration which can change allele frequencies).
3. Totally random mating occurs.
4. Mutations do not occur.
5. Natural selection does not occur.

If these conditions are met, then the gene percentage of a population will not change and evolution does not occur. However, if the gene percentage is not the same generation after generation, then, according to Hardy—Weinberg, evolution has occurred.

Hardy—Weinberg can be used when there are only two alleles involved for one trait. In this case, the percentage of dominant alleles for a trait plus the percentage of recessive alleles for the same trait in a population equal 100% of the possible alleles for the trait.

If the dominant allele is designated p and the recessive allele is designated as q, mathematically, this can be shown as:

$$p + q = 1$$

As a result, the chances of all possible combinations of alleles occurring randomly is

$$(p + q)^2 = 1$$

or simply

$$p^2 + 2pq + q^2 = 1$$

in which p^2 is the predicted frequency of homozygous dominant, $2pq$ is the predicted frequency of heterozygous, and q^2 is the predicted frequency of homozygous recessive.

EXERCISE 1 Observing Hardy—Weinberg Equilibrium

In this exercise, we will use red and white beads to simulate allele frequencies within a population. We will carry this population through several generations to observe how allele frequencies are affected when the Hardy—Weinberg equilibrium conditions are met.

Purpose

To observe the allele frequencies in a population when the Hardy—Weinberg equilibrium conditions are met.

Materials

- Red and white beads
- Plastic cup

Methods

- Place 50 red (allele R) and 50 white (allele r) beads in a plastic cup. As such, both the R and r allele frequencies are 0.50.
- Without looking, remove 2 beads which represents a single offspring from the population.
- Record the genotype (*RR*, *Rr*, or *rr*) in the **Hardy—Weinberg Equilibrium – Generation # 1** table.
- Replace the beads in the plastic cup. Shake the cup and pick another 2 beads.
- Repeat this procedure for a total of 50 times (which represents 1 generation).
- Count the number of R and r alleles and record the results in the **Hardy—Weinberg Equilibrium – Generation # 2** table.

- Record the number of R alleles in the **Frequency of Red Alleles** table in the generation 2 row.
- Make a new population of red and white beads based on the results in the **Hardy—Weinberg Equilibrium – Generation # 2** table.
- Repeat this entire procedure 3 more times.
- Use the **Frequency of Red Alleles** data to prepare the **Frequency of Red Alleles** graph.

EXERCISE 2 Observing the Effect of Natural Selection on Allele Frequencies

In this exercise, we will use red and white beads to simulate allele frequencies within a population. However, in this exercise, we will carry this population through several generations to observe how allele frequencies are affected when natural selection occurs.

Purpose

To observe the allele frequencies in a population when the **Hardy—Weinberg Equilibrium** conditions are not met.

Materials

- Red and white beads
- Plastic cup

Methods

- Place 50 red (allele R) and 50 white (allele r) beads in a plastic cup. As such, both the R and r allele frequencies are 0.50.
- Without looking, remove 2 beads which represents a single offspring from the population.
- Record the genotype (*RR*, *Rr*, or *rr*) in the **Effect of Natural Selection on Allele Frequencies – Generation # 1** table.
- Replace the beads in the plastic cup. Shake the cup and pick another 2 beads.
- Repeat this procedure for a total of 50 times (which represents 1 generation).
- Every rr individual has a lethal condition and is removed from the population.
- Count the number of R and r alleles remaining in the population.
- Prorate the next population based on only the R and r alleles remaining.
- Record this allele data in the **Effect of Natural Selection on Allele Frequencies – Generation # 2** table.
- Record the number of R alleles in the **Frequency of Red Alleles due to Natural Selection** table in the generation 2 row.
- Make a new population of red and white beads based on the results you recorded in the **Effect of Natural Selection on Allele Frequencies – Generation # 2** table.
- Repeat this entire procedure 3 more times.
- Use the **Frequency of Red Alleles due to Natural Selection** data to prepare the **Frequency of Red Alleles due to Natural Selection** graph.

Name:

Exercise 1 Results

1. Record your results for the Hardy—Weinberg Equilibrium exercise.

Hardy—Weinberg Equilibrium – Generation # 1 Red = 50 White = 50									
1		2		3		4		5	
6		7		8		9		10	
11		12		13		14		15	
16		17		18		19		20	
21		22		23		24		25	
26		27		28		29		30	
31		32		33		34		35	
36		37		38		39		40	
41		42		43		44		45	
46		47		48		49		50	

Hardy—Weinberg Equilibrium – Generation # 2 Red = White =									
1		2		3		4		5	
6		7		8		9		10	
11		12		13		14		15	
16		17		18		19		20	
21		22		23		24		25	
26		27		28		29		30	
31		32		33		34		35	
36		37		38		39		40	
41		42		43		44		45	
46		47		48		49		50	

Hardy—Weinberg Equilibrium – Generation # 3 Red = White =									
1		2		3		4		5	
6		7		8		9		10	
11		12		13		14		15	
16		17		18		19		20	
21		22		23		24		25	
26		27		28		29		30	
31		32		33		34		35	
36		37		38		39		40	
41		42		43		44		45	
46		47		48		49		50	

Hardy—Weinberg Equilibrium – Generation # 4 Red = White =									
1		2		3		4		5	
6		7		8		9		10	
11		12		13		14		15	
16		17		18		19		20	
21		22		23		24		25	
26		27		28		29		30	
31		32		33		34		35	
36		37		38		39		40	
41		42		43		44		45	
46		47		48		49		50	

Hardy—Weinberg Equilibrium – Generation # 5 Red = White =

2. Record the allele frequencies for each generation.

Frequency of Red Alleles	
Generation	Frequency
1	P = 0.50
2	p =
3	p =
4	p =
5	p =

3. Graph the **Frequency of Red Alleles**.

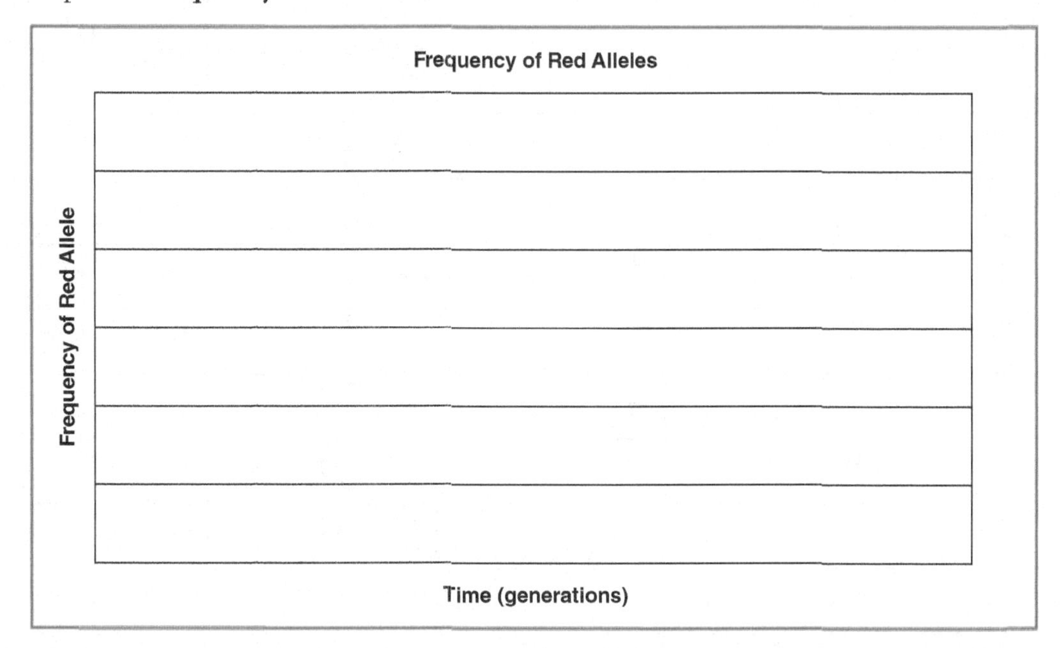

Exercise 2 Results

1. Complete the table below.

Effect of Natural Selection on Allele Frequencies – Generation # 1 Red = 50 White = 50									
1		2		3		4		5	
6		7		8		9		10	
11		12		13		14		15	
16		17		18		19		20	
21		22		23		24		25	
26		27		28		29		30	
31		32		33		34		35	
36		37		38		39		40	
41		42		43		44		45	
46		47		48		49		50	

Effect of Natural Selection on Allele Frequencies – Generation # 2 Red = White =									
1		2		3		4		5	
6		7		8		9		10	
11		12		13		14		15	
16		17		18		19		20	
21		22		23		24		25	
26		27		28		29		30	
31		32		33		34		35	
36		37		38		39		40	
41		42		43		44		45	
46		47		48		49		50	

Effect of Natural Selection on Allele Frequencies – Generation # 3 Red = White =									
1		2		3		4		5	
6		7		8		9		10	
11		12		13		14		15	
16		17		18		19		20	
21		22		23		24		25	
26		27		28		29		30	
31		32		33		34		35	
36		37		38		39		40	
41		42		43		44		45	
46		47		48		49		50	

Effect of Natural Selection on Allele Frequencies – Generation # 4 Red = White =									
1		2		3		4		5	
6		7		8		9		10	
11		12		13		14		15	
16		17		18		19		20	
21		22		23		24		25	
26		27		28		29		30	
31		32		33		34		35	
36		37		38		39		40	
41		42		43		44		45	
46		47		48		49		50	

Effect of Natural Selection on Allele Frequencies – Generation # 5 Red = White =

2. Record the allele frequencies for each trial.

Frequency of Red Alleles due to Natural Selection	
Generation	**Frequency**
1	P = 0.50
2	p =
3	p =
4	p =
5	p =

3. Graph the **Frequency of Red Alleles due to Natural Selection**.

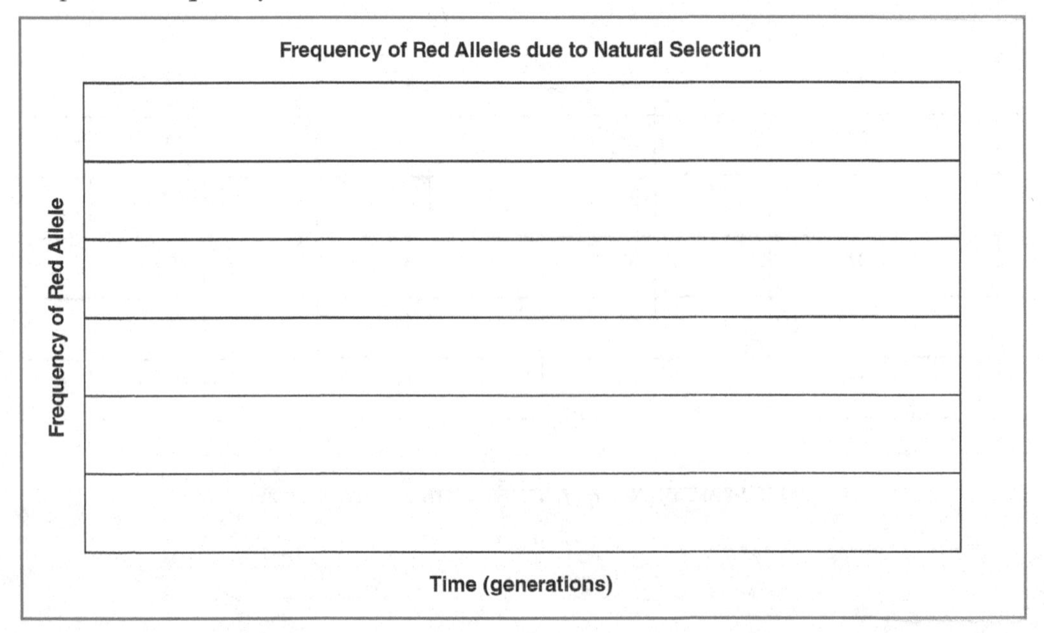

Summary Results

1. Calculate the phenotype frequencies for each snapdragon generation based on the allele frequencies that result from natural selection. Assume that the red allele in snapdragons is incompletely dominant over the white allele.

Allele and Phenotype Frequencies — Snapdragons					
Population	**Allele frequencies**		**Phenotype frequencies**		
	Red (p)	White (q)	Red (p^2)	Pink (2pq)	White (q^2)
Initial population	0.50	0.50.	0.25	0.50	0.25
1st generation	0.60	0.40.			
2nd generation	0.70	0.30.			
3rd generation	0.80	0.20.			

2. Using the phenotypic frequency data from question 1 above, calculate the number of red, pink, and white snapdragons that would be expected in a population of 200 plants.

Phenotypic Frequencies — Snapdragons						
Population	**Phenotype frequencies**			**Number of each phenotype**		
	Red (p^2)	Pink (2pq)	White (q^2)	Red	Pink	White
Initial population	0.25	0.50	0.25	50	100	50
1st generation						
2nd generation						
3rd generation						

CHAPTER 18

Viruses

Photo courtesy of Holly J. Morris

OBJECTIVES

After completing this exercise, you will be able to:

- Describe the anatomical shape and features of a typical virus.
- Describe a typical virus attack.
- Describe how a disease, such as that caused by a virus, can be transmitted in a population.

INTRODUCTION

Viruses are a group of submicroscopic infective agents that some regard as being extremely simple microorganisms and which others regard as being extremely complex molecules. Their actual position is somewhere in the gray area between living and nonliving.

Viruses depend on the host cells that they infect for reproduction. After contacting a host, a virus inserts its genetic material into the host and then effectively takes over control of the host cell. The infected cell produces viral components instead of its usual products. Some viruses may remain dormant inside a host cell for long periods and cause no obvious change in their host cell. However, if the dormant virus becomes stimulated, it can form, assemble, and release new viruses that can infect other cells.

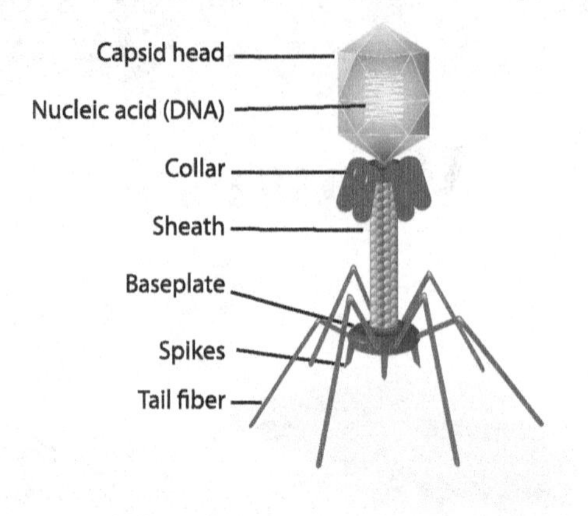

Viruses cause many diseases including the common cold, hepatitis, HIV, warts, yellow fever, herpes, rubella, measles, chickenpox, influenza, polio, rabies, and Ebola. Additionally, some viruses (such as some members of the human papilloma virus group) have been linked to some types of cancer (such as cervical cancer).

Because viruses can transfer genetic material between different hosts, they can be useful in genetic engineering. In fact, viruses carry out natural genetic engineering. For example, a virus during its replication phase may incorporate some genetic material from its host and then transfer this genetic information to a new and/or unrelated host.

EXERCISE 1 Following the Spread of a Viral Based Disease

Viruses cause significant amounts of disease in humans. Viruses can be spread through many routes including respiratory (common cold, influenza), sexually (hepatitis B, HIV), skin contact (warts), and insect vectors (yellow fever). This exercise follows the spread of a viral based disease through a population.

Purpose

To simulate how an HIV infected individual, having multiple sexual encounters, can spread the disease through a population.

Materials

- Empty testing cups.
- Cups containing distilled water.
- One cup containing an unknown solution indicating an infected individual.
- Phenolphthalein solution.

Methods

- Select one of the cups on the counter. All of the cups are filled with distilled water except one cup which contains an unknown solution indicating an "infected" individual.
- Each student will now make a series of "body fluid" exchanges with other students. The procedure for making a "body fluid" exchange is:
    - Select a student in the lab with whom you have not exchanged "body fluid."
    - Pour the contents of your cup into their cup.
    - Swirl the solution to mix it.

 - Your partner pours half of the contents of their cup into your cup.
 - Return to your desk.

- Perform another "body fluid" exchange with a different student. After this second "body fluid" exchange:
  - Pour a small volume of your "body fluid" cup into your testing cup and add a few drops of phenolphthalein.
  - Swirl the mixture.
  - If the solution turns pink, you have been "infected". If the solution remains clear, you have not been "infected".
  - Rinse and dry the testing cup.
  - Indicate on the **Individual Infection Status** table whether you have been "infected" in the "after two exchanges" row.
- Perform two additional "body fluid" exchanges with students with whom you have not previously exchanged "body fluid." After these two exchanges, perform the indicator test a second time and record the results in the "after four exchanges" row in the **Individual Infection Status** table.
- Finally, perform two additional "body fluid" exchanges with students with whom you have not previously exchanged "body fluid." After these two exchanges, perform the indicator test a third time and record the results in the "after six exchanges" row in the **Individual Infection Status** table.
- Summarize the class data in the **Class Disease Simulation Results** table.
- Prepare a graph of the results of your class in the space provided.

Name:

Exercise 1 Results

1. Complete the tables.

Individual Infection Status		
After Exchange	**Infected**	**Not Infected**
2		
4		
6		

Class Disease Simulation Results	
	Number Infected
Start	1
After exchange 2	
After exchange 4	
After exchange 6	

2. Prepare a graph of your results:

3. Infections don't spread as rapidly in real life as they did in our experiment. Why not?

____

____

4. Suggest some methods to slow the rate of infection in the general population.

____

____

CHAPTER 19

Bacteriology

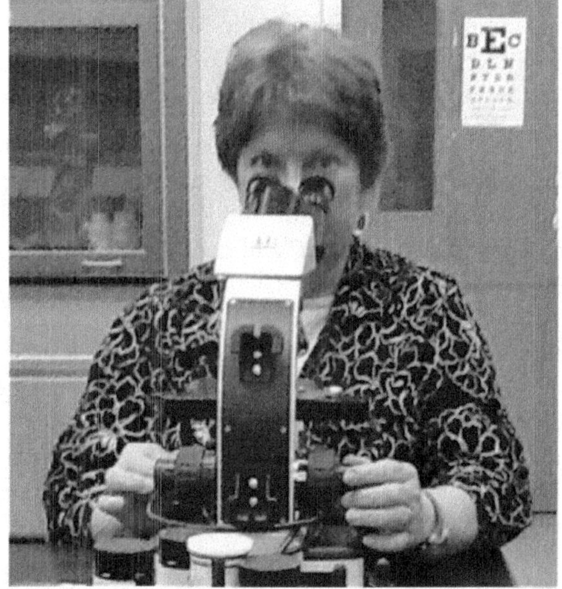
Photo courtesy of Holly J. Morris

OBJECTIVES

After completing this exercise, you will:

- Become familiar with and practice aseptic technique.
- Learn how to describe bacteria colony morphology.
- Learn how to Gram stain sample bacteria and interpret the results.
- Learn how to do streak plating.
- Learn how to make a bacterial lawn.
- Learn how to perform and interpret the results of the effects of antibiotics on bacteria.
- Learn how to perform and interpret the results of the effects of antiseptics and disinfectants on bacteria.
- Learn about bacterial succession.

INTRODUCTION

Prokaryotes are microscopic, single-celled organisms that make up two of the three known domains: Bacteria and Archaea. They are the oldest known life-forms on earth, with fossil records indicating their presence for over 3.5 billion years. They arose when the earth was a very inhospitable place—hot and anoxic with lots of ultraviolet radiation. Many prokaryotes (most of them Archaea) continue to live in the most extreme environments to this day—hot springs, thermal vents, hypersaline ponds, and arctic ice—in environments where there is no oxygen and where there are extremely high pressures, like the bottom of deep oceans. The vast majority of prokaryotes, however, have adapted to non-extreme environments and now dominate the biosphere, being found in every habitat imaginable. Bacterial biomass on earth exceeds that of animals and plants combined.

Prokaryotes, also commonly known as bacteria, are able to survive in such diverse environments because they have developed many different methods of metabolism. For example, anaerobic bacteria function and multiply in the absence of oxygen; some are even "poisoned" by air. Like plants and algae, some bacteria are photosynthetic and can fix inorganic carbon into organic molecules, whereas other bacteria accomplish the same process in the absence of light (chemosynthesis). Most bacterial species can metabolize a wide variety of molecules and thus they contribute to the important process of nutrient cycling, whereby complex organic molecules are broken down into simpler molecules that are then utilized by other organisms. Bacteria can even break down molecules such as pesticides or petroleum that are harmful to humans, and so are used for bioremediation. Others produce helpful products such as antibiotics, and some have been genetically modified to produce important pharmaceuticals, including human insulin.

Bacteria were unknown until the first microscope was invented in the 1680s. Over the next 200 years, people began to understand that bacteria caused infections and that infections could be prevented by sterilizing surgical instruments and washing hands. In the 1880s, the Gram stain was invented and the classification of bacteria by staining characteristics began (and persists today). Some bacteria absorb the purple crystal violet stain and are classified as Gram positive, whereas others absorb the pink counter-stain (safranin) and are classified as Gram negative. They are also described by their different shapes: round (coccus), rod (bacillus), spiral (spirillum), or comma (vibrio). Most exist as single organisms but some form chains (strepto-) or geometric clusters (staphylo-). So, round bacteria in a chain formation are called streptococcus, and cocci in clusters are called staphylococcus. Once aseptic technique was developed, bacteria could be reliably cultured and studied and were further classified by whether they grew in the presence or absence of oxygen, by the types of carbon molecules they could break down, and the by-products they produced from that metabolism. Now the classification of prokaryotes is based on genetics.

Many people think of bacteria as germs and wish to get rid of them. And it is true that some bacteria do cause infections, and those should be controlled. However, the vast majority of prokaryotes are not pathogenic (disease-causing) and, in fact, are helpful and protective. There are millions of bacteria on our skin protecting us from pathogens, and in our intestines producing essential nutrients and helping us digest our food. So keep in mind that antibacterial products kill off good bacteria as well harmful ones, and thus may have unintended consequences.[5]

ASEPTIC TECHNIQUE

Aseptic techniques are any techniques employed to avoid contamination. In a microbiology laboratory it is often necessary to transfer microbes, most commonly bacteria, from one place to another aseptically. There are multiple precautions utilized to maintain the purity of the cultures that are being manipulated. Initially to prevent contamination, all inoculating instrument (loops and needles) should be sterilized prior to use. This is generally achieved through flaming these objects. All growth media should also be sterilized to ensure an axenic culture (pure culture of organism of interest). Typically growth media is sterilized in an autoclave which utilizes steam under pressure.

Aseptic techniques include:

- All work space should be cleaned with Wavicide and free of clutter. All bacterial transfers and dispensing of media should be done as close to the flame as possible to reduce contamination from airborne microbes.
- Test tubes are fitted with loose caps so you should always hold the tube and not the cap. Do not lay the tubes on the table or shake them as the contents may spill. Gently mix broth cultures before transfer by rolling them back and forth between your palms.

- Hold the tube in your non-dominant hand (right-handed people should hold the tube in their left hand) and the inoculating loop in your dominant hand.
- Pass the inoculating loop through the flame of the Bunsen burner. Allow the loop and upper part of the wire to get red hot. This will incinerate any contaminates on the loop. Allow your loop to cool completely before you attempt to pick up cells. A singeing sound indicates you have killed the cells you were attempting to transfer.
- Grasp the cap of the tube between the ring and pinky fingers and the palm of your dominant hand, being careful not to touch your sterile loop to your hand, the tube, or other surface. Do not lay the cap down and be careful not to touch the opening of the cap to your hand. Pass the opening of the tube through the flame a couple of times. Any spills that occur during a bacterial transfer should immediately be cleaned up with Wavicide.
- Insert your sterile loop into the tube and submerge it in the liquid media or touch it to the surface of an agar slant. Be careful not to gouge into the agar if it is a slant culture.

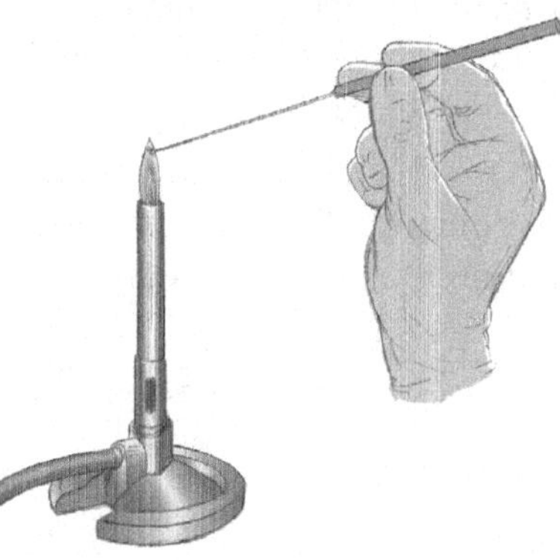

A. Sterilize loop end until red hot. Cool completely before proceeding.

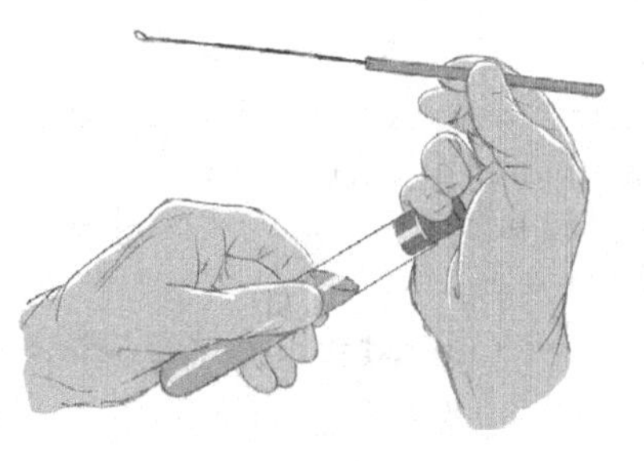

B. Remove stopper from sample tube as shown.

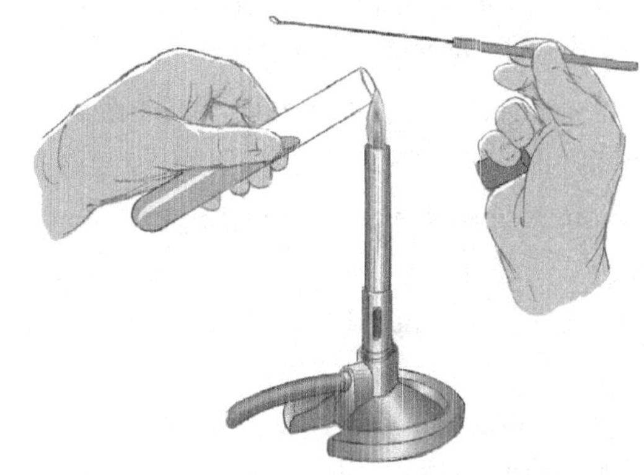

C. Briefly heat tube rim before inserting loop.

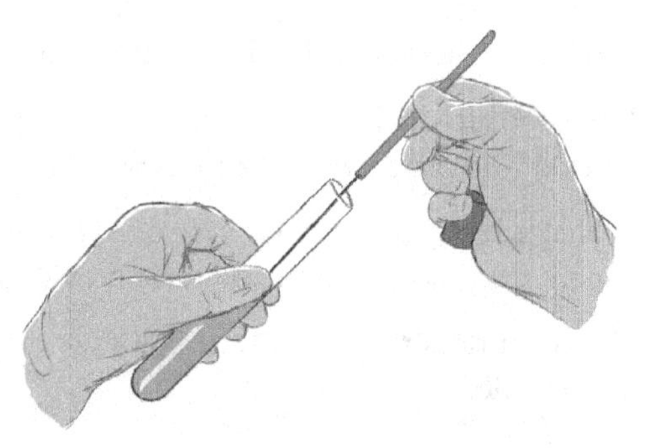

D. Dip loop into sample culture, heat rim again after removing loop, and replace stopper.

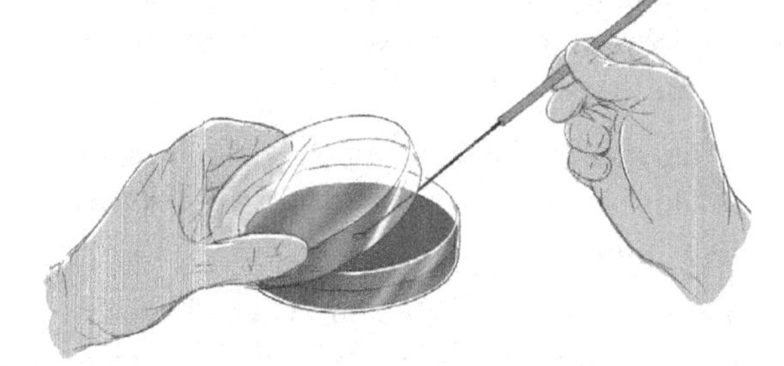

E. Tilt up petri dish cover, lightly wipe streaks across surface of agar medium and replace cover.

Aseptic transfer of bacteria (From Microbes in Health and Disease Lab Manual by Cynthia W. Littlejohn.

- Pass the opening of the tube back through the flame, flame the cap opening gently and then replace the cap onto the tube. Put the tube back into the test tube rack. Work quickly as the longer the tube is open to the environment, the more likely contamination will occur.
- Pick up the tube to be inoculated with your dominant hand. Remove the cap by grasping it between your ring and pinky fingers and your palm. Flame the opening of the tube and insert your loop. Be careful not to contaminate your loop by allowing it to touch and surface outside the tube. DO NOT heat the loop before you insert it into the growth medium.
- To inoculate a broth, insert the loop into the fluid and gently mix to remove bacteria from the loop. To inoculate a slant culture, begin at the bottom of the slant and draw the loop across the agar surface in an s-like pattern being careful not to gouge into the agar surface. To inoculate an agar plate, the cover should be only partially removed to form a clam shell-like opening and the inoculating loop should be drawn across the agar surface in an s-like pattern being careful not to gouge into the agar surface. To inoculate a stab culture, an inoculating needle is used to insert the bacterium directly into the agar.
- Flame the top of the tube and the cap opening. (Agar plates are never flamed but should be only partially opened to allow inoculation.) Replace the cap onto the tube and put it back in the rack. Flame your loop, allowing both the loop and upper part of the wire to get red hot.
- Most bacterial cultures will be incubated at 37°C for 24 to 48 hours. It is important to label each tube or plate with your name, the date, and what bacteria it contains. Plates should be labeled on the bottom (agar side not removable lid) as they will be stored upside down to reduce condensation on the agar surface.
- When you have finished your transfer, clean your work space with Wavicide and wash your hands.[6]

EXERCISE 1 Determining Colony Morphology

The appearance of the organisms grown on solid and in liquid media can be useful in the identification process. In this exercise, you will observe the colonies that develop on a typical medium, such as trypticase soy agar, and evaluate them according to the properties described below.

SIZE: How large are the colonies (compared to typical bacterial colonies)? Examine colonies in the sparsest area of the plate. Distinctions to be made: pinpoint, small, medium, and large.

SHAPE: What is the general appearance of a typical colony? Observe your plate under the stereoscope and compare the colonies to the pictures provided.

MARGIN: What does the edge of the colony look like? Compare the edge of the colonies to the illustrations provided.

ELEVATION: How much does a typical colony stick up above the surface of the plate? Distinctions to be made: flat, raised, convex. Observe your plate from the side of the plate. Use the stereoscope if necessary.

OPACITY: Observe how much light is able to pass through the colony. Hold the plate obliquely to the overhead light. Alternatively, place the plate over the writing of this page. Be careful not to contaminate your notebook!

- Transparent colonies allow all of the light to pass through them.
- Translucent colonies allow most of the light to pass through them; some light is blocked.
- Opaque colonies block most of the light.

COLOR: Look for color in the colony or the medium. The color must be distinct from the color of the uninnoculated medium. If the bacteria produce an intracellular pigment that is retained by the cell, the colony becomes colored. If the bacteria produce an extracellular pigment that is excreted by the cell, the medium becomes colored. If the organisms are translucent, the colored media will show through the colonies, and the colonies will appear to have the same color as the media. White colonies or translucent colonies that take on the color of the straw-colored media are termed "non-pigmented."[7]

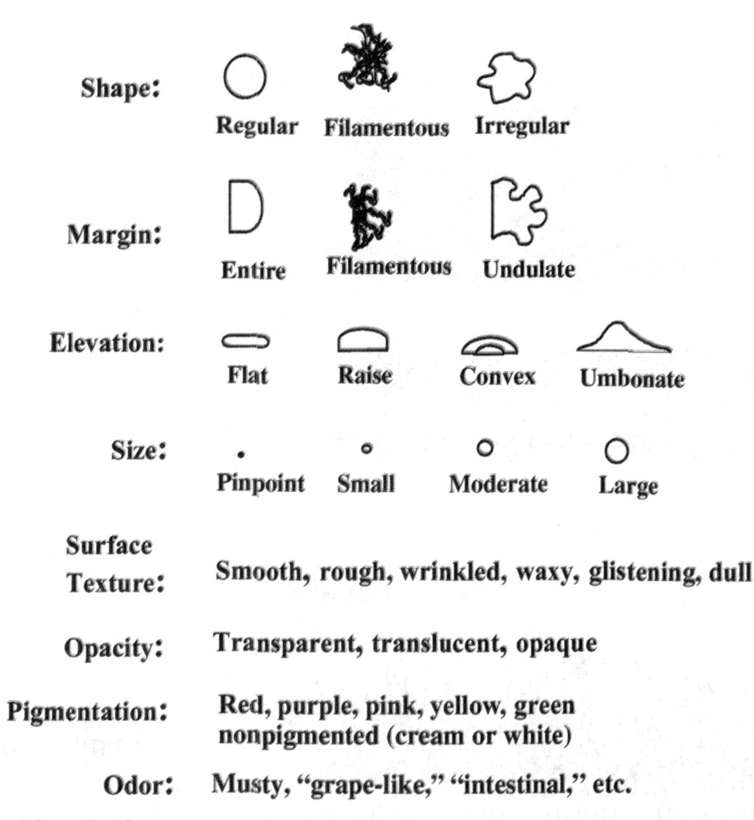

Bacterial Colony Morphology (From Fundamentals of Microbiology for Allied Health by Kathleen Dannelly, Angela K. Chamberlain and William M. Chamberlain. Copyright © 2009 by Kendall Hunt Publishing Company. Reprinted by permission.)

Purpose

To learn how to characterize bacterial colonies by their morphology.

Materials

- Petri plates containing pure cultures of various types of bacteria
- Ruler
- Magnifying glass
- Stereoscope

Methods

- Take 3 different Petri plates containing pure cultures of various types of bacteria.
- Determine the characteristics of each of the various bacteria using the information above as a guide.
- Record the results in the table below.

EXERCISE 2 Identifying Bacteria by Gram Staining

Gram staining is a method of differentiating bacterial species into two large groups: Gram-positive (G^+) and Gram-negative (G^-). The name comes from the developer of the technique, Hans Christian Gram, a Danish bacteriologist. As such, the Gram stain is usually the first step in the identification of bacteria.

G^+ and G^- bacteria are distinguished from each other by differences in their cell walls. These differences affect many aspects of the cell, including the way the cell takes up and retains stains. G^+ bacteria take up the crystal violet, which is then fixed in the cell with the iodine mordant forming a crystal-violet iodine complex which remains in the cell even after decolorizing and causes them to be stained purple. G^- bacteria do not retain this complex when decolorized which causes them to be stained pink.

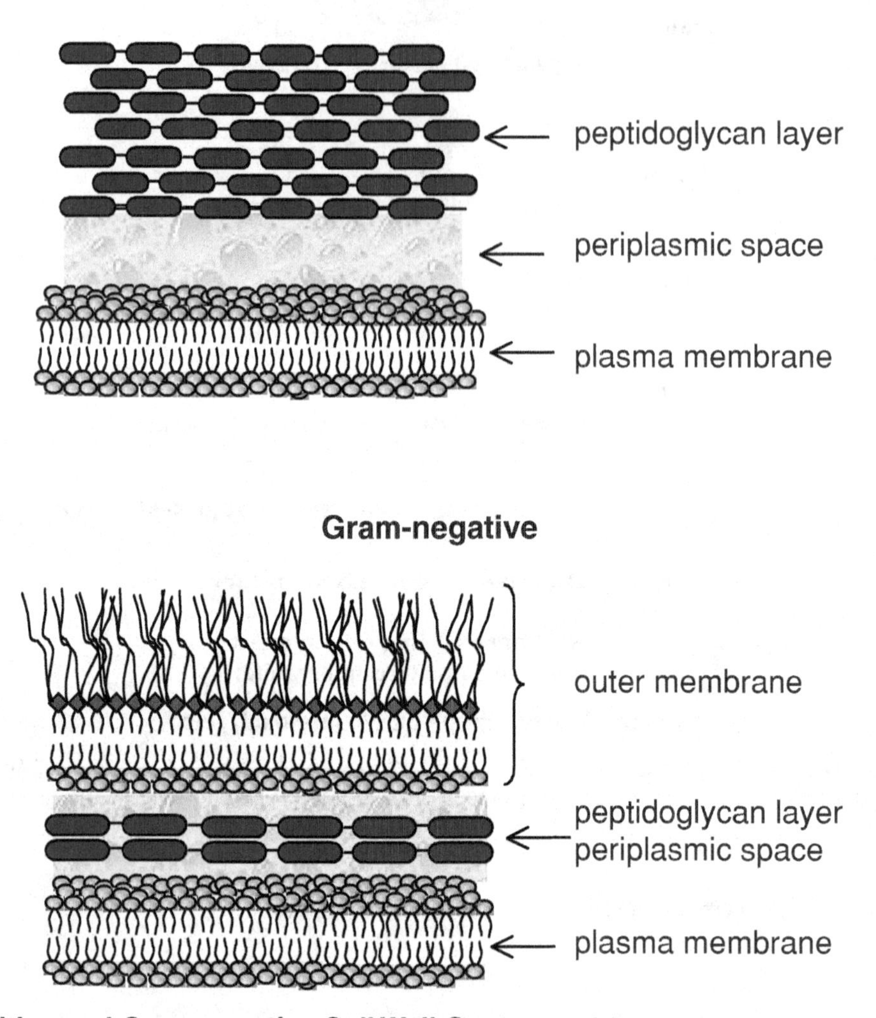

Gram positive and Gram negative Cell Wall Components. (From Fundamentals of Microbiology for Allied Health by Kathleen Dannelly, Angela K. Chamberlain and William M. Chamberlain. Copyright © 2009 by Kendall Hunt Publishing Company. Reprinted by permission.)

Below is an example of a G^+ and G^- bacterial stain:

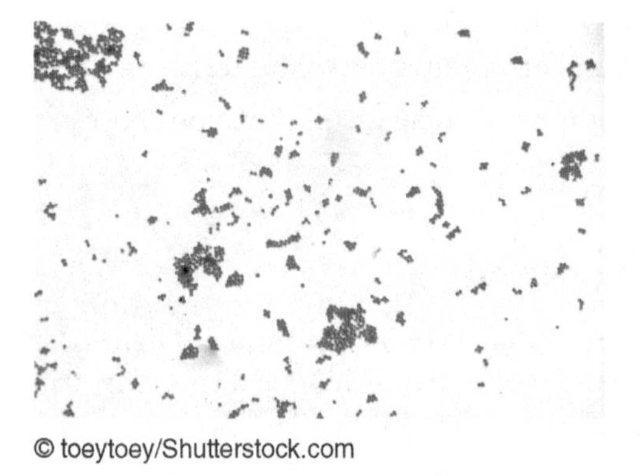

© toeytoey/Shutterstock.com

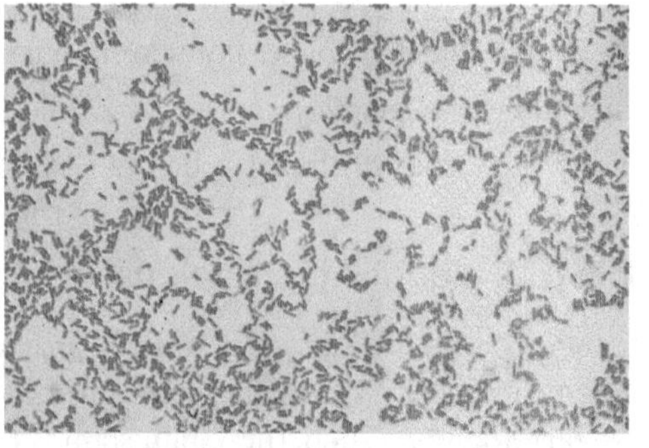

© toeytoey/Shutterstock.com

Purpose

To learn how to Gram stain bacteria and interpret the results.

Materials

- Bacteria broths or slants
- Microscope slides
- Pencil
- China marking pencil
- Ethanol
- Water bottle
- Clothespin
- Gram stain reagents:
    - Crystal violet
    - Gram's iodine
    - Safranin
- Compound light microscope
- Immersion oil
- Staining rack
- Bibulous paper
- Inoculating loop

Methods

- Using a china marking pencil, make a circle about the size of a dime on a clean microscope slide.
- Using aseptic technique, takes a loopful of broth, place it within the circle, and then using the loop, evenly cover the entire area of the circle. If using a slant, put a drop of water in the circle, add a loopful of bacteria, mix the water and the bacteria, and then using the loop, evenly cover the entire area of the circle.
- Prepare a different smear from each of the cultures provided.
- Air dry and heat fix the smears.
- Place the slide on the staining rack over the sink.
- Cover the slide with crystal violet and leave on for 1 minute.
- Hold the slide at a 45° angle with a clothespin and gently rinse the slide with water for a few seconds.
- Cover the slide with iodine and leave on for 1 minute.
- Rinse with water.
- Hold the slide at 45° with a clothespin and decolorize with 95% ethyl alcohol by dripping the alcohol on the slide until no purple washes off the slide.
- Rinse with water.
- Cover the slide with safranin for 1 minute.
- Rinse with water.
- Blot dry with bibulous paper and air dry.
- Examine the slide under oil immersion. See Use of an Oil Imerrsion Lens below.
- Record your observations of the Gram stain characteristics (G^+ or G^-; shape) for the cultures provided on the **Gram Staining Results** table.

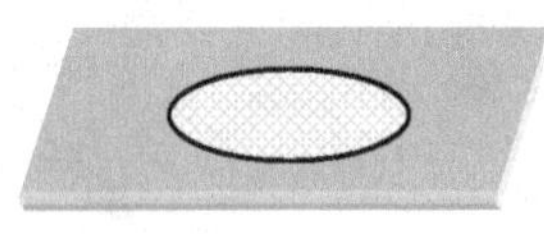

1. Start with a bacterial smear.

2. Cover the smear with crystal violet stain for 1 minute.

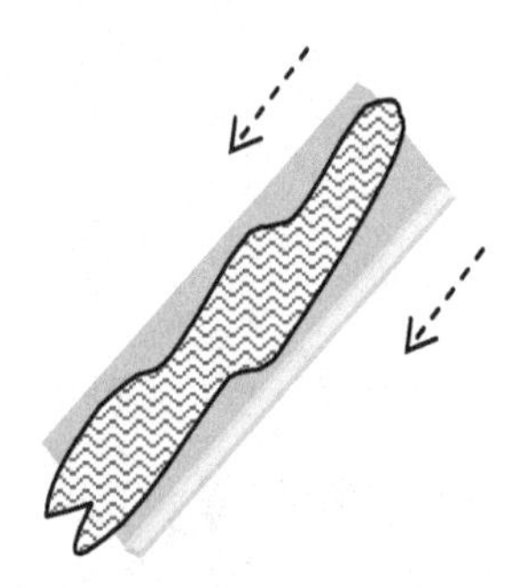

3. Gently and thoroughly rinse the slide with water.

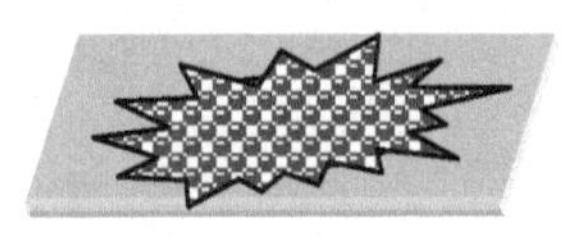

4. Cover the smear with Gram's iodine for 1 minute.

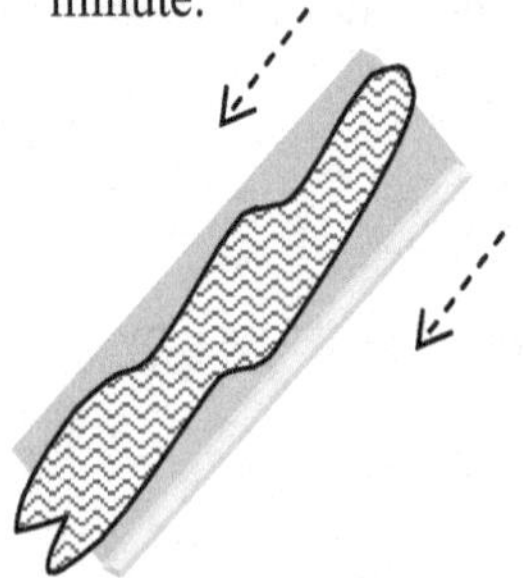

5. Gently and thoroughly rinse the slide with water.

6. Add decolorizer drop wise until runoff is clear (10–20 seconds. Do not over decolorize.

7. Gently and thoroughly rinse the slide with water.

8. Cover the smear with safranin stain for 1 minute.

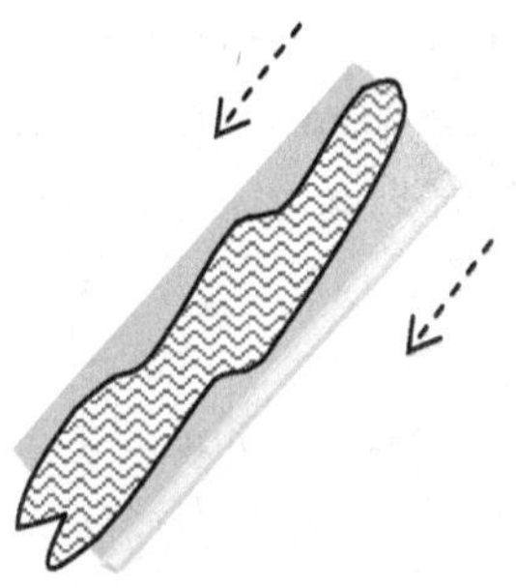

9. Gently & thoroughly rinse the slide with water.

10. Gently blot the slide dry with paper towels.

11. Observe stain under oil immersion.

Gram Stain Procedure (From Fundamentals of Microbiology for Allied Health by Kathleen Dannelly, Angela K. Chamberlain and William M. Chamberlain. Copyright © 2009 by Kendall Hunt Publishing Company. Reprinted by permission.)

EXERCISE 3 Observing Bacteria

Bacteria are essentially invisible, existing beneath the capabilities of human vision. Because the ability to magnify and visualize bacteria is an important facet of their study, we need to use the oil immersion lens of our compound light microscopes to be able to view them.[8]

Use of an Oil Immersion Lens

After you have focused the compound light microscope with the high power objective lens (40x), rotate the objectives halfway between the high power objective lens (40x) and the oil immersion objective lens (100x). Place a drop of immersion oil on the slide where the light is passing through and rotate the oil immersion objective lens into place. Use the fine adjustment focus knob as needed.

Caring for an Oil Immersion Lens

- Oil must not get on the high power objective lens (40x). Once oil is placed on the slide, the high power lens must not be rotated into place. If oil does get on the high power lens, please wipe it off immediately.
- A blurry or smudged image in high power may indicate the presence of oil. Try to wipe the lens clean with lens paper or ask the instructor for assistance.
- If the oil drop is too small, it may not form a complete layer between the oil immersion lens and the slide. The edge of the field of view may appear fuzzy. Rotate the oil immersion lens away and add more oil.
- When the oil immersion lens is turned into place, rock it back and forth to remove any air bubbles from the oil. Bubbles will distort the magnified image.
- Use lens paper to remove oil from the oil immersion lens after each use.
- Since oil is placed directly on heat-fixed bacterial specimens, it is possible for some of the cells (particularly in a thick smear) to lose adherence to the slide. If you see cells floating in your field of view, it may be necessary to make a new slide. Wiping oil off the slide may pull the majority of the cells off the glass and the slide may no longer be useful for microscopic study.[9]

Purpose

To observe bacteria using the oil immersion objective lens.

Materials

- Prepared bacteria slides.
- Compound light microscope.
- Immersion oil.

Methods

- Obtain prepared bacteria slides and focus them with the oil immersion lens.
- Make a detailed drawing of 2 different types of bacteria.

EXERCISE 4 Environmental Culturing

Microbes exist in mixed populations in nature. In order to study a specific bacteria, the bacteria must be isolated from the other types of bacteria. The *T streak plate* method is a technique to prepare isolated bacterial colonies. An isolated bacterial colony could then be used to grow a pure culture (a culture containing only one kind of microbe).

Purpose

To learn the technique of streak plating as a method to isolate pure cultures.

Materials

- Petri plates containing nutrient agar
- China marking pencil
- Sterile cotton swabs
- Test tubes containing sterile water

Methods

- Each person obtains 2 Petri plates, 2 sterile cotton swabs, and a test tube of sterile water.
- Sample 2 areas on campus that you want to check for the presence of bacteria.
- Label the bottom half of each Petri plate with your name, date, and area sampled.
- Draw a "T" on the bottom of the plate to define the 3 areas for the dilution of the organism. See T streak diagram below.
- If the area to be sampled is a liquid, dab the sterile cotton swab in the liquid. If the area to be sampled is a dry area, dip the sterile cotton swab into the sterile water and then dab the area with the sterile cotton swab.
- Slightly open the Petri plate and starting at the edge of the agar, streak the primary inoculation area (the area above the "T" crossbar). Because the agar is very fragile, gently streak the surface of the agar and do careful not to gouge the agar. Make as many streaks as possible but do not overlap previous streaks.
- Rotate the plate a quarter turn.
- Streak the secondary inoculation area (the area to the right of the "T" vertical) by going back into the primary inoculation area once with your swab. Streak this area in the same manner as the primary inoculation area covering as much agar as possible without overlapping the previous streaks.
- Rotate the plate another quarter turn.
- Streak the tertiary inoculation area (the area to the left of the "T" vertical) by going back into the secondary inoculation area once with your swab. Streak this area in the same manner as the primary inoculation area covering as much agar as possible without overlapping the previous streaks.
- Invert the Petri plates and tape them together.
- Incubate the Petri plates at room temperature for 7 days.
- After 7 days, examine the Petri plates and determine whether the Petri plates contain pure or impure cultures, colony types, and relative abundance of each bacterial type.
- Record your observations in the **Types and Abundance of Colonies Found at Selected Environments** table.
- Remove the tape from the Petri plates and place them in the fume hood for disposal.

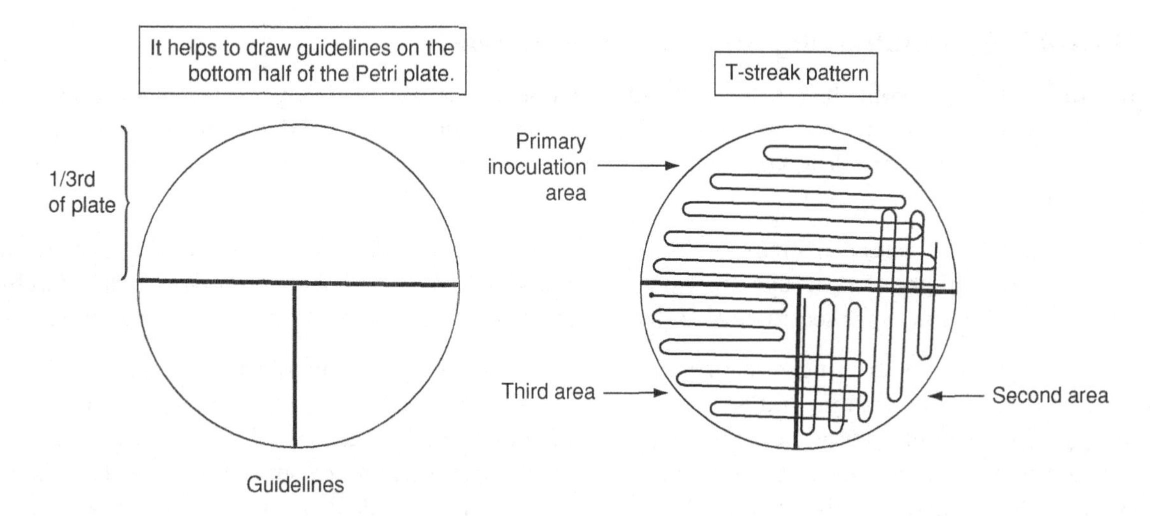

Streaking a Plate with the T Streak Technique. (From Fundamentals of Microbiology Laboratory Manual by Marlene Demers. Copyright © 2009 by Kendall Hunt Publishing Company. Reprinted by permission.)

EXERCISE 5 Determining Ecological Succession in Milk

All of the groups of organisms living and interacting with each other in a specific environment are called a community. Members of a community interact in a variety of ways. The most common interaction is the predator – prey relationship. However, there are other types of interactions within the populations living in the community including interactions within the bacterial populations. As one bacterial type grows, it can change the environment which then enables other bacterial species to grow. This process is called ecological succession. In this experiment, you will observe ecological succession in two types of milk: plain milk and chocolate milk.

Purpose

To determine the ecological succession of bacteria in milk.

Materials

- Samples of plain and chocolate milk: refrigerated and aged for 1, 4, and 8 days
- pH strips
- Materials to perform Gram staining on the samples
- Compound light microscope
- Immersion oil

Methods

- Observe the various plain and chocolate milk samples for color, consistency, and odor. Additionally, the pH of each sample is also obtained.
- Prepares slides of each of the various plain and chocolate milk samples for Gram staining.
- Gram stain each of the slides and examine them to determine the morphology of the bacteria present in each of the samples.
- Record all of the results in the **Ecological Succession of Bacteria in Milk** table.

EXERCISE 6 Determining Antibiotic Resistance

When the body is overcome by disease, chemotherapeutic agents (antibiotics and synthetic drugs) may be needed. Unlike disinfectants, however, antibiotics must act within the host and usually within the pathogen. The ideal antimicrobial drug kills the harmful organism without damaging the host, that is, it is selectively toxic.

Different microbial species and strains of the same species have different sensitivities to microbial agents. Often, the susceptibility of a pathogen to a specific antibiotic may change during therapy. As such, a physician may need to know the sensitivities of the disease organism before treatment can be started.

The earliest methods of utilizing agar media in the determination of antimicrobial resistance were developed and described by Schmith and Reymann in the 1940's to determine the minimum inhibitory concentration (MIC) of sulphapyridine for the treatment of gonorrhea. In this method, the antimicrobial agent was added directly to an agar medium prior to inoculation. By noting any growth on the treated agar, resistance or susceptibility to the antimicrobial agent in question could be determined. In 1945, A. R. Frisk used this method to determine the MIC of penicillin for the treatment of *Streptococcus pneumoniae*. However, performing the agar dilutions was time consuming and labor intensive. In the late 1950s, a simplified and widely used method of testing the effectiveness of an antimicrobial compound against a specific culture was developed by University of Washington professors, Kirby and Bauer. In this Kirby--Bauer method (disk diffusion method), which is still used today, a Petri dish containing an agar medium is inoculated with the test organism. Next, a filter-paper disk impregnated with a known concentration of a chemotherapeutic agent is placed on the agar. The chemical will diffuse through the agar and, if it is inhibitory to the isolate, a clear zone (zone of inhibition) will be seen around the disk. The size of the zone of inhibition is determined by measuring its diameter and comparing this value to known values. The size of the zone of inhibition depends upon the relative susceptibility of the microbe, as well as, the diffusion rate of the chemical. Therefore, a wider zone does not always indicate greater antimicrobial activity. The results are reported as sensitive, intermediate, or resistant.

The test is simple, inexpensive, and is often used when more sophisticated facilities are not available. In today's exercise you will test the relative susceptibility/resistance of some common bacteria to several different antibiotics.[10]

Purpose

To learn the techniques to determine the sensitivity or resistance of pathogenic bacteria to various antibiotic compounds.

Materials

- Petri plates containing nutrient agar (4)
- Broth cultures of various bacteria (4)
- Sterile cotton swabs
- Various types of antibiotic disks
- Alcohol
- Bunsen Burner
- Ruler
- Tape
- Forceps
- Susceptibility Data Sheet

Methods

- Each team obtains four Petri plates containing nutrient agar.
- Use a sterile cotton swab to create a bacterial lawn on each Petri plate with a different bacterial culture.
- Sterilize the forceps by dipping it in alcohol and flaming it before handling each sterile antibiotic disc.
- Take one of antibiotic discs from each of the antibiotic dispensers and then carefully place the disc on the agar in the Petri plate as shown below.

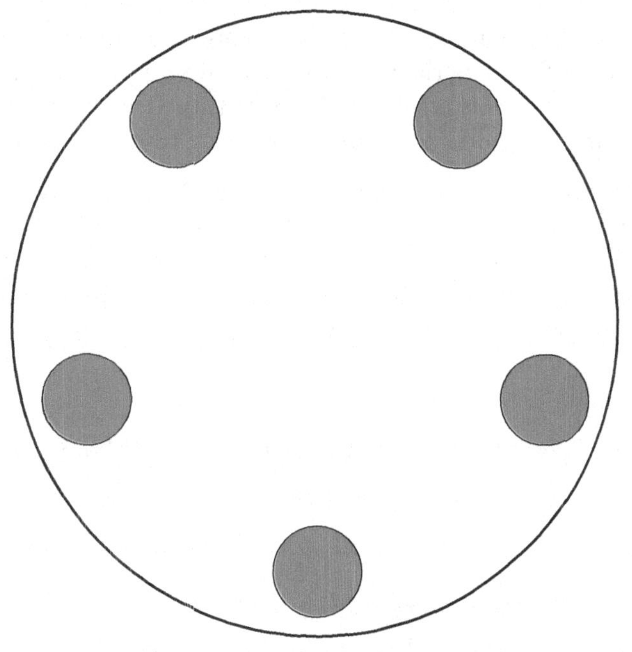

© Kendall Hunt Publishing

- Place a disc for all of the antibiotics on each of the Petri plates.
- Label the Petri plates with your name, date, type of bacteria, and location of each of the discs. Tape the four Petri plates together.
- Invert the Petri plates and incubate them for 7 days at room temperature.
- After 7 days, measure the diameter of the zone of inhibition to the nearest millimeter for each disc on all four Petri plates. If there is no zone of inhibition, record the diameter as 0.

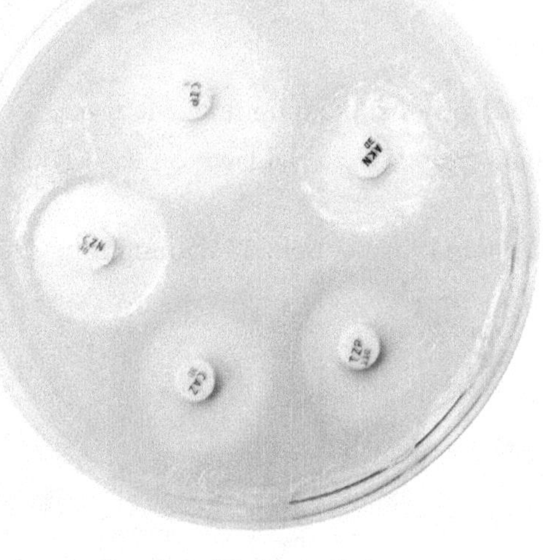

© grebcha/Shutterstock.com

- Record your zone of inhibition results in the **Controlling Bacteria with Antibiotics** table.
- Remove the tape from the Petri plates and place them in the fume hood for disposal.
- Use the Susceptibility Data Sheet handout to interpret your data.
- After interpreting your data, record your antibiotic susceptibility results in the **Controlling Bacteria with Antibiotics** table.

EXERCISE 7 Determining Antiseptic & Disinfectant Resistance

Many chemical substances are toxic to bacteria. These agents either inhibit microbial growth (bacteriostatic) or kill the microorganisms (bactericidal). Most inhibitory agents are antiseptics and can be used topically on living tissues. Disinfectants usually are lethal to pathogenic forms and are used on inanimate surfaces. These substances are usually too harsh for use on living tissues.

In this exercise, we will investigate the effects of various antiseptics and disinfectants on bacterial growth. A sterile disc is impregnated with the substance to be tested. The disc is then placed on a Petri dish that has been covered with a bacterial lawn. As the substances diffuse into the medium, bacterial growth is inhibited. A distinct zone is usually formed at a region where the concentration of the agent inhibits the growth of the organism. By measuring this zone we can quantitatively compare the effects of the different agents.[20]

Purpose

To learn the techniques to determine the sensitivity or resistance of pathogenic bacteria to various antiseptic and disinfectants.

Materials

- Petri plates containing nutrient agar (4)
- Broth cultures of various bacteria (4)
- Sterile cotton swabs
- Sterile filter paper discs
- Various types of antiseptics and disinfectants:
- Alcohol
- Bunsen Burner
- Ruler
- Tape
- Forceps

Methods

- Each team obtains four Petri plates containing nutrient agar.
- Use a sterile cotton swab to create a bacterial lawn on each Petri plate with a different bacterial culture.
- Sterilize the forceps by dipping it in alcohol and flaming it before handling each sterile filter paper disc.

- Touch the sterile disc to one of the antiseptics/disinfectants and then carefully place the disc on the agar in the Petri plate as shown in the drawing below.

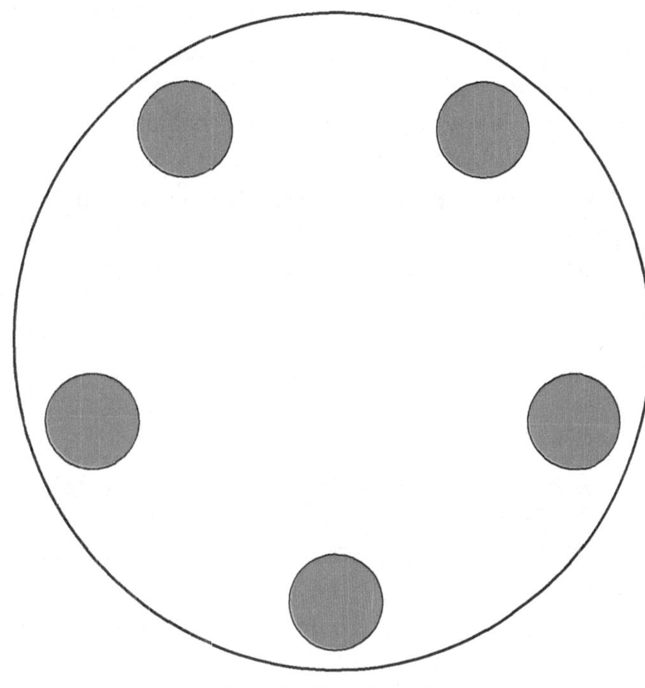

© Kendall Hunt Publishing

- Place a disc for all of the antiseptics/disinfectants on each of the Petri plates.
- Label the Petri plates with your name, date, type of bacteria, and location of each of the discs. Tape the four Petri plates together.
- Invert the Petri plates and incubate them for 7 days at room temperature.
- After 7 days, measure the diameter of the zone of inhibition to the nearest millimeter for each disc on all 4 Petri plates. If there is no zone of inhibition, record the diameter as 0.

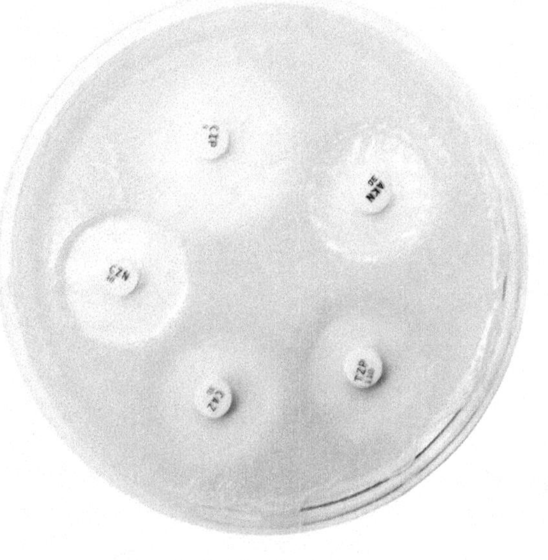

© grebcha/Shutterstock.com

- Record your zone of inhibition results in the **Controlling Bacteria with Antiseptics and Disinfectants** table.
- Remove the tape from the Petri plates and place them in the fume hood for disposal.
- Measure the diameter of the zone of inhibition and interpret the results using these parameters:
    - 0, record "R" (resistant) in the RSI column.
    - Less than 10 mm, record "I" (intermediate) in the RSI column.
    - 10 mm or more, record "S" (susceptible) in the RSI column.
- Record your susceptibility results in the **Controlling Bacteria with Antiseptics and Disinfectants** table.

Name:

Exercise 1 Results

1. Complete the table with your observations of the bacterial colonies examined.

Characteristics of Bacterial Colonies						
Name of Bacteria	Size	Shape	Margin	Surface	Opacity	Color
1						
2						
3						

2. What are the most common colony shapes, margin and surface characteristics of the species you observed?

3. Is colony morphology a reliable way to identify a bacterial species?

Exercise 2 Results

1. Complete the table with the results of your Gram Staining.

Gram Staining Results	
Name of Bacteria	Gram Stain Characteristics
1	
2	
3	
4	

2. What factors can modify the expected results of the staining procedure?

____

____

Exercise 3 Results

1. Make **detailed drawings** of the individual cells from 2 different bacterial species as you observed them with the oil immersion lens.

Name:	Name:
Total magnification:	Total magnification:

Exercise 4 Results

1. Complete the table below.

Types and Abundance of Colonies Found at Selected Environments	
Environments	**Colony Types & Abundance**
1	
2	
3	

2. What was your hypothesis regarding the presence of bacteria in the different environments you sample?

____

____

2. Did the plates differ in the number and diversity of bacterial colonies? Explain.

______

______

3. Did your predictions match your hypothesis? Explain.

______

______

4. What factors might be responsible for your results?

______

______

Exercise 5 Results

1. Complete the table.

Ecological Succession of Bacteria in Milk		
Age/Type of Milk	**Environmental Characteristics (pH, color, consistency, odor)**	**Organisms Present (Morphology & Gram Stain Results)**
Refrigerated plain		
24 hour plain		
4 day plain		
8 day plain		
Refrigerated chocolate		
24 hour chocolate		
4 day chocolate		
8 day chocolate		

2. Describe the changing sequence of organisms and the corresponding environmental conditions during succession in plain milk.

3. Describe the changing sequence of organisms and the corresponding environmental conditions during succession in chocolate milk.

4. Compare the ecological succession in plain milk with that of chocolate milk. Propose any reasons for any differences.

Exercise 6 Results

1. Complete the table with your results.

Controlling Bacteria with Antibiotics														
Antibiotic														
Bacteria Name	**Zone**	**RSI**	**Zone**	**RSI**	**Zone**	**RSI**	**Zone**	**RSI**	**Zone**	**RSI**	**Zone**	**RSI**	**Zone**	**RSI**
1														
2														
3														
4														

2. Were any bacteria very sensitive to all of the antibiotics? Explain

____

____

3. Were the results different between G^+ and G^- bacteria? Explain.

____

____

4. List an antibiotic that you would prescribe to control each of the bacterial species you tested.

Controlling Bacterial Growth with Antibiotics		
Bacteria Name	**Antibiotic to control growth**	**Explain your reasoning**
1		
2		
3		
4		

Exercise 7 Results

1. Complete the table with your results.

Controlling Bacteria with Antiseptics & Disinfectants												
Substance Name												
Antiseptic or Disinfectant?												
Bacteria	**Zone**	**RSI**	**Zone**	**RSI**	**Zone**	**RSI**	**Zone**	**RSI**	**Zone**	**RSI**	**Zone**	**RSI**
1												
2												
3												
4												

2. What are the most appropriate situations to use an *antiseptic*?

3. Which *antiseptic* was the most effective in controlling the growth of bacteria?

4. Were the results different between G^+ and G^- bacteria for *antiseptics*? Explain.

5. What are the most appropriate situations to use a *disinfectant*?

6. Which *disinfectant* was most effective in controlling the growth of bacteria?

7. Were the results different between G^+ and G^- bacteria for *disinfectants*? Explain.

CHAPTER 20

Photosynthesis

Photo courtesy of Holly J. Morris

OBJECTIVES

After completing this exercise, you will be able to:

- Describe the importance of photosynthesis.
- Describe the various types of photosynthesis.
- Describe the generalized microanatomy of a leaf.
- Describe the role of the stomata and guard cells.
- Understand the role played by the chloroplasts in photosynthesis.
- Understand the function of the various pigments in photosynthesis.

INTRODUCTION

Photosynthesis is the process used by plants, some protists, and cyanobacteria to transform light energy into chemical energy that can be used for cellular processes. This chemical energy is stored in the form of glucose. The equation for this process is:

$$6CO_2 + 6H_2O + \text{light} \rightarrow C_6H_{12}O_6 + 6O_2$$

Photosynthesis generally occurs in the plant's leaves and specifically, occurs in the chloroplasts. The carbon dioxide needed for photosynthesis enters the plant's leaves through the stomata, pores in the leaves. Water is absorbed through the roots and travels up to the leaves. Light energy is absorbed by several pigments such as carotenes and xanthophylls, but the primary pigment involved in photosynthesis is chlorophyll, which is found primarily in the chloroplasts. Oxygen is a by-product of photosynthesis and is either used for cell respiration or exits the leaves through the stomata.

Chloroplasts, the actual site of photosynthesis, contain:

- Outer and inner membranes that serve as protective coverings.
- Thylakoids, where chlorophyll is found, are arranged in stacks called grana.
- Stroma, a dense fluid, is the site of the conversion of carbon dioxide to sugar.

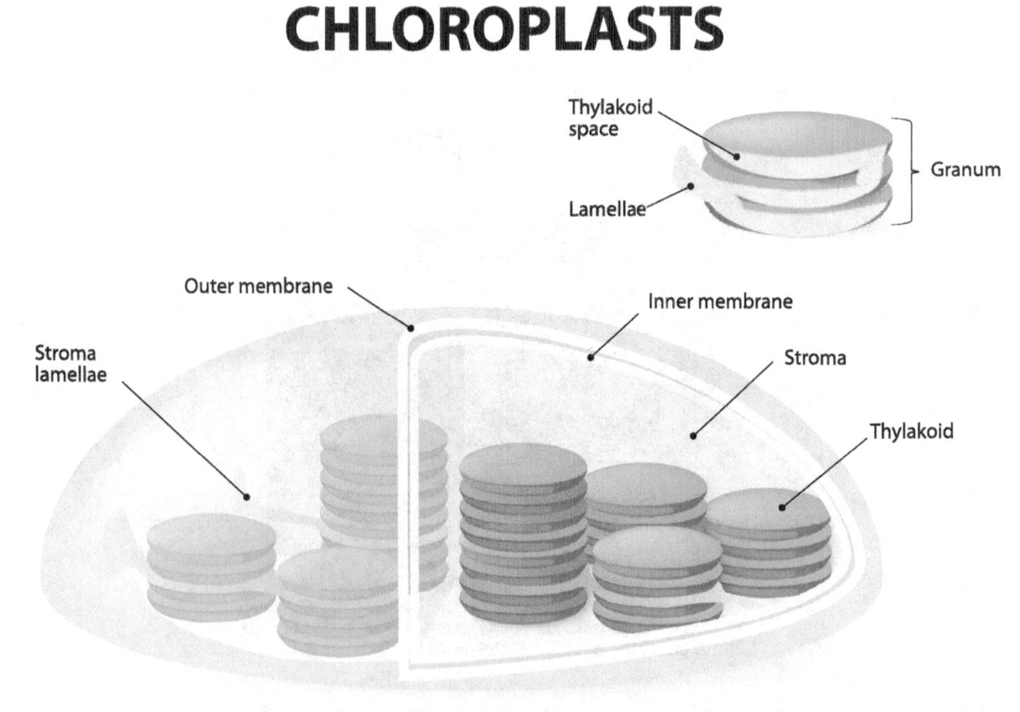

© Designua/Shutterstock.com

Photosynthesis occurs in two stages—the light dependent stage and the light independent stage (Calvin cycle).

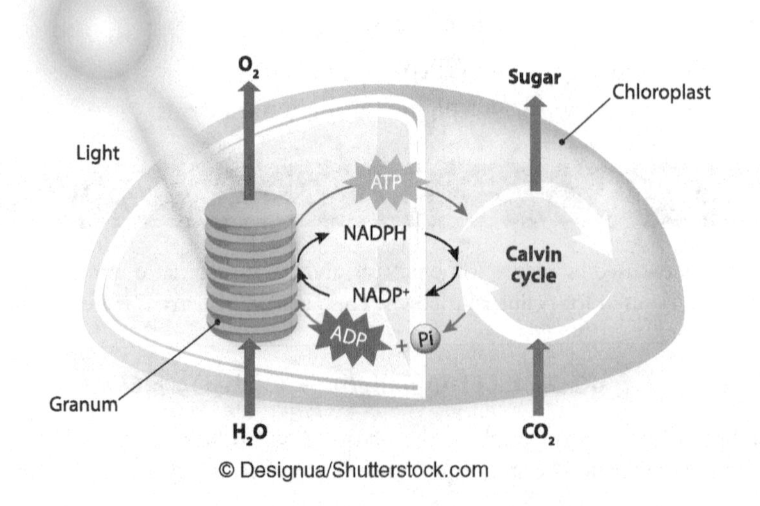

© Designua/Shutterstock.com

The light dependent reaction occurs in the thylakoid stacks of the grana. Light energy is converted to chemical energy in the form of adenosine triphosphate (ATP) and nicotinamide adenine dinucleotide phosphate (NADPH). After light energy is absorbed by the various pigments, chlorophyll a starts the process that produces ATP, NADPH, and oxygen, by splitting water. The ATP and NADPH produced in the light cycle, power the Calvin cycle, which occurs in the stroma, and using the ATP and NADPH, carbon dioxide is converted to sugar.

EXERCISE 1 Separating and Analyzing Photosynthetic Pigments

The chloroplasts contain the photosynthetic pigments that absorb light energy. The principal photosynthetic pigment is chlorophyll a, which absorbs blue-violet and red light and plays the central role in the conversion of light energy into chemical energy. Another type of chlorophyll, chlorophyll b, absorbs mainly blue and orange light but only plays an indirect role in photosynthesis by transferring the light energy it absorbs to chlorophyll a. In addition to the chlorophylls, there are other pigments, the carotenes and xanthophylls that function to increase the overall absorbance of light energy by absorbing light that is not absorbed by the chlorophylls. The carotenes and xanthophylls mainly absorb blue-green light and then transfer the light energy they absorb to chlorophyll a. Plant leaves appear to be green because green and yellow light are not effectively absorbed by the photosynthetic pigments in plants and are reflected.

To be able to identify the various pigments in a plant's leaves, the pigments need to be separated. Paper chromatography is a technique than can be used to separate the photosynthetic pigments of a plant's leaves. Paper chromatography has two phases---a stationary phase (an absorbent paper) and a mobile phase (a solution that travels up the stationary phase carrying the various photosynthetic pigments with it). The various pigments will separate according to their relative attraction to the components of the stationary phase as the mobile phase passes over it.

The relative rate of migration (R_f value) for each pigment is the ratio of the distance traveled by the pigment divided by the distance travelled by the solvent front. This value should be consistent when the chromatography is carried out the same way with the same conditions.

Purpose

To use paper chromatography to determine he absorption spectrum for spinach.

Materials

- Spinach leaves
- Mortar and pestle
- Acetone
- Graduated cylinder
- Sea sand
- Beaker
- Scissors
- Chromatography paper
- Chromatography set-up (see below)
- Chromatography solvent
- Disposable pipettes
- Pencil
- Toothpicks

- Metric ruler
- Spectrophotometer
- Cuvettes

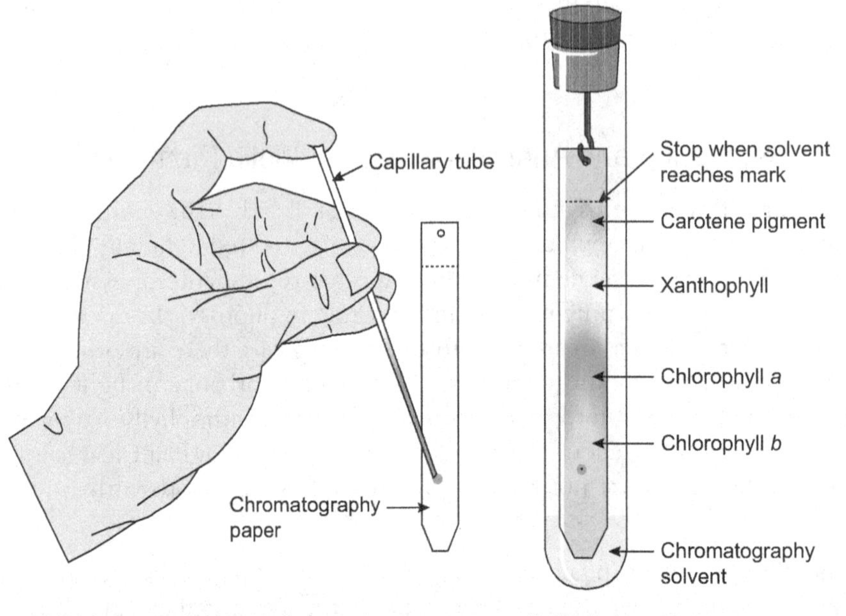

A. Applying spot of chlorophyll pigment

B. Final chromatogram

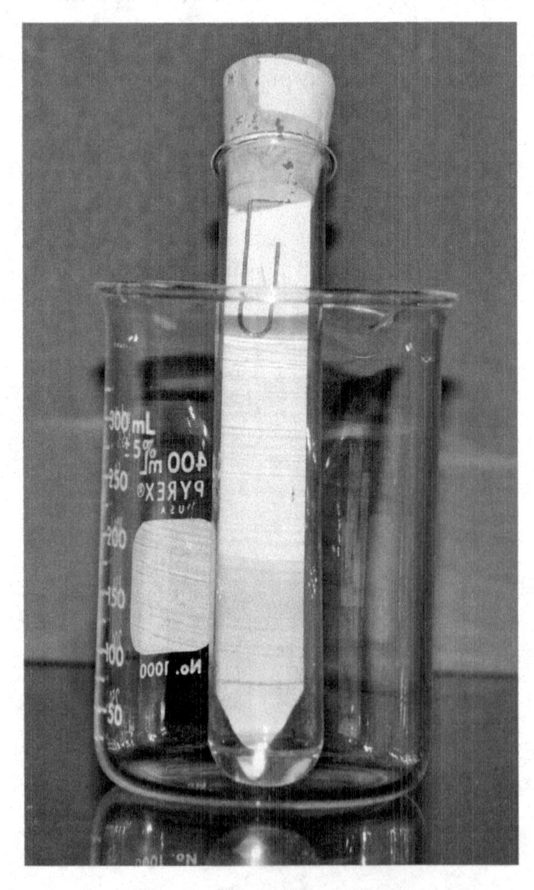

Chromatography set-up

(Photo courtesy of Holly J. Morris and David T. Moat)

Methods

- Obtain two or three spinach leaves and place them in a mortar and pestle with a pinch of sea sand and 10 mL of acetone.
- Grind the spinach leaves with the pestle until they have the consistency of a liquid. Pour the pigment extract solution into a beaker.
- Cut a piece of chromatography paper so that it fits into the chromatography set-up as shown above. Handle the chromatography paper only by edges. Do not touch the front or the back of the chromatography paper to prevent depositing oil and dirt from your hands on the chromatography paper and thereby contaminating it.
- Cut the bottom of the chromatography paper to a point.
- Draw a faint pencil line across the chromatography paper about 2 cm from the bottom.
- Using a toothpick, paint a thin strip of the pigment extract solution over the pencil line. After the solution has dried, repeat the application 20 to 25 times.
- Put approximately 2 mL of chromatography solvent in the bottom of the chromatography set up and then place the tip of the chromatography paper into the solvent. Be certain that the pigment extract solution on the chromatography paper does not touch the chromatography solution.
- Remove the chromatography paper from the chromatography set up when the solvent line approaches the top of the chromatography paper. Below is a chromatogram showing the distances traveled by the various pigments in a paper chromatography exercise.

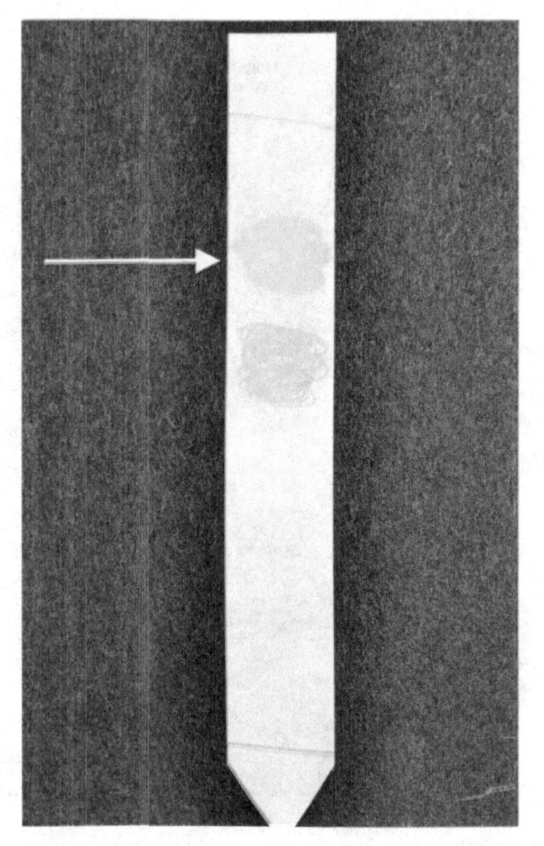

(Photo courtesy of Holly J. Morris and David T. Moat)

- Quickly mark the solvent line before the solvent line evaporates.
- Measure the distance that the solvent front and each of the solution components traveled and use this data to calculate the R_f values in the **R_f Values** table.
- Place the chlorophyll a bands (blue-green) from the chromatography paper into a small beaker and then add 10 mL of acetone.

- Stir the mixture to dissolve the chlorophyll a.
- Place the mixture containing the dissolved chlorophyll a into a cuvette. Fill another cuvette with pure acetone and use this cuvette as the "blank."
- Turn on the spectrophotometer and measure the absorbance of chlorophyll a from 380 nm to 720 nm in increments of 20 nm.
- Appropriately dispose of all chemical used in this exercise.
- Record your results in the **Absorption Spectrum of Chlorophyll a** table.

Note: Pour all of the liquids used in this exercise in a fume hood.

EXERCISE 2 Microanatomy of a Leaf

The function of a leaf is to collect sunlight for photosynthesis. The leaf structure consists of an upper and lower epidermis that secretes a waxy layer called the cuticle. The cuticle helps prevent water from escaping from the leaf, and protects the internal structures from insects, bacteria, and other invaders. Stomata, found primarily in the lower epidermis, are openings surrounded by two guard cells that allow for gas exchange in and out of the leaf. The next layer from the top is the palisade mesophyll that consists of elongated cells. Most of the cell's chlorophyll is found in this layer and this is where most of the plants photosynthesis occurs. The spongy mesophyll consists of loosely packed cells that hold both the raw materials for, and the products of, photosynthesis.

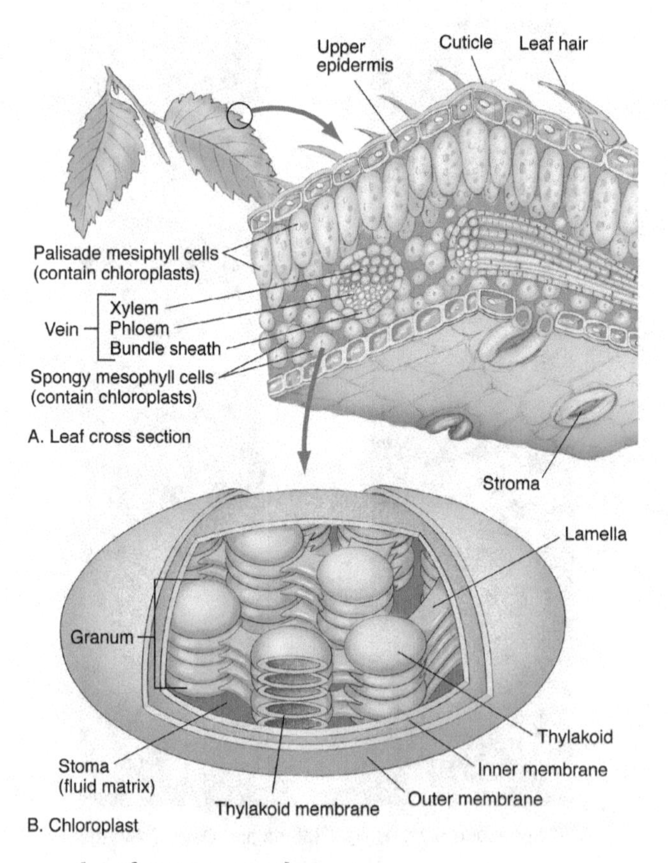

Leaf cross section (© Kendall Hunt Publishing)

Plants have developed three metabolic pathways to fix carbon dioxide for photosynthesis—C3, C4, and CAM carbon fixation. The major difference between the three types of photosynthesis is how the Calvin cycle is carried out.

C_3 fixation is the optimal type of photosynthesis for plants in moist conditions while the optimal environment for C_4 plants is warm, sunny, and dry conditions. The optimal condition for CAM plants is desert-like conditions. Because of the differences in their environments, each of these pathways has evolved different leaf structures to accommodate their environments.

Purpose

To become familiar with the internal structure of C_3 and C_4 leaves and also with stomata and their guard cells.

Materials

- Compound light microscope
- Prepared cross section slide of a C3 leaf
- Prepared cross section slide of a C4 leaf
- *Tradescantia* leaf
- Scalpel
- Clean slides
- Cover slips

Methods

- Obtain a prepared cross section slide of a C3 leaf and observe the cross section of the leaf with a compound light microscope. Make a detailed drawing of the cross section of the leaf in the space provided.
- Obtain a prepared cross section slide of a C4 leaf and observe the cross section of the leaf with a compound light microscope. Make a detailed drawing of the cross section of the leaf in the space provided.
- Obtain a *Tradescantia* leaf. Peel the epidermal layer on the lower side of the leaf of this C_3 plant. Use a scalpel to help separate the epidermal layer. Make a wet mount and observe the stoma and guard cells with a compound light microscope and make a detailed drawing of these structures in the space provided.

EXERCISE 3 Photosynthesis in *Elodea*

Photosynthesis may be summarized as follows:

$$\text{Carbon dioxide} + \text{Water} \xrightarrow[\text{Chlorophyll}]{\text{Light}} \text{Glucose} + \text{Oxygen}$$

Carbon dioxide is the source of carbon which is converted into carbohydrates during the light independent stage (Calvin Cycle) of photosynthesis.

Purpose

To demonstrate the uptake of carbon dioxide during the light independent reactions of photosynthesis.

Materials

- *Elodea* sprigs
- Bromothymol blue solution

- Two test tubes
- Test tube rack
- Light source
- Straw

Methods

- Fill both test tubes about ¾ full with water and place them in a test tube rack.
- Add several drops of bromothymol blue to both test tubes.
- Using the straw, blow into one of the test tubes until the bromothymol blue turns yellow. The yellow color indicates that the solution has become acidic because of the addition of carbon dioxide.
- Place several *Elodea* sprigs in the now yellow test tube.
- Place both test tubes in direct sunlight or under a bright light for 20–30 minutes.
- Note the color change in the test tube containing the *Elodea* sprigs which occurs because the *Elodea* is using up the carbon dioxide as it undergoes photosynthesis.

EXERCISE 4 Oxygen Production in *Elodea*

Photosynthesis may be summarized as follows:

$$\text{Carbon dioxide} + \text{Water} \xrightarrow[\text{Chlorophyll}]{\text{Light}} \text{Glucose} + \text{Oxygen}$$

Water is the source of the electrons in the light dependent stage of photosynthesis that generates the ATP and NADPH needed to power the subsequent light independent stage. Oxygen is produced as a by-product of these light dependent reactions.

Purpose

To demonstrate the production of oxygen as a by-product of the light dependent reactions in photosynthesis.

Materials

- *Elodea* sprigs
- Test tube
- Large beaker
- Funnel
- Test tube holder
- China marking pencil
- Light source

Methods

- Place several *Elodea* springs in a large beaker ¾ filled with water.
- Place the funnel over the *Elodea*.
- Invert a filled test tube and place it over the funnel stem. Use a test tube holder to hold the test tube in place.
- Mark the water level on the test tube with the China marking pencil.
- Place the set up in direct sunlight or under a bright light for 20–30 minutes.

- Observe the oxygen bubbles that are produced by the *Elodea* as a result of photosynthesis.
- After 20–30 minutes, mark the water level on the test tube indicating the amount of oxygen produced.

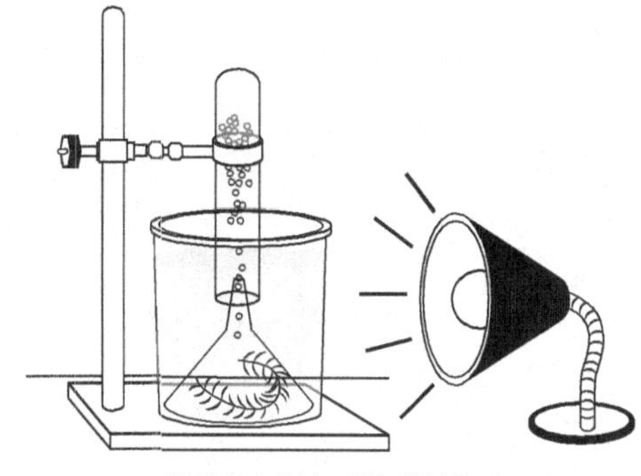

Name:

Exercise 1 Results

1. Calculate the R_f value for each pigment:

R_f Values			
Pigment	**Color**	**Distance Traveled**	**R_f Value**
Chromatography solvent			
Carotene			
Xanthophyll			
Chlorophyll a			
Chlorophyll b			

2. Complete the table with your absorption spectrum data.

Absorption Spectrum of Chlorophyll a					
Wavelength	**Absorbance**	**Wavelen gth**	**Absorbance**	**Wavelength**	**Absorbance**
380 nm		500 nm		620 nm	
400 nm		520 nm		640 nm	
420 nm		540 nm		660 nm	
440 nm		560 nm		680 nm	
460 nm		580 nm		700 nm	
480 nm		600 nm		720 nm	

3. Prepare the spectral absorbance curve of chlorophyll a

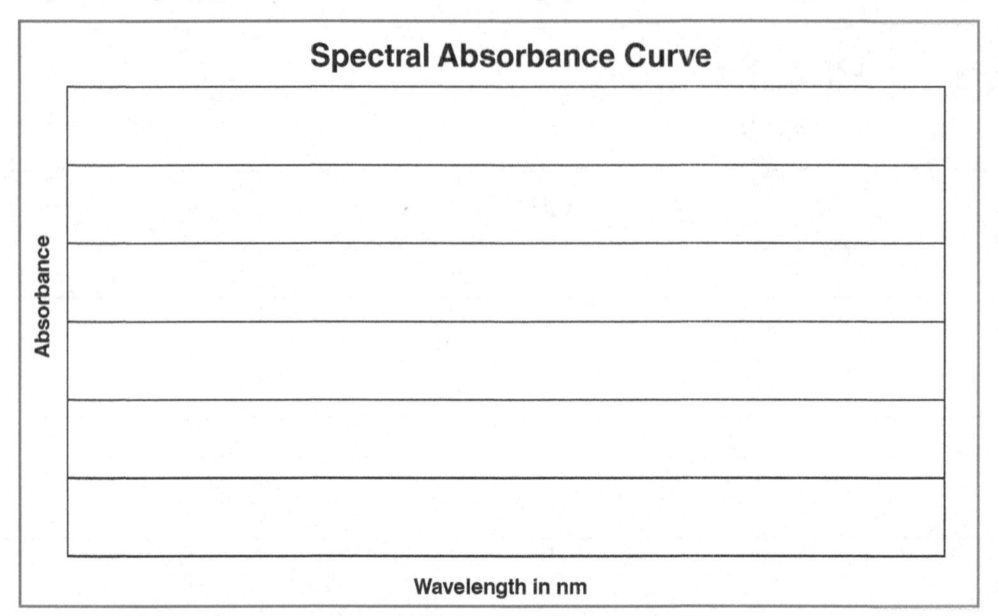

4. What is the basic theory of paper chromatography?

____

____

5. What is significance of the R_f value?

____

____

6. What is the purpose of the accessory pigments in photosynthesis?

____

____

7. The yellow xanthophyll and yellow carotene are also present in the spinach extract. Why did the extract appear green?

____

____

8. Which of the pigments moved the farthest from the point of origin? Why?

____

____

9. At what wavelength(s) does chlorophyll a absorb the most energy? How can you tell? What is the significance of this observation?

____

____

10. Briefly describe the major differences between the light dependent and light independent phases of photosynthesis.

____

____

____

Exercise 2 Results

1. Male a **detailed drawing** of a C3 and a C4 leaf cross section and label the following structures (if present): cuticle, epidermis, palisade parenchyma, and spongy parenchyma.

C3 Plant:	**C4 Plant:**
Total Magnification	Total magnification:

2. What is/are the main difference(s) between the leaves of C3 and C4 plants?

____

____

3. What is/are the main difference(s) between C3 and C4 photosynthesis?

____

____

4. Make a **detailed drawing** of an open and closed *Tradescantia* stomata.

Open Stomata	**Closed Stomata**
Total Magnification	Total magnification:

5. How do guard cells regulate the opening of the stomata?

Exercise 3 Results

1. Describe the results of the uptake of carbon dioxide exercise.

2. What is the significance of the color change in the test tube containing the bromothymol blue?

3. If you were to wrap the test tube containing the bromothymol blue in aluminum foil, what would happen?

Exercise 4 Results

1. How do you know that oxygen was produced?

2. Where did the oxygen that was produced come from?

3. How could you design an experiment to measure the rate of oxygen production?

CHAPTER 21

Classification

Photo courtesy of Holly J. Morris

OBJECTIVES

After completing this exercise, you will be able to:

- Define taxonomy
- Understand and use a dichotomous key
- Construct a dichotomous key
- Understand and use a cladogram
- Construct a cladogram

INTRODUCTION

Humans have a need to classify objects into related groups. In ancient times it would have been important, at the very least, to be able to identify plants and animals that were beneficial or dangerous. More recently, categorizing organisms became structured, following common rules. The science of classifying organisms is called **taxonomy**, which means, essentially, "arrangement system."

The person most recognized for developing a structured system of classification is Carl Linnaeus, a Swedish botanist, who published his first edition of *Systema Naturae* in 1735. His two most important contributions to taxonomy were the development of hierarchical categories, and the system of binomial nomenclature for naming organisms, i.e., two-part names. Linnaeus' original hierarchy contained kingdom, class, order, genus, and species. Over time, modifications have occurred, adding domain, phylum, and family.

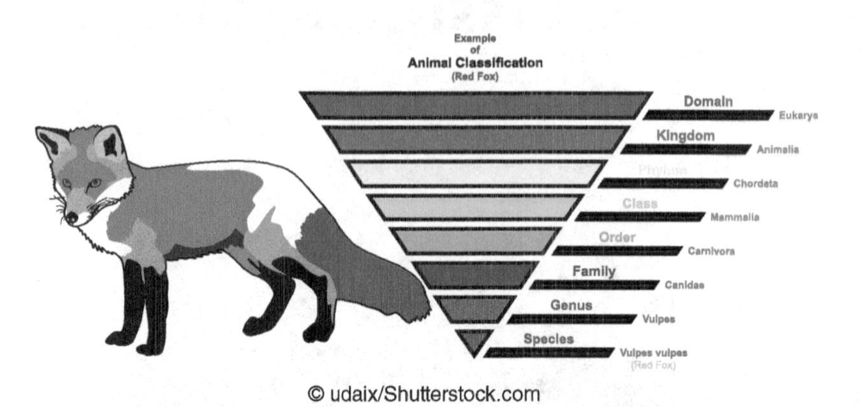

© udaix/Shutterstock.com

The system of binomial nomenclature, which is still used today, is the formal naming system for all living organisms, and uses the genus and species of each organism. The first part of the name, the genus is capitalized and either italicized or underlined. The second part of the name is the species and is lower case and either italicized or underlined, for example *Homo sapiens*.

Classification was originally based on morphology, that is, physical characteristics. Organisms were grouped by visible similarities starting with kingdom (e.g., animal) and getting more and more specific at each level of the hierarchy. This was a good start but classification solely on the basis of appearance was soon recognized as inadequate. As laboratory techniques became more sophisticated, the evolutionary relationship between groups of organisms was easier to discern. Using molecular biology, similarities and dissimilarities were easier to recognize, and organisms that had been grouped together on the basis of a morphological feature were being reclassified on the basis of molecular features. This led to a new field called **systematics**, which evaluates the evolutionary relationship between organisms. An evolutionary relationship can be drawn as a tree and is called a **phylogeny**. The image below is a representative phylogeny.

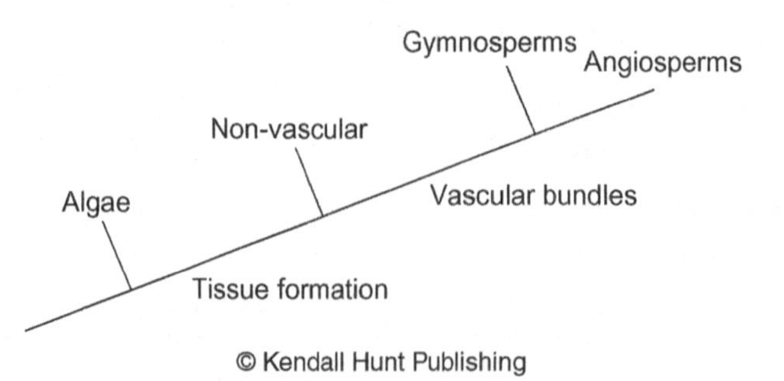

© Kendall Hunt Publishing

EXERCISE 1 Using and Constructing a Dichotomous Key

Using a Dichotomous Key

In this exercise, you will use and then construct a dichotomous key. A dichotomous key is a type of taxonomic key that gives you two choices, and at each level you choose one of the choices. Dichotomous keys are constructed so that there is one, and only one, choice at each level. For example, if you were looking at leaves on a tree and the dichotomous key gave you the choice between compound leaves and simple leaves, there would be only one choice.

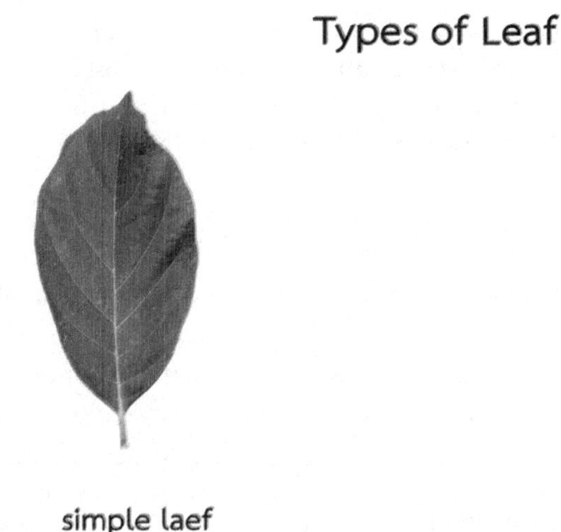

Dichotomous keys are clear and systematic. They must be followed in order, with no skipping of any steps. Keys usually start with large encompassing features and get more detailed with subsequent steps.

Constructing a Dichotomous Key

Constructing a dichotomous key takes patience and care. Here are some suggestions for steps.

1. Observe the features of the organisms and make notes about both common features and distinctive features.
2. Organize the organisms into two tentative groups based on a mutually exclusive characteristic. Rearrange groups, using a new feature, if necessary.
3. At each step, make sure you write down the dichotomous feature.
4. Subdivide each of those groups into two groups, again using a mutually exclusive characteristic.
5. Continue until you have separated all of your organisms.
6. Complete your key and doublecheck it to make sure each step is mutually exclusive.

Purpose

To become familiar with the use of dichotomous keys and the methodology to construct a dichotomous key.

Methods

- Use a dichotomous key to identify the bords.
- Develop a dichotomous key to identify the various types of candies listed.

Go to "Exercise 1 Results" to use and then create a dichotomous key.

EXERCISE 2 Constructing a Cladogram

When different organisms have many common homologous structures (i.e., structures that share the same embryological origin), their presence indicates that the organisms had a common ancestor. However, these structures may be modified from the ancestral trait (i.e., derived shared characteristics) and be functionally different. For example, the wings of a bat and the forelimb of a human are homologous structures but they perform different functions.

A cladogram is a diagram connecting organisms that shows their different derived shared characteristics. The more shared derived structures that two organisms share, the closer is their evolutionary relationship and the more recently they had a common ancestor. Close relationships are shown by a recent split from the supporting branch. The closer the split between the two organisms, the closer is their relationship.

To construct a cladogram:

- Determine the derived-shared characteristics that will be compared. In the following example, morphological traits are compared.
- Determine the number of derived shared characteristics for each of the groups of organisms being analyzed. Building a table, as shown below, provides a method to organize your thoughts.

Derived Shared Characteristics	Fish	Amphibian	Birds	Mammals
Vertebrae	x	x	x	x
Tetrapod		x	x	x
Amniotic eggs			x	x
Hair				x

- Build a Venn diagram containing the organisms and their derived traits as shown below.

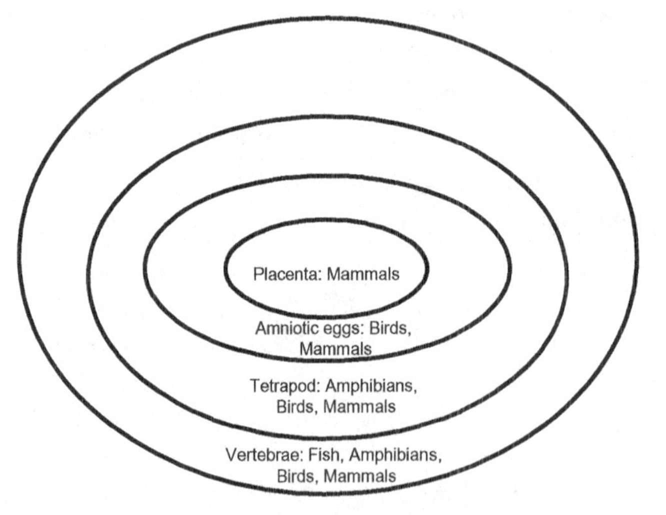

- Convert your Venn Diagram into the cladogram. Indicate shared derived traits after the branch points.

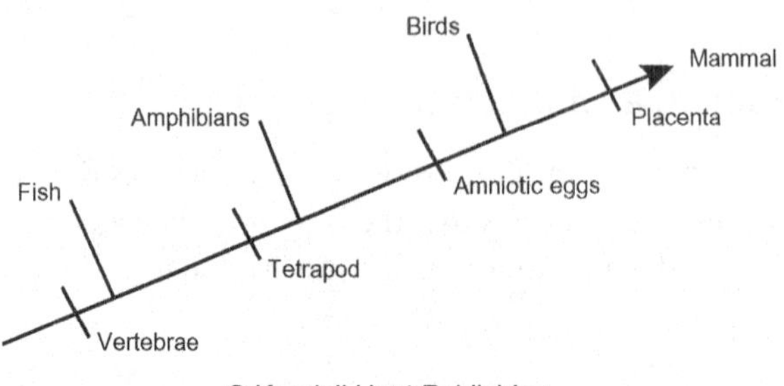

Purpose

To become familiar with the methodology to develop a cladogram.

Methods

- Develop a cladogram using the data provided.

Go to "Exercise 2 Results" to construct a cladogram.

Notes

Name:

Exercise 1 Results

1. Use the dichotomous key shown below to identify the species of bords. Write their complete scientific name (genus + species) in the blank line below each of the bords.

Bords belong to the genus *Bordo* and can be divided into eight species that are generally located in specific regions of the world.

Drawings courtesy of Holly J. Morris

1. Has no comb ....go to 2
....Has comb ....go to 3

2. Has pointed tail ....Arcturus
....Has bushy tail ....Telsius

3. Comb has no feathers ....go to 4
....Comb has feathers ....go to 5

4. Has pointed tail ....Kiberius
....Has bushy tail ....Malcorius

5. Has pointed tail ....Regula
....Has bushy tail...go to 6

6. Has downward beak ....Talaxius
....Beak points upward ....go to 7

7. Feathers are red and yellow ....Cardassius
....Feathers are purple ....Pentaris

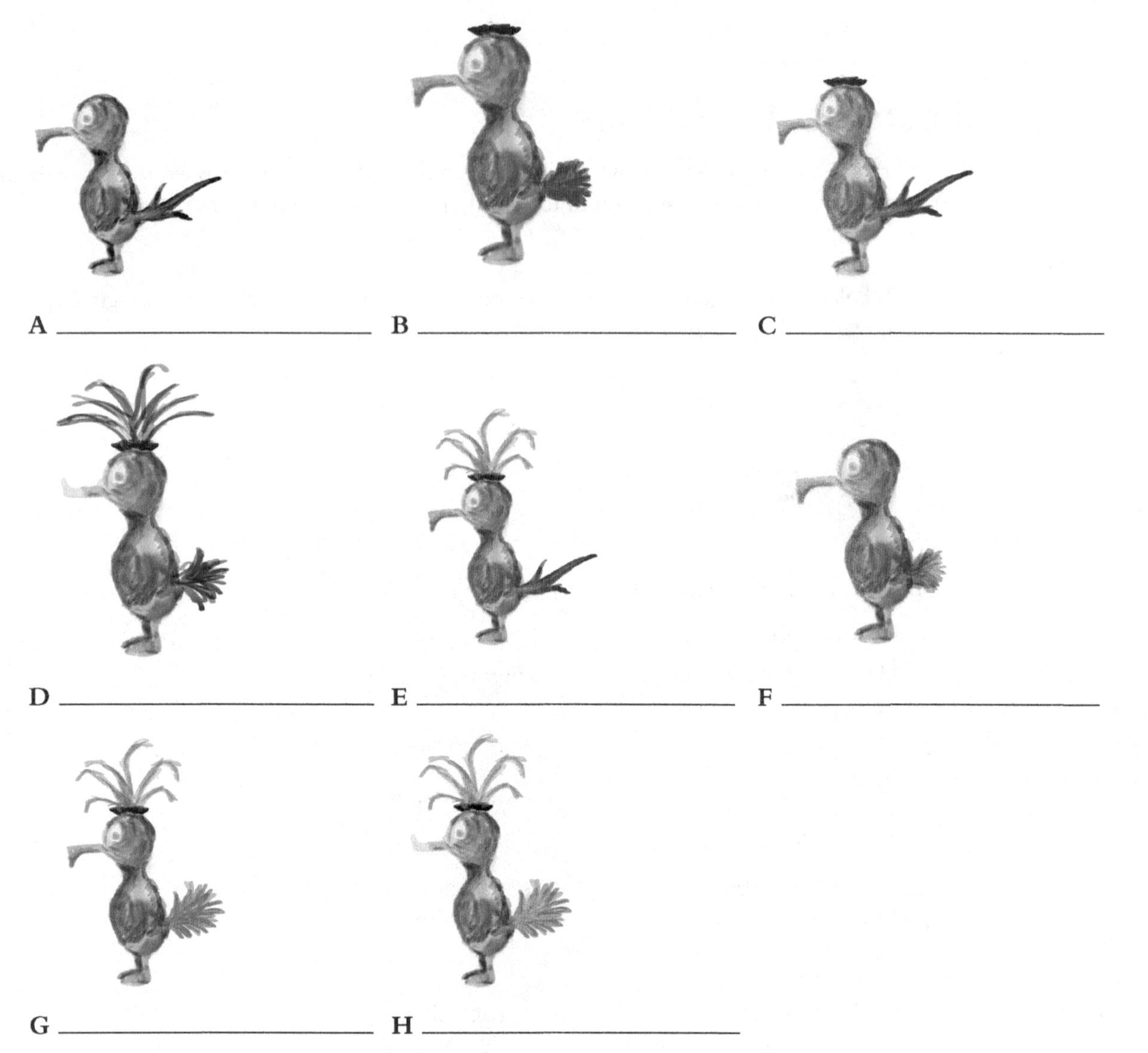

Drawings courtesy of Holly J. Morris

2. In the space provided, create a dichotomous key to identify the 10 types of Hershey products listed below:

A. Kit Kat Bars

B. York Peppermint Pattie

C. Almond Joy Bars

D. Hershey's Kisses

E. Hershey's Milk Chocolate Bar

F. Jolly Rancher

H. Reese's Peanut Butter Cups

I. Milk Chocolate Whoppers

J. Reese's Pieces

K. Cadbury Crème Eggs

1. a. ____

b. ____

2. a. ______________________________
   b. ______________________________
3. a. ______________________________
   b. ______________________________
4. a. ______________________________
   b. ______________________________
5. a. ______________________________
   b. ______________________________
6. a. ______________________________
   b. ______________________________
7. a. ______________________________
   b. ______________________________
8. a. ______________________________
   b. ______________________________
9. a. ______________________________
   b. ______________________________
10. a. ______________________________
    b. ______________________________
11. a. ______________________________
    b. ______________________________
12. a. ______________________________
    b. ______________________________

Exercise 2 Results

1. Complete the table below to display the relationships between the organisms listed.

	Derived Shared Characteristics					
	Placenta	**Limbs**	**Hair**	**Segmented**	**Jaws**	**Multicellular**
Catfish						
Earthworm						
Sponge						
Horse						

Gecko						
Paramecia						
Kangaroo						

2. Develop a simple cladogram that includes the animals listed above. Be sure to include the nodes and clades showing their relationships based on their derived shared characteristics.

CHAPTER 22

Protists

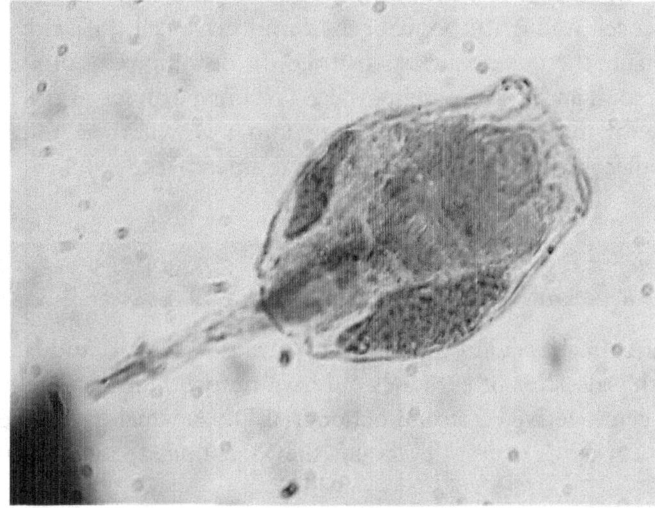

Photo courtesy of Holly J. Morris and David T. Moat

OBJECTIVES

After completing this exercise, you will be able to:

- Discuss the diversity within protists.
- Discuss the phylogenetic relationships between the various groups of protists.
- Identify representatives of each major group of protists.
- Identify the distinguishing features of each major group of protists.
- Discuss the ecological role and economic importance of each major group of protists.

INTRODUCTION

The first eukaryotic organisms to appear on Earth were the protists. Since protists are composed of *eukaryotic cells* there are clear distinguishing features that separate them from the prokaryotes, such as a distinct nucleus and membrane-bound organelles in the cytoplasm. Protists also have cytoskeletal components that provide support and motility (flagella and cilia). Protists use mitosis for asexual reproduction and many can also reproduce sexually by the fusion of "gametes." Most protists are *unicellular*, but there are a few species of *multicellular* algae. There are a variety of feeding strategies used by protists; many are *autotrophic* (photosynthetic) and many are *heterotrophic*.

Traditionally protists were grouped into one kingdom, the Kingdom Protista. Today with the use of molecular systematics it is widely accepted that the organisms placed in the Kingdom Protista represent several different, not too closely related lineages, making many biologists create several different kingdoms out of what used to be Kingdom Protista. Some biologists even believe that members in this group can be separated out into 30 different kingdoms.

Originally the Kingdom Protista contained organisms that biologists could not place with the plants, fungi, or animals. So basically the protist group became a hodgepodge of many unrelated species, up to around 65,000 species, many of which are more closely related to higher eukaryotes than to other protists (based on more recent molecular systematics. Protists that were photosynthetic, and more plantlike, were all grouped together and called algae. While all algae share some basic features, algae as a group or clade is not a monophyletic taxon. The animal-like protists were called protozoa. They were all grouped together because of their lack of a cell wall (recall animal cells do not have cell walls either) and the fact that they are all heterotrophic. Finally, there were protists that fed on decomposing organic material; they are the slime molds and water molds and are more funguslike. For our purposes in the lab we will use the term *protist* as a generic name for any organism in this "kingdom"; we will then study the major evolutionary clades as indicated by molecular systematics and structural differences.

EXCAVATA

Excavates are unicellular animal parasites that move by means of flagella and lack mitochondria in their cells. It is thought that the ancestors of excavates did have mitochondria and they have been secondarily lost (they have nuclear genes derived from mitochondrial DNA which suggest that mitochondria were once present). We will study two groups of excavates, the Metamonada and the Euglenozoa.

1. *Metamonada*

   (a) *Diplomonadida*

   Diplomonad cells possess two nuclei and have multiple freely beating flagella. Diplomonads also lack endoplasmic reticulum and Golgi complex. The best known diplomonad is the parasite *Giardia lamblia*, which infects the gastrointestinal tract of mammals and causes diarrhea and abdominal cramps. *Giardia* is found in freshwater streams and lakes. Many hikers contract *Giardia* when they drink unfiltered stream or lake water.

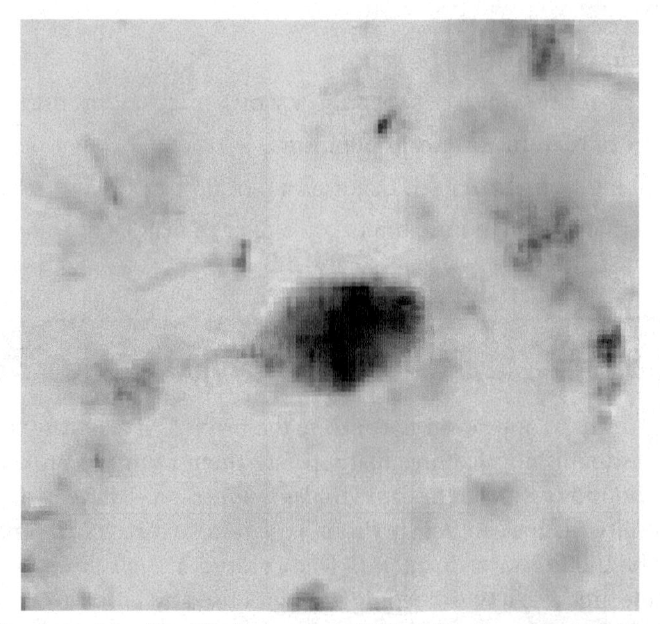

Giardia lamblia, 400× (Photo courtesy of Holly J. Morris and David T. Moat)

(b) *Parabasala*

Parabasala species have multiple freely beating flagella and a single flagellum buried in a cytoplasmic fold called an *undulating membrane*.The buried flagellum allows the cell to move through highly viscous fluids. The most common members of this clade are the trichomondas. Many urogenital infections are caused by *Trichomonas vaginalis*, which is often sexually transmitted (though not transmitted exclusively by sexual means).

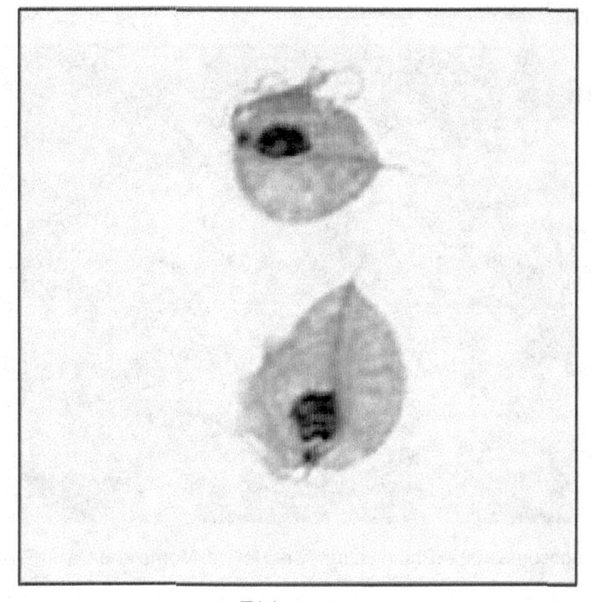

Trichomonas
(http://www.cdc.gov/std/trichomonas/stdfact-trichomoniasis.htm)

2. *Euglenozoa*

(a) *Euglenids*

The euglenid group is mainly freshwater, photosynthetic protists. Many euglenoids can absorb organic compounds through their cell membrane, so they can be autotrophic and heterotrophic at the same time. A few euglenoids are not photosynthetic and are simply heterotrophs. Euglenoids have a variety of cytoplasmic organelles including chloroplasts and contractile vacuoles. Most of the photosynthetic euglenoids have an *eyespot*, a cluster of carotenoid pigments that functions in light detection. Many members of this group belong to the genus *Euglena*.

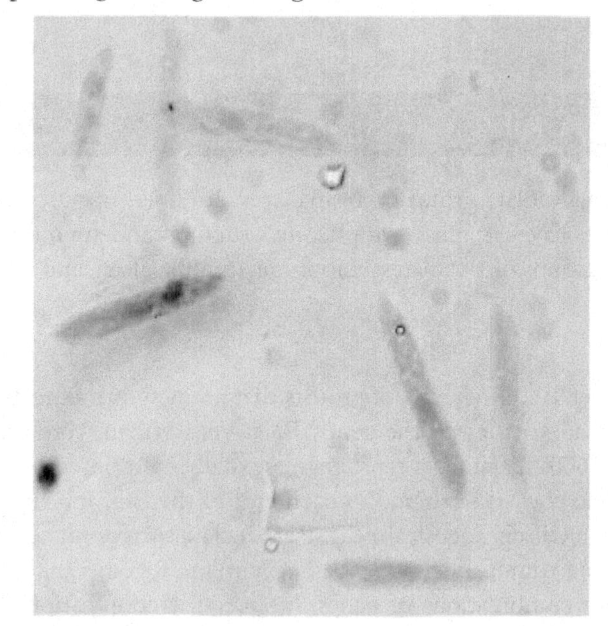

Euglena 400x (Photo courtesy of Holly J. Morris and David T. Moat)

(b) *Kinetoplastids*

Kinetoplastids are animal parasites that have a single mitochondrion, which contains a large DNA/protein deposit called a *kinetoplast*. Most kinetoplastids have two flagella; sometimes one of the flagella can form an undulating membrane. The most common kinetoplastids are the trypanosomes which are responsible for diseases that inflict millions of humans each year. Common trypanosome diseases include African sleeping sickness, Chagas disease, and leishmaniasis.

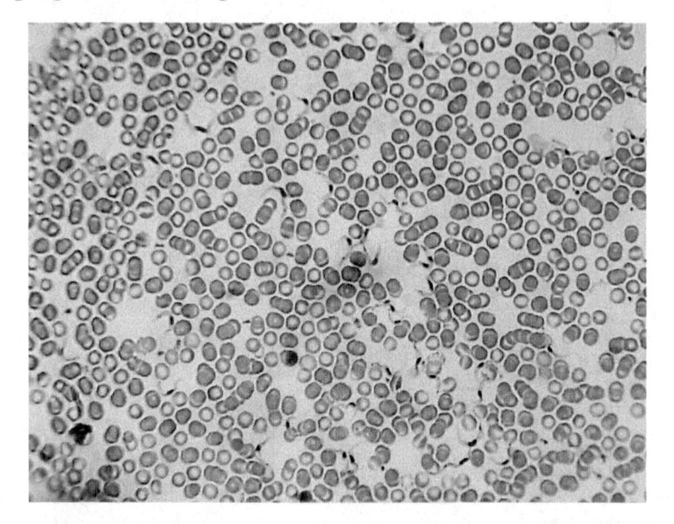

Trypanosoma 400x (Photo courtesy of Holly J. Morris and David T. Moat)

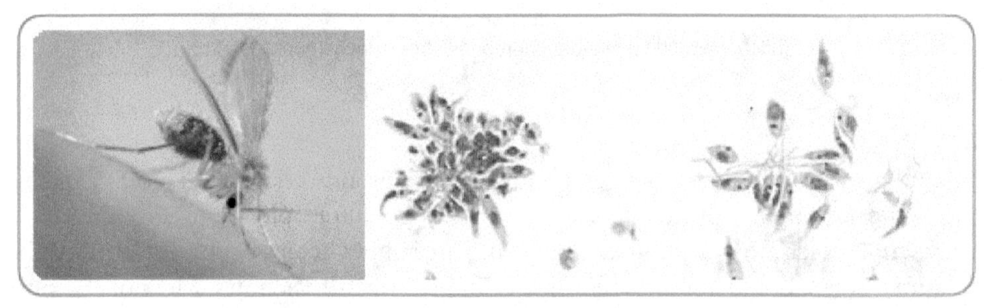

Leishmania
(http://www.cdc.gov/parasites/leishmaniasis/index.html)

CHROMALVEOLATA

All members of the Chromalveolata group have small vesicles lined up just below their cell membrane, called *alveoli*. Most alveolates have complex cytoplasmic structures and are motile, using cilia or flagella to move. We will study three groups of alveolates: ciliophora, dinoflagellata, and apicomplexa.

1. *Alveolata*

(a) *Ciliophora*

Ciliophora includes over 10,000 species of protists that are heterotrophic and move by means of *cilia*. The most common ciliates belong to the genus *Paramecium* which were first observed by Anton van Leeuwenhoek in the 1600s. Paramecia reproduce asexually by a process called *fission*. During fission the DNA is copied, one copy is separated to each end of the cell, and the cell pinches in half transversely creating two genetically identical cells. Paramecia also reproduce sexually through a process called *conjugation*. During conjugation two paramecia merge together and swap DNA, then split back apart carrying different combinations of genes. Notice that conjugation does not actually produce more cells, just genetic recombinants. Conjugation is typically used when environmental conditions become harsh. Genetic recombinants may be able to better withstand environmental changes and have a better chance at continuing the population or species.

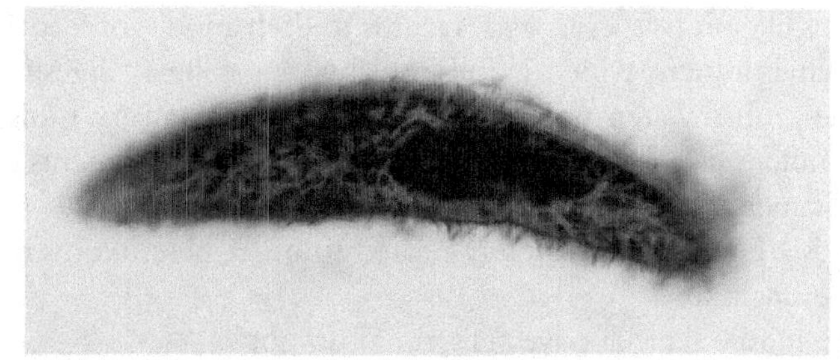

Paramecium 400x (Photo courtesy of Holly J. Morris and David T. Moat)

(b) *Dinoflagellata*

Most dinoflagellates are unicellular autotrophs that comprise much of the *phytoplankton* in marine waters. Most have a shell formed from *cellulose plates* and move by means of flagella. The flagella often lie in grooves on the cellulose plates, which make the cells spin when the flagella are beating (*dinos* = spinning). Seasonal population blooms of dinoflagellates color the water red, orange, or brown. This phenomenon is called the *red tide*. Red tide dinoflagellates may produce toxins that kill fish and any animals that consume the infected fish.

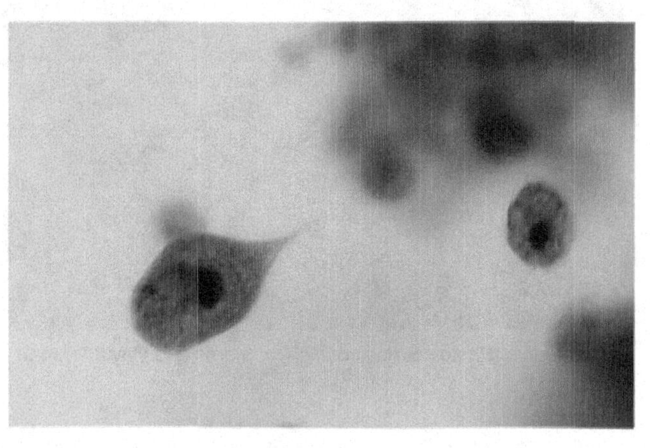

Dinoflagellates 400x (Photo courtesy of Holly J. Morris and David T. Moat)

(c) *Apicomplexa*

Apicomplexans are nonmotile animal parasites. The name of this group comes from the *apical complex*, a group of organelles at one end of the cell that functions for invasion of host cells and attachment. Malaria is caused by the apicomplexan *Plasmodium*.

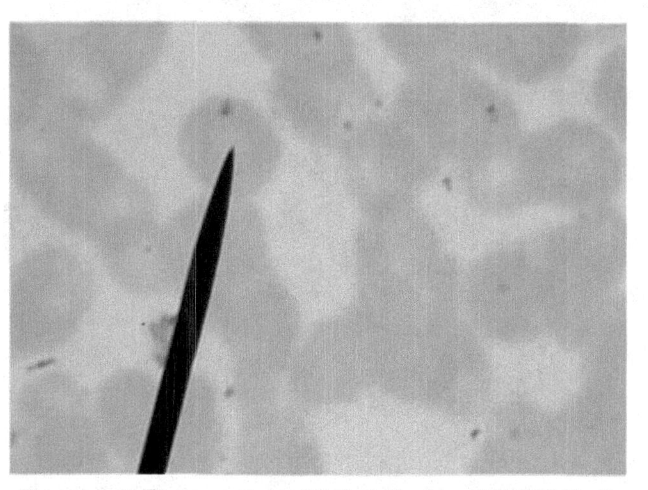

Plasmodium (Photo courtesy of Holly J. Morris and David T. Moat)

2. *Stramenopila*

(a) *Oomycota*

Oomycetes are fungus like protists, even with regards to their modes of feeding. They secrete digestive enzymes into their environment which is followed by extracellular digestion, and the nutrients are then absorbed by the protist. Water molds live in freshwater or in very moist terrestrial habitats. The structure of a water mold is like that of a fungus, having nonmotile filaments called *hyphae* which form a large network of filaments called a *mycelium*. Despite their similarities to fungi, DNA analysis indicates that water molds are much more closely related to other heterokonts than to fungi. In addition, other basic differences include:

- Water molds have motile reproductive cells, fungi do not.
- Water molds have cellulose cell walls, fungi have cell walls made of chitin.
- Water mold cells are diploid, fungal cells are haploid.

Saprolegnia is a common freshwater oomycete.

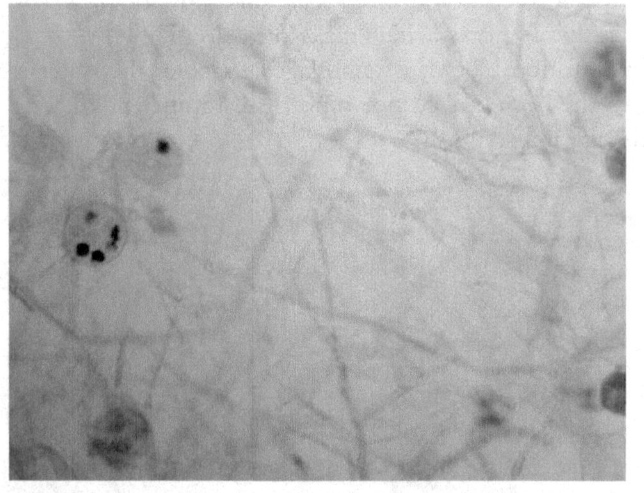

Saprolegnia (Photo courtesy of Holly J. Morris and David T. Moat)

(b) *Bacillariophyta*

Diatoms are a diverse group of unicellular, photosynthetic heterokonts. They are surrounded by a shell of *silica* (glass). Diatoms are important components of phytoplankton in both marine and freshwater habitats. Over 35,000 extinct species of diatoms have been described from fossils. Grinding fossilized shells into a powder creates *diatomaceous earth*, which can be used as an abrasive powder, a filtering agent (like in pool filters), and a pesticide.

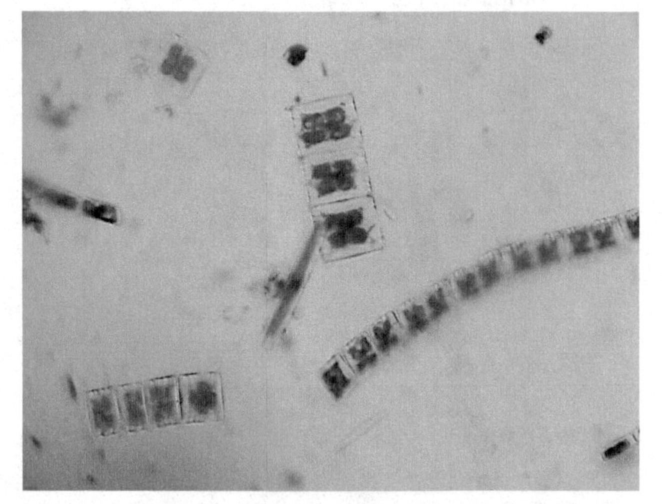

Diatoms 400× (Photo courtesy of Holly J. Morris and David T. Moat)

(c) *Chrysophyta*

Chrysophytes are the golden algae. Many golden algae live in colonies, with each cell in the colony having two flagella. The golden algae are photosynthetic and are important components of the phytoplankton in marine waters. *Synura* is a common genus of mainly freshwater golden algae.

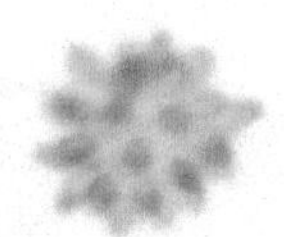

Synura (From Investigating Biology: The Diversity of Life Lab by Paul Florence and Annisa Florence. Copyright © 2013 by Paul Florence and Annisa Florence. Reprinted by permission of Kendall Hunt Publishing Company)

(d) *Phaeophyta*

Phaeophytes are the brown algae. They are photosynthetic and range from microscopic planktonic algae to giant kelps. Most brown algae inhabit cool coastal waters. Kelps are the largest and most complex protists. Phaeophytes obtain their brown color from a pigment called *fucoxanthin*. Their cells and tissues are arranged into specific structures such as *blades* (like leaves), *holdfasts* (like roots), and hollow gas-filled *bladders* for buoyancy. Cell walls of brown algae contain cellulose and a mucilaginous carbohydrate called *alginic acid*. When extracted, then called *algin*, alginic acid is used as a thickening agent in ice cream, pudding, salad dressing, jellybeans, cosmetics, among other products.

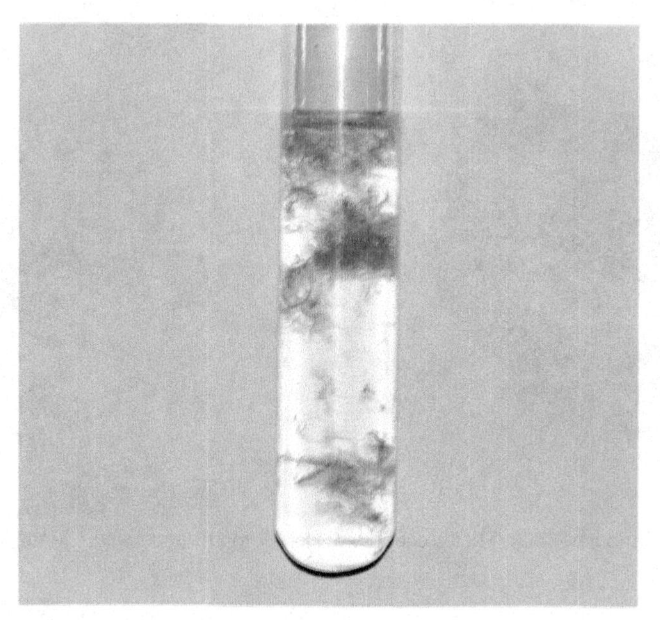

Brown algae (Photo courtesy of Holly J. Morris and David T. Moat)

RHIZARIA

Members of the Rhizaria group are unicellular and move by means of slender extensions of their cytoplasm called *pseudopodia* (pseudopods). These protists are amoeboid (like an amoeba) in form, but their pseudopodia are slender, stiff, and filamentous and are supported intracellularly by parts of the cytoskeleton. Many of the cercozoa produce hard outer shells called *tests*. The two main types of cercozoans are heterotrophs, namely the Radiolaria and the Foraminifera (forams).

1. *Radiolaria*

   Radiolaria are characterized by very slender, long pseudopods called *axopodia*. Prey organisms stick to the axopodia and are then pulled in and digested for food. Radiolarians form internal shells made of *silica*. Radiolarians live in marine habitats.

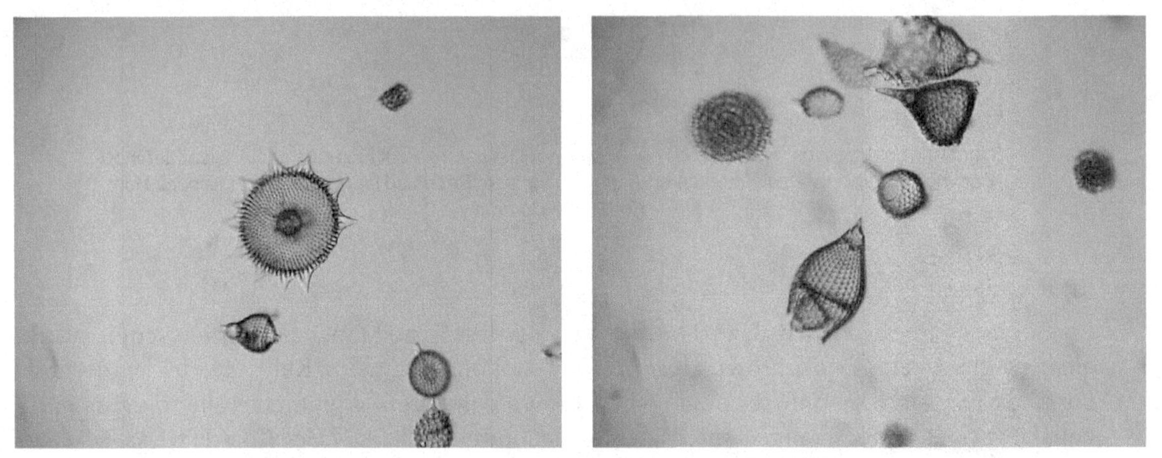

Radiolaria 400x (Photos courtesy of Holly J. Morris and David T. Moat)

2. *Forams*

   Forams are marine cercozoans that secrete tests (shells) of *calcium carbonate*. Marine sediments are typically packed with the tests of dead forams.

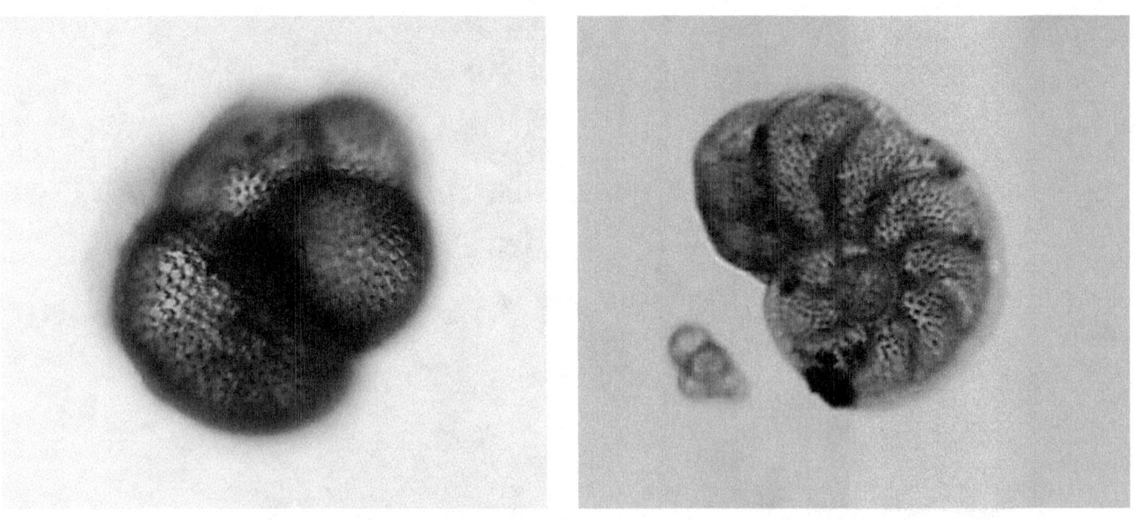

Forams 400x (Photos courtesy of Holly J. Morris and David T. Moat)

ARCHEOPLASTIDA

The Archaeplastida clade includes the red algae and green algae, which are protists, and the land plants that evolved from a group of green algae.

1. *Rhodophyta*

   This group is the red algae, which are mostly small marine seaweeds (a few are found in freshwater). Most red algae are autotrophs (photosynthetic) that attach to sandy or rocky substrates. Many red algae are *multicellular*. They contain pigments called *phycobilins*, which mask the green chlorophylls and make the algae appear reddish-brown (and a variety of other colors). The cell walls of red algae cells contain a mucilaginous carbohydrate which is extracted and used as *agar* in biology laboratories.

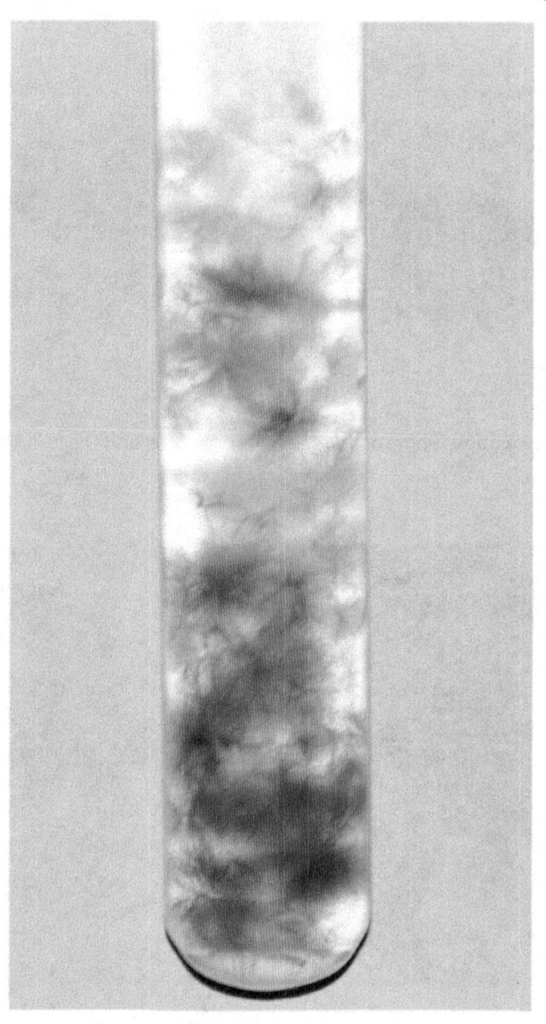

Red algae (Photo courtesy of Holly J. Morris and David T. Moat)

2. *Chlorophyta*

   This group includes the green algae. Green algae are autotrophic (photosynthetic) and can occur as *solitary* cells, as *colonies*, or as long *filaments* of cells. Some green algae were the ancestors of land plants.

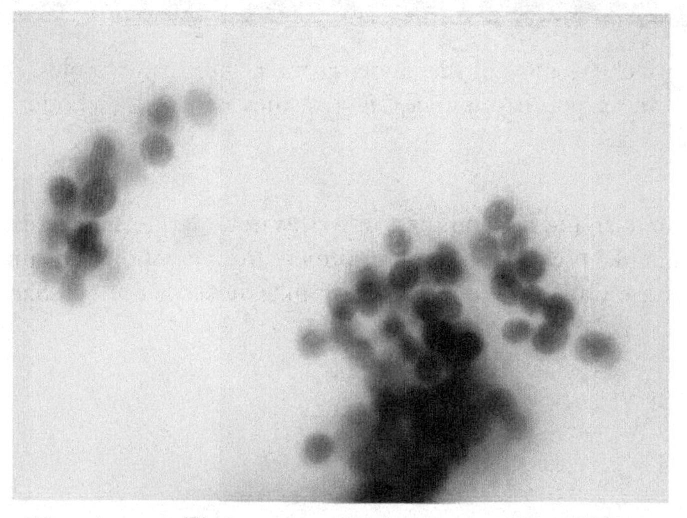

Chlamydomonas (Photo courtesy of Holly J. Morris and David T. Moat)

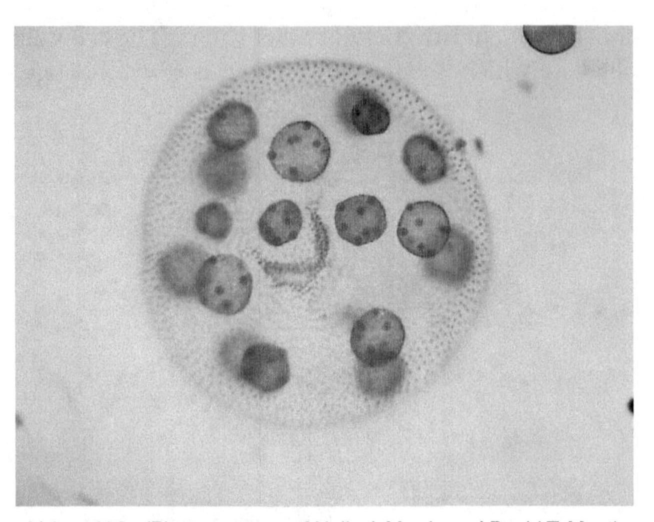

Volvox 100× (Photo courtesy of Holly J. Morris and David T. Moat)

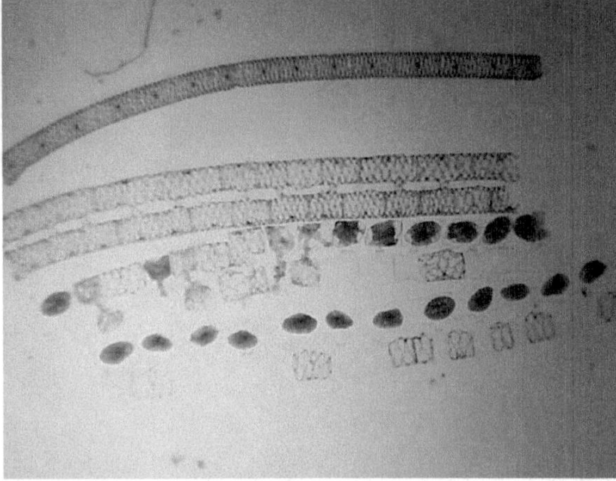

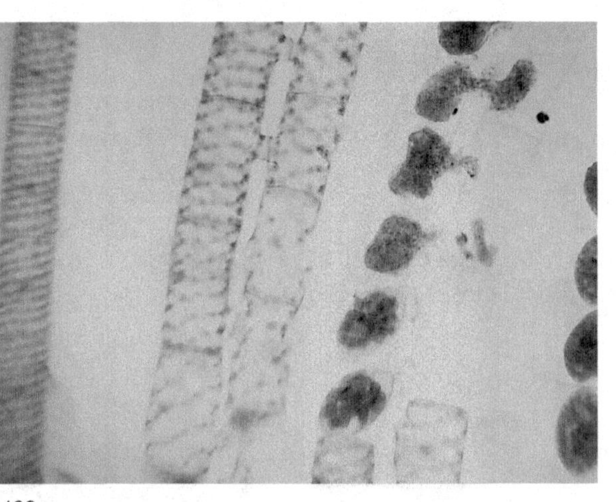

Spirogyra 40×
(Photos courtesy of Holly J. Morris and David T. Moat)

100×

AMOEBOZOA

The Amoebozoa clade includes most of the amoebas, the cellular slime molds, and the plasmodial slime molds. All members of this group have *pseudopodia* that they use for locomotion and feeding (at least at some stage in their life cycle).

1. *Amoebas*

   Amoebas are unicellular and are common in freshwater, marine, and moist terrestrial soils. These amoebas have a lobelike pseudopod that can project from any point around the cell. The pseudopodia of amoebas are not supported internally like those of the cercozoa. Amoebas reproduce entirely asexually.

Amoeba 100× (Photo courtesy of Holly J. Morris and David T. Moat)

2. *Slime Molds*

Slime molds are heterotrophic organisms that exist as amoeboid individuals at some point in their life cycle, and as more complex forms during the remainder of their life cycle. When reproducing, slime molds develop a funguslike stalk called a *fruiting body*; it produces spores by asexual or sexual reproduction. *Cellular slime molds* exist as single cells, either individually or together in a cellular mass. *Plasmodial slime molds* exist as a large mass of cells whose cytoplasm has fused. This mass is called the *plasmodium* (do not confuse this with the apicomplexan genus *Plasmodium*).

Plasmodial slime mold (*Physarum*) (Photo courtesy of Holly J. Morris and David T. Moat)

OPISTOKONTS

The Opisthokonts clade includes choanoflagellates, fungi, and animals. The choanoflagellates are protist ancestors of fungi and animals. We will not be studying any specimens from this clade in this lab; it is included here just to complete the clades to which protists belong.[13]

EXERCISE 1 Observing Protists

The term "protist" refers to a diverse group of eukaryotic, mostly unicellular, organisms that cannot be classified as plants, animals, or fungi. They do not form a clade. Rather, they are a paraphyletic collection of organisms that are grouped together for convenience. Protists include organisms such as amoebae, paramecia, diatoms, dinoflagellates, red algae, and slime molds.

Purpose

To become familiar with the various types of protists.

Materials

- Compound light microscope
- Magnifying glass
- Various protist exhibits on display
- Various protist cultures
- Various protist prepared slides
- Disposable pipettes
- Slides
- Cover slips

Methods

- Examine the various protist exhibits on display.
- Make wet mounts of the various protist cultures and examine the wet mounts with a compound light microscope.
- Examine the various protist prepared slides with a compound light microscope.
- Record your observations in the spaces provided.

Name:

Exercise 1 Results

Excavata

1. Make **detailed drawings** of *Euglena* and *Trypanosoma* as you observed them.

Euglenozoa – Euglenids	**Euglenoza - Kinetoplastids**
Euglena	***Trypanosoma***
Total Magnification:	Total Magnification:

Chromalveolata

1. Make **detailed drawings** of *Paramecium* and *Dinoflagellates* as you observed them.

Alveolata – Ciliophora	**Alveolata - Dinoflagellata**
Paramecium	***Dinoflagellate***
Total Magnification:	Total Magnification:

2. Make a **detailed drawing** of a *Plasmodium* as you observed it.

Alveolata –Apicomplexa

Plasmodium
Total Magnification:

3. Make **detailed drawings** of Diatoms and *Diatomaceous Earth* as you observed them.

Stramenopila – Bacillariophyta Stramenopila - Bacillariophyta

Diatoms	***Diatomaceous Earth***
Total Magnification:	Total Magnification:

4. Make a **detailed drawing** of brown algae as you observed it.

Stramenopila – Phaeophyta

Brown Algae
Total Magnification:

Rhizaria

1. Make **detailed drawings** of *Radiolaria* and *Foraminifera* as you observed them.

Radiolaria	Foraminifera
Radiolaria	**Foraminifera**
Total Magnification:	Total Magnification:

Archeoplastida

1. Make a **detailed drawing** of Rhodophyta as you observed it.

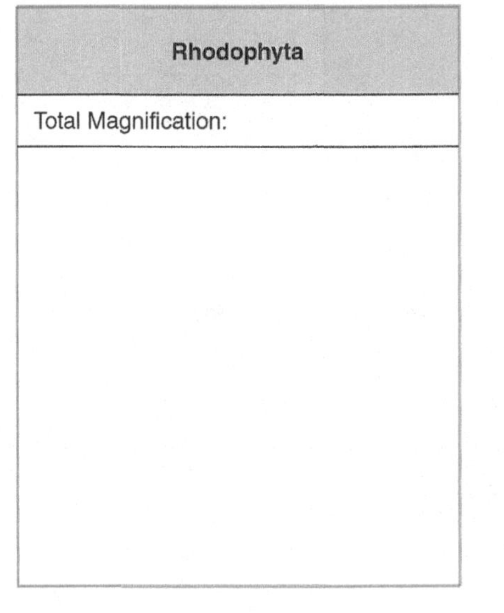

Rhodophyta
Rhodophyta
Total Magnification:

2. Make **detailed drawings** of Chlorophyta as you observed them.

Chlorophyta

Chlamydomonas
Total Magnification:

Chlorophyta

Volvox
Total Magnification:

Chlorophyta

Spirogyra
Total Magnification:

Amoebozoa

1. Make a **detailed drawing** of an amoeba and a *Physarum* culture as you observed them.

Amoeba

Amoeba
Total Magnification:

Plasmodial Slime Mold

Physarum
Total Magnification:

Additional Questions

1. Complete the **Comparison of Protists** table.

Group	Example	Characteristics (see below)			Ecologic Role (see below)
		Size	Energy Source	Type of Movement	
Euglenozoa	*Trypanosoma*				
Alveolata	*Paramecia*				
	Dinoflagellates				
	Diatoms				
Stramenopila	Brown Algae				
Foramenifera					
Radiolarians					
Amoebas	*Amoeba*				
Slime Molds	*Physarum*				
Rhodophyta	Red Algae				
Chlorophyta	*Spirogyra*				

Characteristics

1. Size (microscopic, macroscopic, or both);
2. Energy source (autotroph, heterotroph, or mixotroph);
3. Type of movement (pseudopod, threadlike pseudopod, flagella, cilia, or none).

Ecologic Role

1. Primary producer, food for others, parasite, or none.

2. Scientists are concerned that the depletion of the ozone layer will result in a decline in the populations of marine algae such as diatoms and dinoflagellates. If these populations would collapse, discuss the likely results of this collapse considering the ecologic roles of these organisms.

____

____

____

____

____

____

Fungi and Lichens

Photo courtesy of Holly J. Morris

OBJECTIVES

After completing this exercise, you will be able to:

- Describe the characteristics of the major groups of fungi: Zygomycetes, Ascomycetes, Basidiomycetes, and Deuteromycetes.
- Characterize the general life cycle of each of the major groups of fungi.
- List the characteristics of lichens.
- Describe the symbiotic relationship between the lichen partners.

INTRODUCTION

As you continue your study of the Eukarya, the heterotrophic fungi include five phyla with 100,000+ identified species. However, it is likely there are between one and one-half million species on this planet. The significant ecological role of the fungi as a group is the *decomposition* of wastes and dead organic matter, particularly woody plants. Extracellular enzymes secreted by a fungus onto a substrate perform digestion of organic matter into soluble molecules that it then can absorb into the fungal body. Fungi can be saprobes, parasites, pathogens, or mutualistic partners. Common names of fungi you have perhaps heard of include mushrooms, toadstools, Athlete's foot, puffballs, yeast, and molds.

The fungus body is sometimes unicellular but most often filamentous. A massive filamentous fungal body is called a *mycelium* (pl., mycelia). A single filament is called a *hypha* (pl., hyphae). Hyphae are either multicellular with each cell possessing one or more nuclei separated by *septa* (cross walls) or multinucleate, *coenocytic* strands without septa. The cell walls of fungi usually possess *chitin*, a modified polysaccharide with nitrogen-containing sugar. Chitin exists in the exoskeletons of arthropods and is resistant to hydrolysis.

Spores are the mode of reproduction for most fungal species, either asexually or sexually. A spore will produce a mycelium after it germinates. Fungal spores are everywhere. Slices of bread uneaten after a few days may have a blue-green mold growing after spores in the air land on this favorable substrate. If you look at the mold under the stereo microscope, you will see hyphae with upright, finger-like tips. On these tips or conidiophores are the asexual spores known as *conidia*. Another kind of mold frequently found on aging fruit or bread shows white threads or hyphae with tiny black knobs under the scope. These tiny asexual knobs held on upright hyphae are the *sporangia*. In the sexual cycles of fungi, a "marriage" or *plasmogamy* of genetically distinct strains yields a *diploid zygote* following the fusion of the two genetically distinct nuclei (+ and –). Meiosis in the diploid zygote nucleus will produce haploid spores. These spores germinate to produce new mycelia. The particular mode of production of the diploid zygote in fungi has led to taxonomic groups or phyla.

We will examine three of these phyla: the Zygomycota, the Ascomycota, and the Basidiomycota. One phylum of fungi, the aquatic Chytridiomycota, produces flagellated spores and gametes found in no other group of fungi. Scientists consider Chytrids the earliest group of fungi arisen from an ancient protistan. One additional group of fungi, the Glomeromycota, is very important in the ecosystems of the world. These fungi are the mycorrhizae that form mutualistic partnerships with the majority of plants, enabling the roots of these plants to receive minerals and nutrients from the surrounding soil habitat. At one time, researchers grouped the fungi not known to reproduce sexually into a phylum known as the Deuteromycota. Today they have reassigned these fungi, but they often still refer to a fungus they have not observed to have a sexual cycle as a deuteromycete.[16]

I. ZYGOMYCETES (PHYLUM ZYGOMYCOTA)

Zygomycetes (1000 species) are fungi within a clade that forms zygosporangia, which are the diploid sporangia created when two haploid cells fuse (see photo).

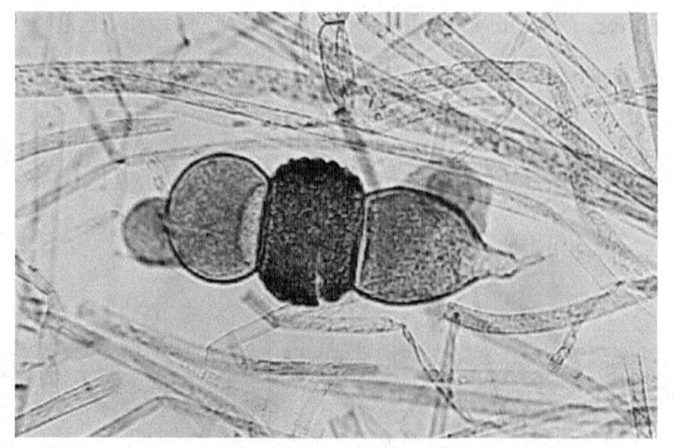

Zygospore 100× (Photo courtesy of Holly J. Morris and David T. Moat)

The cell wall of zygomycetes is different than most other fungi. The cell walls of most fungi contains chitin as the structural polysaccharide. Zygomycetes contain chitosan, which is a variation of chitin.

Some zygomycota act as decomposers (host is dead) or parasites (host is living), and are mostly terrestrial. One example of a genus of zygomycetes is Rhizopus, which includes the type of fungus that forms black bread mold.

Molds found growing on aging fruits and breads are usually members of the conjugating zygomycetes. The hyphae are *coenocytic or nonseptate*. *Sporangia* produce asexual spores at the tips of *sporangiophores*. The horizontal hyphae that run across the substrate are the *stolons*. *Rhizoids* penetrate the substrate anchoring the mold, secrete the digestive enzymes, and absorb the nutrients.[16]

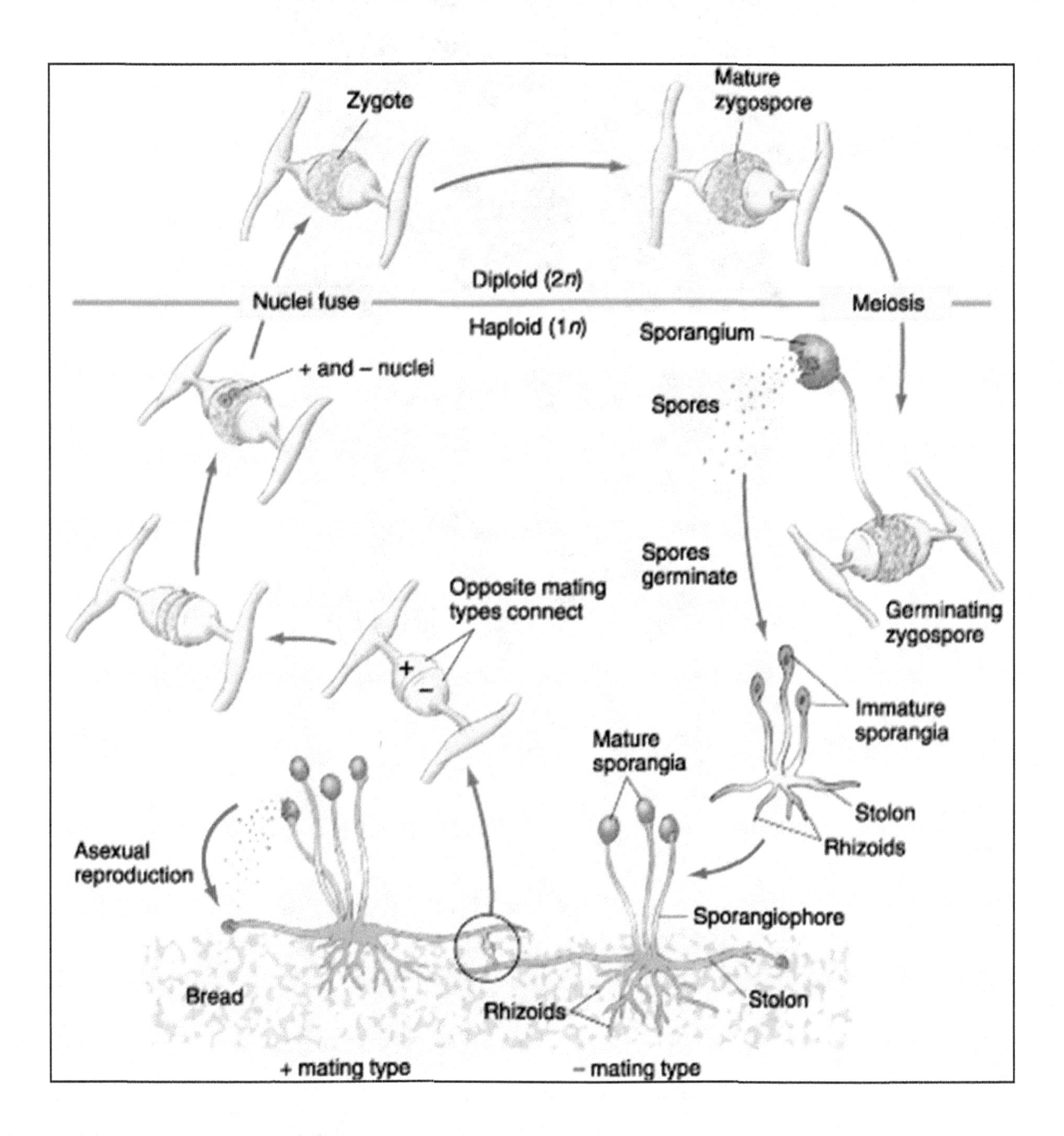

Note the fusion or plasmogamy when the genetically distinct stolons of the mating types come in contact. The lateral buds from each hypha are called *progametangia*. Nuclei migrate into the tips, and septa form to create the *gametangia*. The + and – nuclei fuse to form diploid zygote nuclei in a *zygospore*, which has a thick-walled rough covering. This is the only diploid stage of the cycle. Meiosis will occur under favorable environmental conditions and produce one sporangium with haploid spores. The new diverse spores can each produce a new mycelium just as the asexual sporangiospores produce upon their germination. Biologists often refer to sexual mating of this type as conjugation.[16]

II. ASCOMYCETES (PHYLUM ASCOMYCOTA)

© ArgenLant/Shutterstock.com

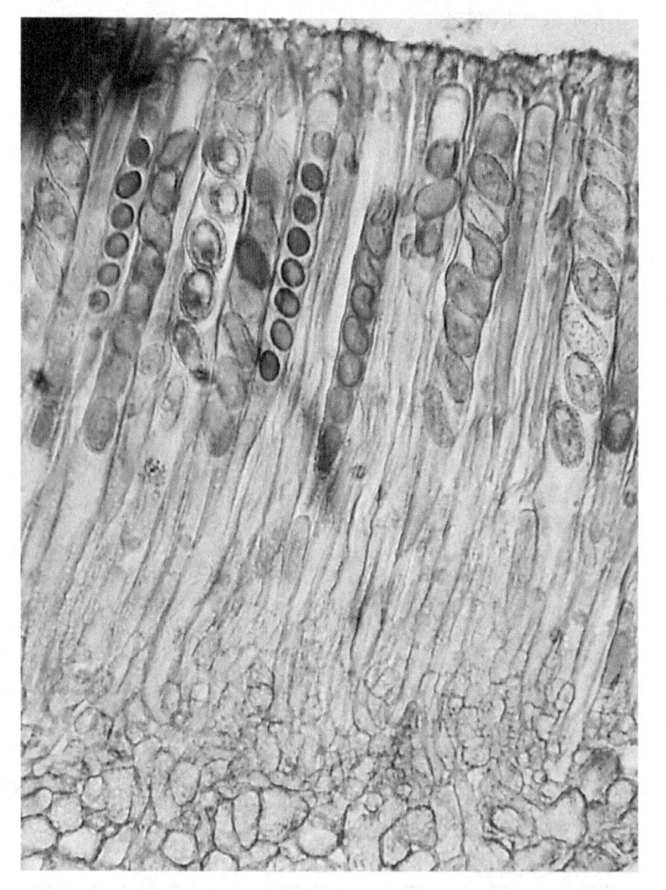

Photo courtesy of Holly J. Morris and David T. Moat

Biologists often refer to the ascomycetes as the *sac* or *cup* fungi. These fungi produce a sac-like structure that produces the sexual ascospores. Hyphae are *septate* with a haploid nucleus in each cell of the primary mycelium. Once mating occurs, a secondary mycelium is produced with each cell possessing two unfused, genetically distinct nuclei (n + n). This is called *dikaryotic.* The growth of the secondary mycelium produces a fruiting body called the ascocarp.[16]

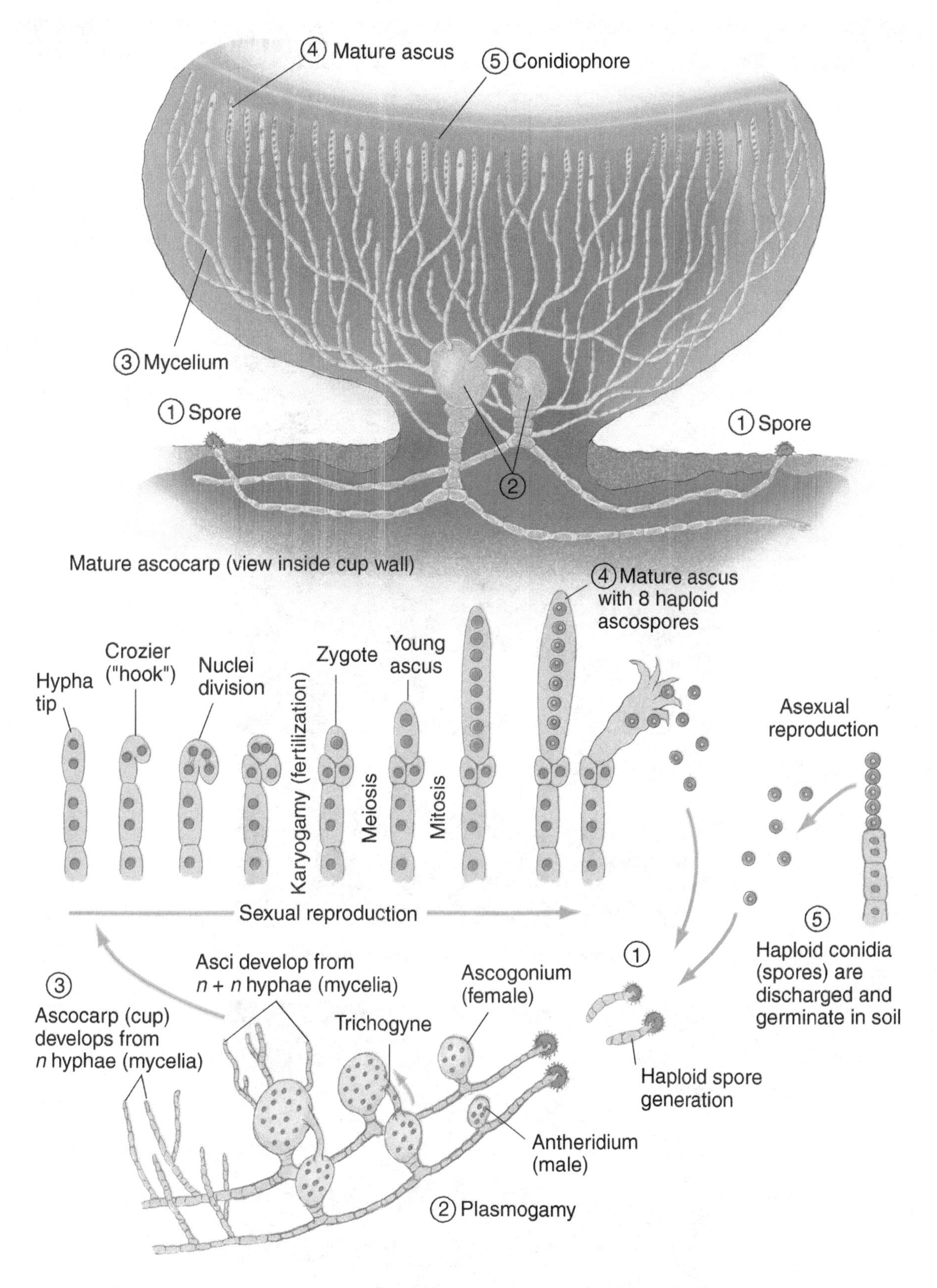

④ Mature ascus
⑤ Conidiophore
③ Mycelium
① Spore
① Spore
②
Mature ascocarp (view inside cup wall)
④ Mature ascus with 8 haploid ascospores
Hypha tip
Crozier ("hook")
Nuclei division
Karyogamy (fertilization)
Zygote
Meiosis
Young ascus
Mitosis
Asexual reproduction
Sexual reproduction
①
⑤
Haploid conidia (spores) are discharged and germinate in soil
③
Ascocarp (cup) develops from n hyphae (mycelia)
Asci develop from n + n hyphae (mycelia)
Ascogonium (female)
Trichogyne
Haploid spore generation
Antheridium (male)
② Plasmogamy

III. BASIDIOMYCETES (PHYLUM BASIDIOMYCOTA)

Shelf Fungus (Basidiomycota) (Photo courtesy of Holly J. Morris)

Basidiomycota (Photo courtesy of Holly J. Morris)

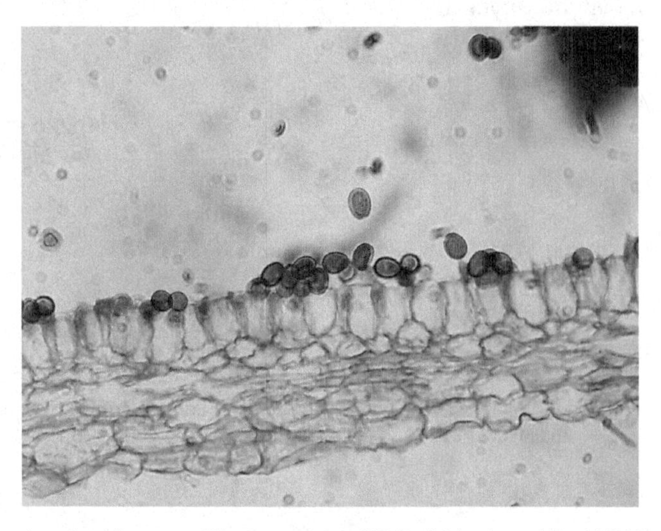

Coprinus basidiospores (Photo courtesy of Holly J. Morris and David T. Moat)

Most of the fungi that you find on a stroll through a woodland are members of this group. The fruiting bodies are usually the only visible stage of the life cycle. A mushroom is a good example of this kind of fruiting body or basidiocarp. The hyphae fuse to form the dikaryotic mycelium which will produce the basidiocarp is likely underground or under leaves and fallen trees. Scratch under some decaying leaves and look for these whitish threads when you go on a hike. You may have a pile of such leaves in your yard in a mulch pile. You should find some hyphae here as well.

The basidiospores that are germinating on the soil each represent a different mating type. Notice that the primary mycelium produced from each spore has one haploid nucleus per cell. Following the fusion of two of these primary mycelia, a new secondary mycelium forms. Note that these cells are *dikaryotic* with a nucleus from each parent strain. This condition is denoted as (n + n). The secondary, dikaryotic mycelium rapidly grows to produce the characteristic basidiocarp for the species. In other words, the different species or kinds of mushrooms and other members of the group each have their own distinctive look just as different kinds of tree species have their own distinctive leaves.

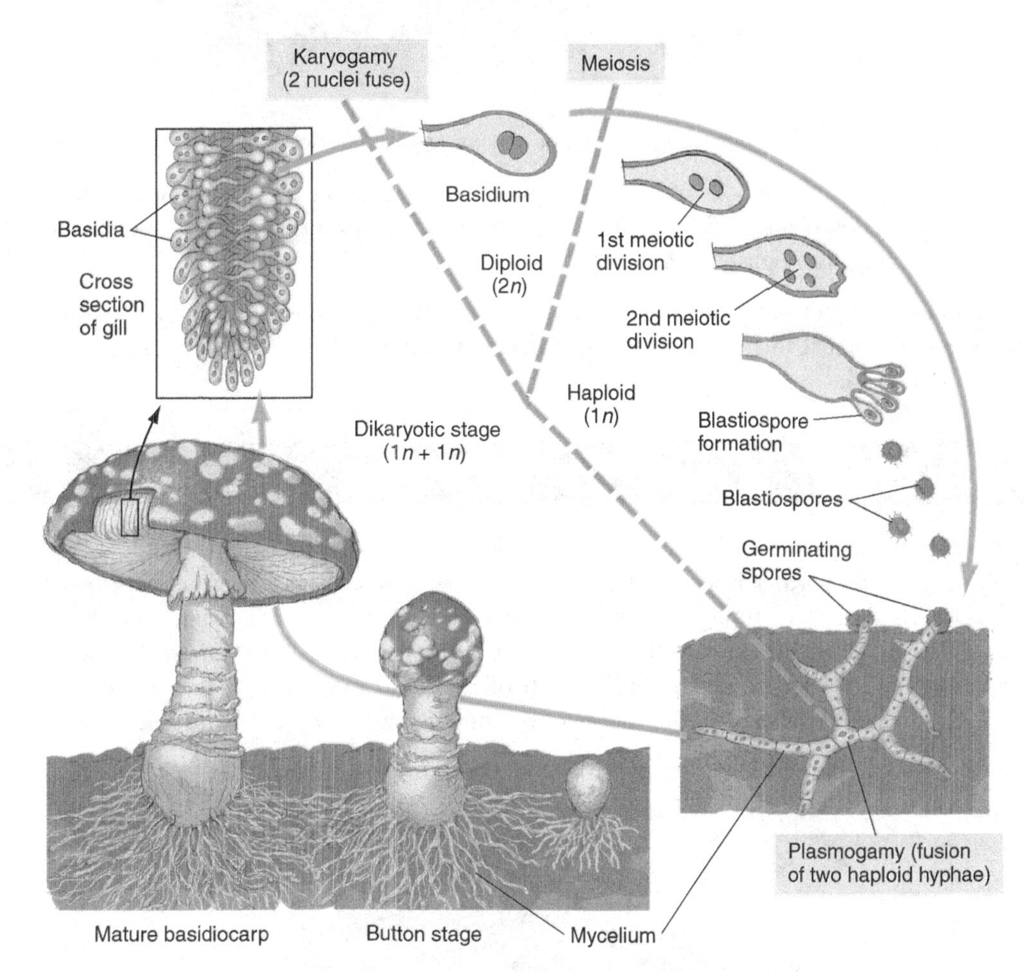

The formation of the secondary mycelium from the two mating strains is like a "marriage" in a sense. The fruiting body or mushroom forms from the underground mycelium when environmental conditions are met. Under the *cap* of the mushroom are *gills or pores.* Along the surfaces of these structures are the terminal club-shaped *basidia.* Each basidium is the site of karyogamy or nuclear fusion. A new diploid zygote form followed by meiosis to produce four haploid diverse *basidiospores.* These are not contained as they were in the sac or ascus, but they are extruded exocytically to the outside. Each of the four basidiospores will briefly hang from the tip of the basidium by cytoplasmic strands. One mushroom can produce close to a billion basidiospores. The wind disperses these sexual spores in most cases. Sometimes the basidiocarp will secrete chemicals that have a strong odor to attract flies. Stinkhorns are an example of fungi that smell like rotten meat. The flies come to the fungi seeking a meal, and the basidiospores stick to their legs for transport to other locations.[16]

IV. PENICILLIUM (PHYLUM DEUTEROMYCOTA)

Photo courtesy of Holly J. Morris and David T. Moat

Penicillium is a mold commonly found growing on decaying fruit, such as oranges. This mold produces penicillin, the well-known antibiotic. The antibiotic characteristic of this mold was discovered in 1929 by Sir Alexander Fleming, an English researcher.[1]

Deuteromycota (imperfect fungi) do not fit into the commonly established taxonomic classifications of fungi because their sexual form of reproduction has never been observed. Their only known form of reproduction is asexual.

V. LICHENS

A lichen consists of two partners who live mutualistically. The fungal partner is a member of the Ascomycota, and the photosynthetic partner is a cyanobacteria species or a green algal species. The growth pattern of the lichen is largely due to the fungus and results in characteristic types: *crustose*, *foliose*, and *fruticose*. You can find lichens on rocks, tree bark, the ground, old cemetery monuments, and even on the old metal of discarded vehicles. Lichens can cause the breakdown of rock and metal due to the secretion of an acid. Lichens are considered "pioneers" in the succession of new plant life on rocks and soil surfaces.[16]

Pictures of the three types of lichens (crustose, foliose, and fruticose) follow.

Crustose (Photo courtesy of Holly J. Morris and David T. Moat)

Foliose (Photo courtesy of Holly J. Morris and David T. Moat)

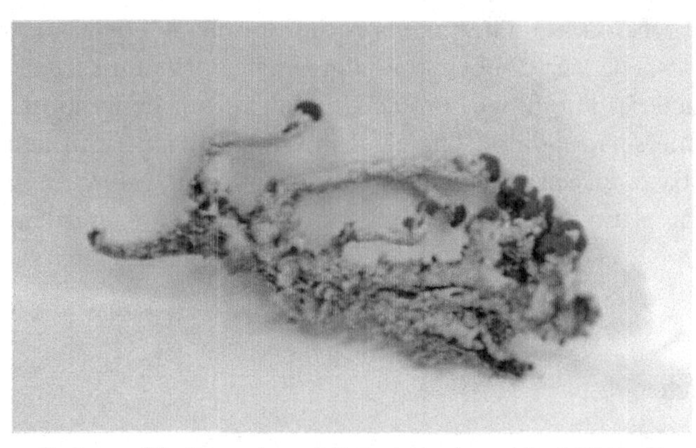

Fruticose (Photo courtesy of Holly J. Morris and David T. Moat)

EXERCISE 1 Observing Fungi

Fungi form a diverse group of organisms that are classified within their own kingdom. Although fungi grow in the ground, they contain no chlorophyll and are therefore, heterotrophic, not autotrophic. They obtain nutrients by secreting digestive enzymes into their surrounds and absorbing nutrients. As a result, fungi act as decomposers within their ecosystems.

Purpose

To observe the macroanatomy and microanatomy of the various types of fungi.

Materials

- Compound light microscope
- Dissecting microscope
- Fresh mushrooms
- Scalpel
- *Rhizopus* display
- Petri plates
- Magnifying glass
- Prepared slides of *Rhizopus*, *Peziza*, *Coprinus*, and *Penicillium*.

Methods

- Examine the *Rhizopus* display with a magnifying glass.
- Examine the prepared *Rhizopus* zygosporangia slides.
- Examine the morels on display.
- Examine the prepared *Peziza* asci slides.
- Obtain a fresh mushroom and examine its parts including the stalk (stipe), cap (pileus), and the gills.
- Examine the prepared *Coprinus* basidia slides.
- Examine prepared *Penicillium* slides
- Record your observations in the space provided.

EXERCISE 2 Observing Lichens

Lichens are a combination of two types of organisms, a fungus and an alga, or sometimes a cyanobacterium, which form a symbiotic relationship. The fungus is the dominant organism, with the algal cells living among the fungal filaments. The alga can be either a green alga or cyanobacteria. Some lichens have both green algae and cyanobacteria. The fungus protects the alga from desiccation and intense sunlight, and absorbs nutrients from the surroundings. The alga provides nutrients through photosynthesis and, for lichens containing cyanobacteria, can provide nitrogen compounds, through nitrogen fixation, to the fungus and other surrounding plants. The appearance of the lichen is different than the appearance of either partner, individually.

Purpose

To observe the various lichen displays.

Materials

- Lichen displays.

Methods

- Examine the Lichens on display.
- Record your observations in the space provided.

Name:

Exercise 1 Results

1. Make **detailed drawings** of the *Rhizopus* specimen on display and of the zygosporangium prepared slide at 400×.

***Rhizopus* specimen**	**Zygosporangium**
Total magnification:	Total magnification:

2. Make a **detailed drawing** of the *Peziza* asci prepared slide at 400x.

***Peziza* asci with ascospores**
Total magnification:

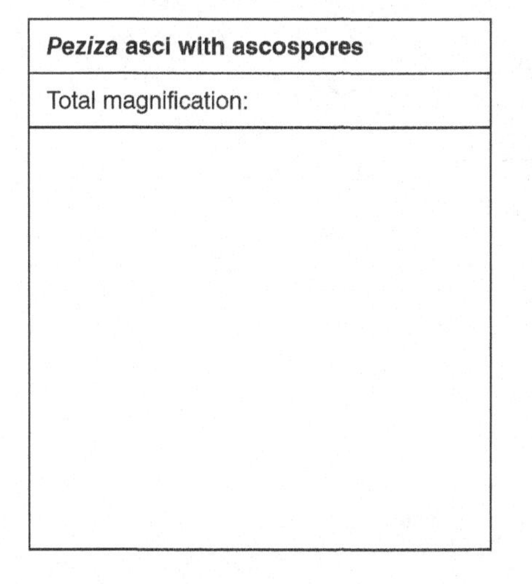

3. Make a **detailed drawing** of the *Coprinus* basidia prepared slide at 400x.

***Coprinus* basidia with basidiospores**
Total magnification:

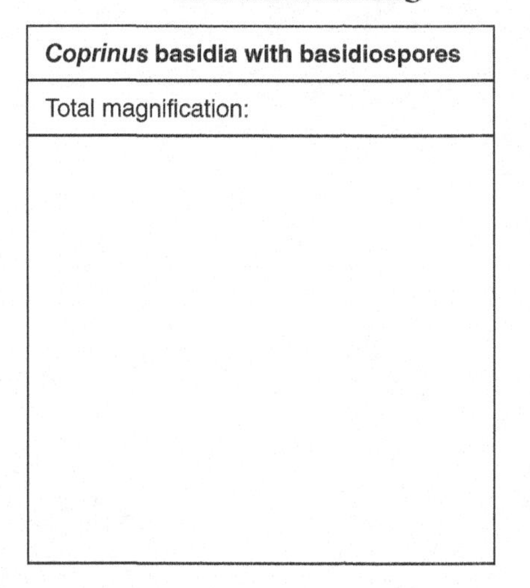

4. Make a **detailed drawing** of the *Penicillium* prepared slide at 400×.

Penicillium
Total magnification:

5. Imagine an ecosystem without any fungi. How would the ecosystem be changed?

____

____

Exercise 2 Results

1. Make **detailed drawings** of the three types of Lichens on display.

Name: Crustose	**Name: Foliose**
Total Magnification:	Total Magnification:

Name: Fruticose
Total Magnification:

CHAPTER 24

Nonvascular and Seedless Vascular Plants

Photo courtesy of Holly J. Morris

OBJECTIVES

After completing this exercise, you will be able to:

- Describe the process of alternation of generations.
- Describe the characteristic features of nonvascular plants and seedless vascular plants.
- Discuss the ecologic role played by nonvascular and seedless vascular plants.
- Understand the difference between homospory and heterospory.
- Discuss the economic importance of nonvascular and seedless vascular plants.

INTRODUCTION

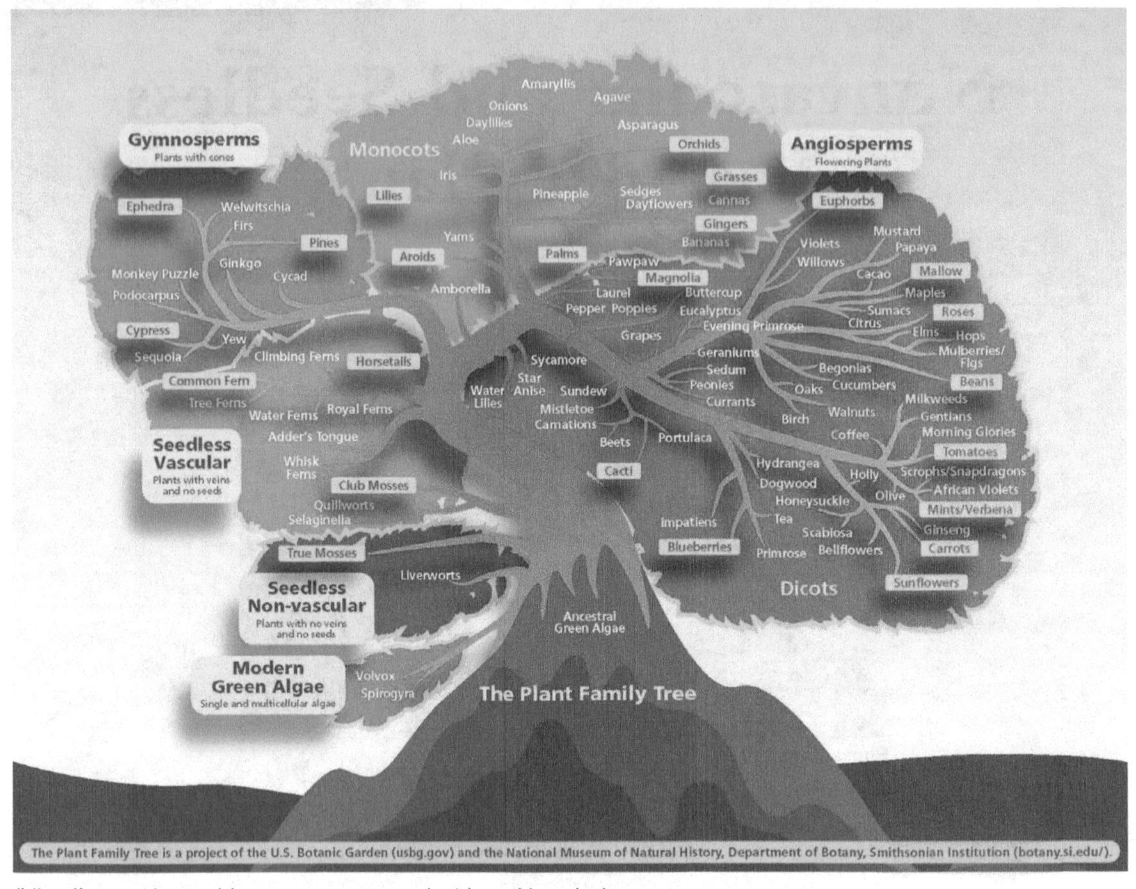

(https://www.usbg.gov/classroom-resources-plant-based-learning)

NONVASCULAR PLANTS

This exercise begins our look at land plants. We start with a group called *avascular* plants, meaning that they do not possess vascular tissue. Vascular tissue is similar to your arteries and veins, and is used to conduct water, minerals, and nutrients around the plant. Since the avascular plants do not have this type of tissue, they are limited to moist habitats. Also, the plant body of an avascular plant is fairly small compared to that of a vascular plant. Avascular plants do not have true organs, such as roots, stems, or leaves. There are approximately 24,000 species of avascular plants in the world.

Plants contain chlorophyll a, are autotrophs, and have cellulose cell walls. The life cycle of a land plant is referred to as *alternation of generations*. There is a $1n$ stage of the life cycle, the *gametophyte* plant, and a $2n$ stage of the life cycle, the *sporophyte* plant. The gametophyte is the gamete-producing generation of the life cycle. Cells in structures called *antheridia* (singular = *antheridium*) go through mitosis to produce sperm cells, and cells in *archegonia* (singular = *archegonium*) go through mitosis to produce egg cells. The sporophyte is the spore producing generation. Once a zygote is formed from fertilization, it develops into a young sporophyte, which then matures into the sporophyte plant. Diploid cells in *sporangia* undergo meiosis to produce $1n$ spores. The spores released from the sporophyte plant germinate and develop into the gametophyte generation. When studying groups of plants, it is important to recognize which of these generations is most common (predominant).[14]

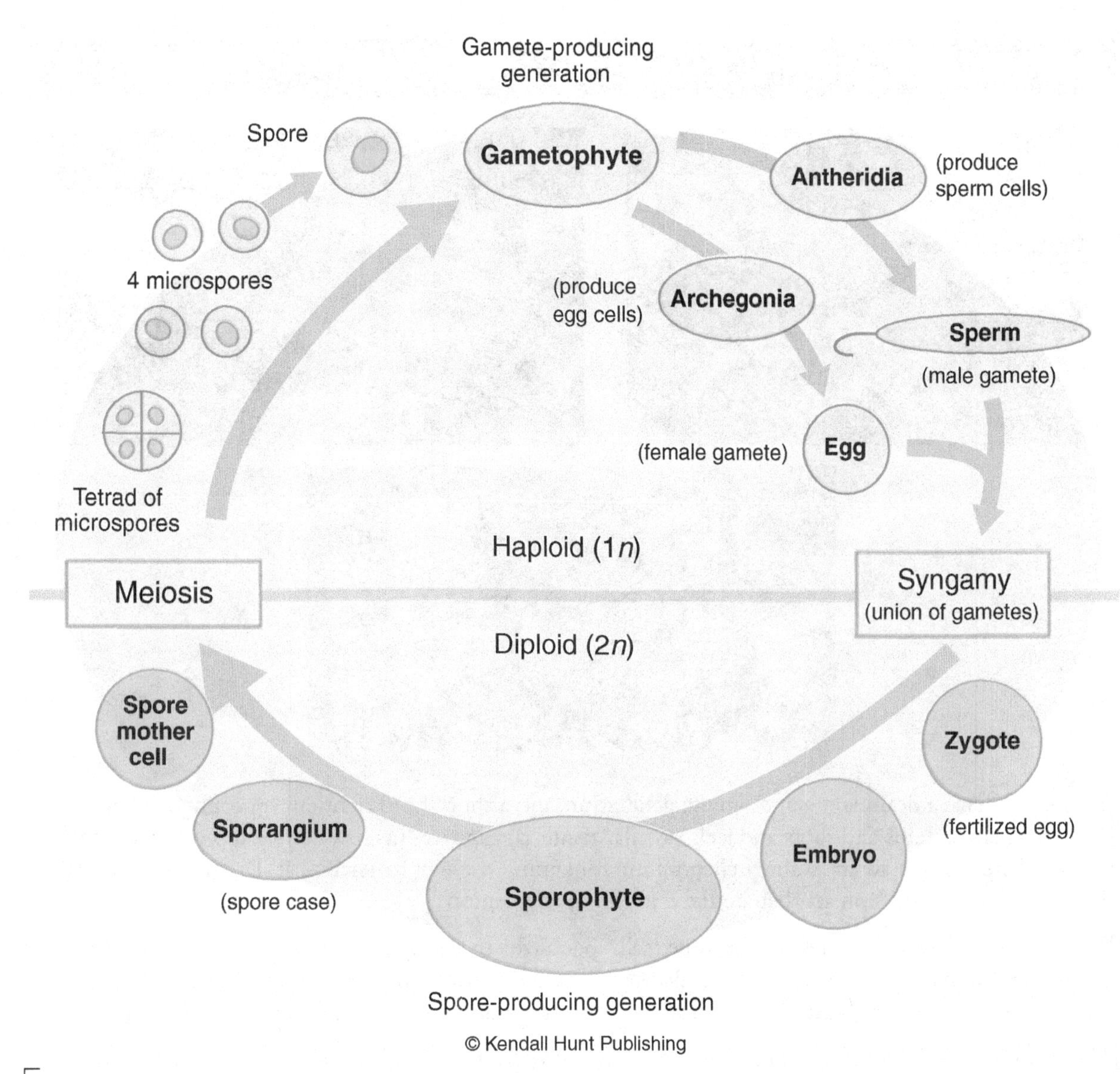

In all avascular plants, the gametophyte is the predominant generation. All plants provide some basic benefits to their environment. Plants fix carbon dioxide and release oxygen, stabilize soil and reduce soil erosion, and degrade rocks into soil.

Nonvascular plants are not economically important with the exception of sphagnum moss (a Bryophyte) which has horticultural uses and additionally stores a significant amount of carbon. However, it is burned as a fuel in some countries releasing carbon into the atmosphere.[14]

Nonvascular plants include three phyla:

- Phylum Bryophyta (true mosses)
- Phylum Hepatophyta (liverworts)
- Phylum Anthocerophyta (hornworts)

PHYLUM BRYOPHYTA

Bryophytes have a dominant gametophyte generation, and a short-lived transient sporophyte. They do not have true leaves, stems and roots and lack vascular tissue. Bryophytes need water for reproduction because the flagellated sperm swim to the archegonium containing the egg to fertilize it. Bryophytes also have a thin waxy cuticle and stomata that are fixed in the open position.

Because these mosses lack true vascular tissue (cells specialized to transport materials), the gametophyte *thallus* (plant body) remains close to the ground. Because there are separate male and female gametophyte plants, a gametophyte plant will have either an archegonia or an antheridia at the tip of the central stem-like projection.

The sporophyte develops as a *capsule*, which grows away from the archegonium on a stalk-like structure called the *seta*. The capsule contains a sporangium composed of diploid cells. These diploid cells undergo meiosis to form haploid spores. When the capsule opens, the spores are released and distributed.[14]

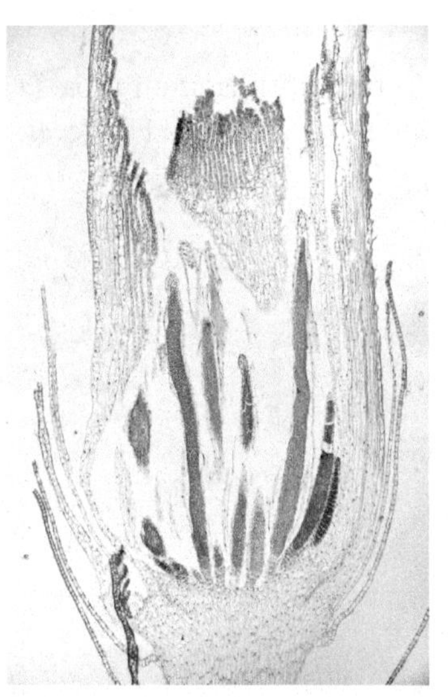

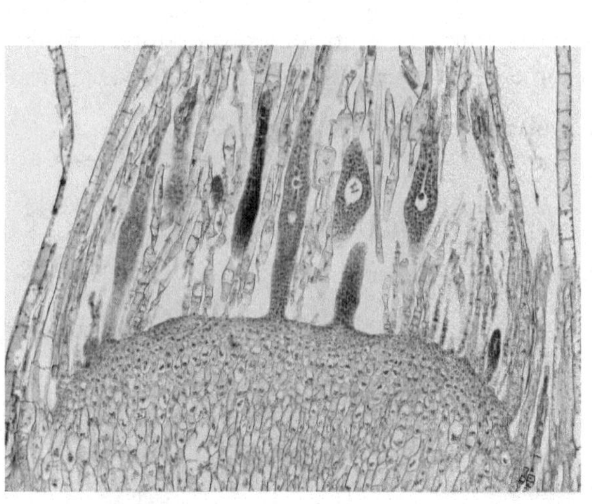

The typical life cycle of a true moss is depicted below.

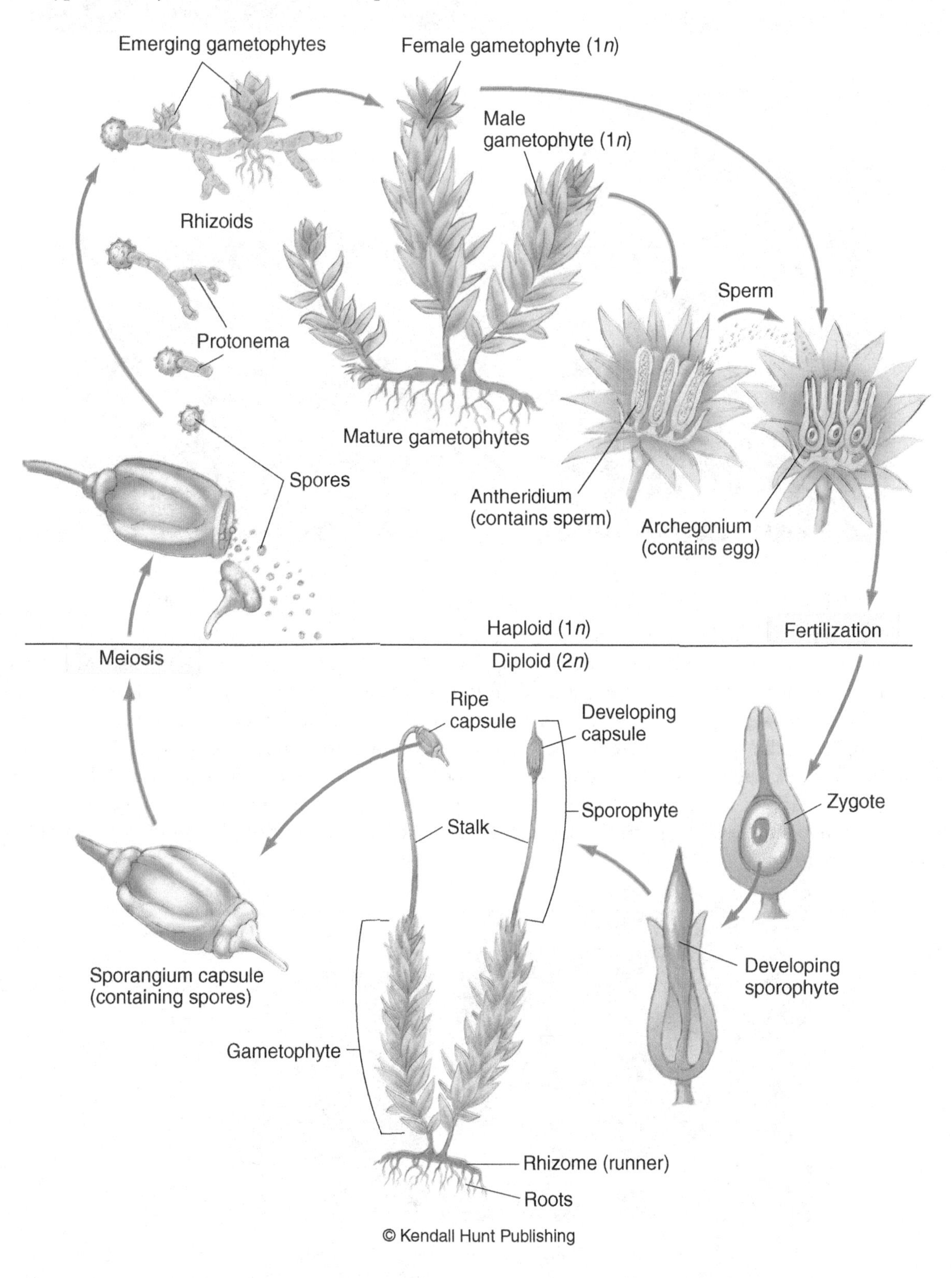

PHYLUM HEPATOPHYTA

Marchantia (© dabjola/Shutterstock.com)

(© AlessandroZocc/Shutterstock.com)

Hepatophytes are similar in most respect to the true mosses but look somewhat different.

The gametophyte of a liverwort may be both male and female, and have both archegonia and antheridia. These reproductive structures in liverworts grow up from the surface of the plant body on stalks. Antheridia are found on *antheridiophores*, and archegonia are found on *archegoniophores*. Antheridia will appear similar to those in the mosses, wherein large areas of darkly stained material will open to the top surface of the antheridiophore. The sperm cells are flagellated and will swim to the archegonium where fertilization occurs. Archegonia will also appear as they did in the mosses. However, the archegonia will develop on the bottom surface of the archegoniophore. Once fertilization has occurred, the sporophyte develops from the archeognium. Once again the sporophyte generation is marked by the presence of a capsule containing spores.[14]

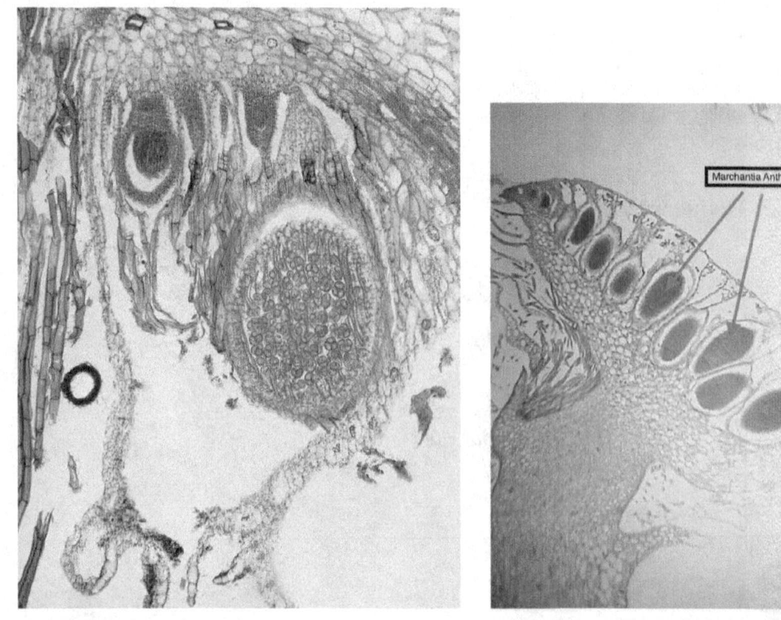

Marchantia archegonia
(© Jubal Harshaw/Shutterstock.com)

Marchantia antheridia, 40×
(Photo courtesy of Holly J. Morris and David T. Moat)

Liverworts can reproduce asexually. Cup shaped structures on the upper surface of the thallus contain groups of cells (gemma) that are capable of breaking off and producing a new plant.

Gemmae Cups (© Henrik Larsson/Shutterstock.com)

The typical life cycle of a liverwort is depicted below.[14]

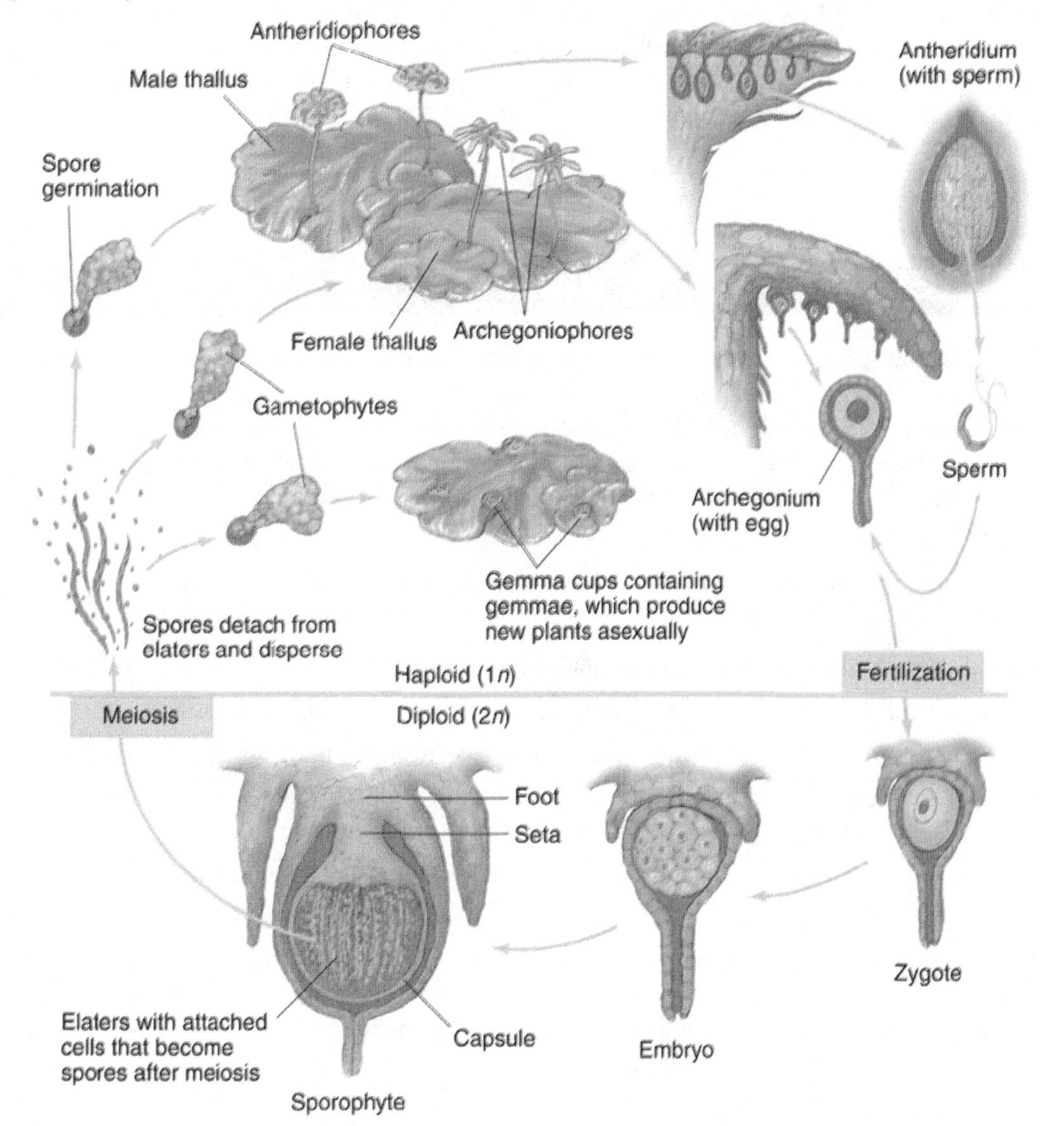

PHYLUM ANTHOCEROPHYTA

Hornwort (© Svetlana Klaise/Shutterstock.com)

Anthocerotophyta, commonly called hornworts, have tall narrow sporophytes emerging from the gametophytes. The appearance of the sporophyte is like that of a tapered horn, thus the name. Hornworts appear to be a transition plant, poised between simpler bryophytes and more advanced ferns. The hornwort sporophyte is embedded in the gametophyte, like other bryophytes, but continues to grow and remain photosynthetic, which sets it apart from simpler bryophytes.

SEEDLESS VASCULAR PLANTS

Scientists have now proven that nonvascular bryophytes are the oldest terrestrial plants, evolving from green algae and colonizing land approximately 470 million years ago. Two early lineages occurred in the evolution of plants, one leading to the nonvascular bryophytes and the other leading to vascular plants. Successful colonization of the terrestrial habitat was largely due to the development of vascular tissue, *xylem* and *phloem*. The earliest vascular plants, which date back to over 400 million years ago during the Silurian Period, were spore producers.

In addition to the presence of vascular tissue, other adaptive characteristics for early terrestrial life in these plants are:

1. Larger, dominant sporophyte with branching stems, leaves, and roots.
2. Lignified vascular tissue for support in addition to transport of water and nutrients.
3. Development of megaphyll or a larger leaf with surface area to support greater food output.
4. Sporophyll or protective, modified leaf to enclose sporangia and strobilus, the cone-shaped structure of sporophylls that concentrated the spores for greater protection and survival.
5. Evolution of heterospores that led to the development of separate gametophytes in the higher vascular plants.

Current evidence supports two clades of vascular seedless plants: the club mosses and ferns. Club mosses are the lycophytes. The fern group or pteridophytes includes some early vascular plants once considered separate phyla or more primitive than ferns. Evidence today includes these whisk ferns and horsetails in the same phylum Pterophyta with the ferns.

During the Carboniferous Period, these were the dominant land plants. The early forests looked quite different from the forests of today. Many of these plants grew to heights of 40 meters. When the climate changed from warm to cold, these giants toppled and eventually become compressed forming *coal* deposits. The carbon compounds or fossil fuel in this coal is stored energy produced millions of years ago. Today, most of the surviving species of the vascular seedless plants are small; however, there are some tropical tree ferns that reach 60 feet.[16]

PHYLUM LYCOPHYTA

© Kuttelvaserova Stuchelova/Shutterstock.com

© Noppharat4569/Shutterstock.com

The 1200 species of club mosses are small in size unlike the tree-like species that existed during the Carboniferous Period. A "tree lycopod" would be about 100 feet or taller than those in the photos. The plants possess tiny scaly leaves (*microphylls*) with one vascular strand only, both horizontal, underground *rhizomes* as well as aerial branches, and *roots*. The cone-shaped strobili consist of sporangia-bearing sporophylls.[16]

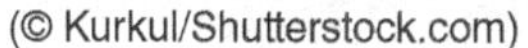

(© Kurkul/Shutterstock.com)

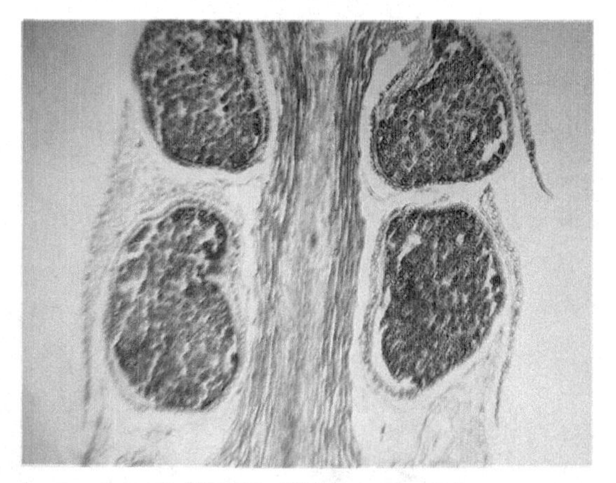

Lycopodium strobilus 40x (Photo courtesy of Holly J. Morris and David T. Moat)

The *Lycopodium* life cycle is depicted below.

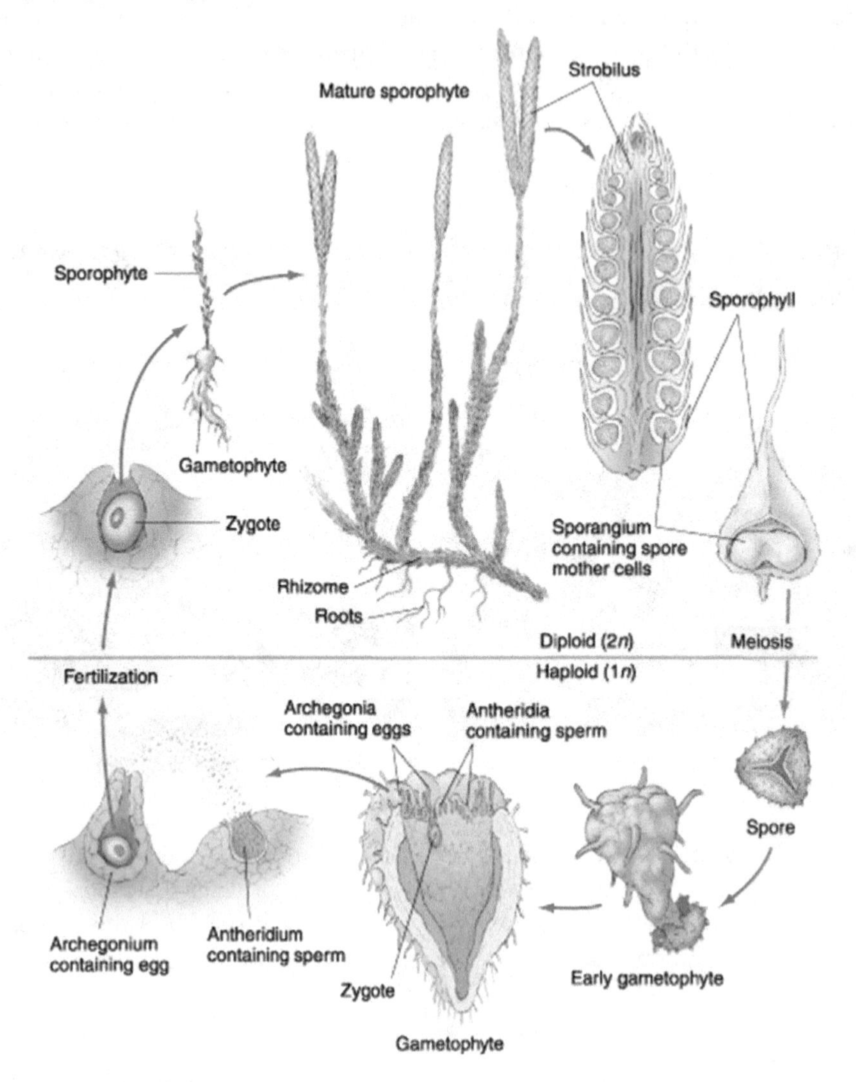

Selaginella or spike moss is a significant Lycophyte because it is heterosporous. All higher plants are also heterosporous, which means that there are separate spores for a separate male gametophyte (*microspore*) and for a separate female gametophyte (*megaspore*).[16]

(© 7pic/Shutterstock.com)

Selaginella strobilus 40× (Photo courtesy of Holly J. Morris and David T. Moat)

The *Selaginella* life cycle is depicted below.

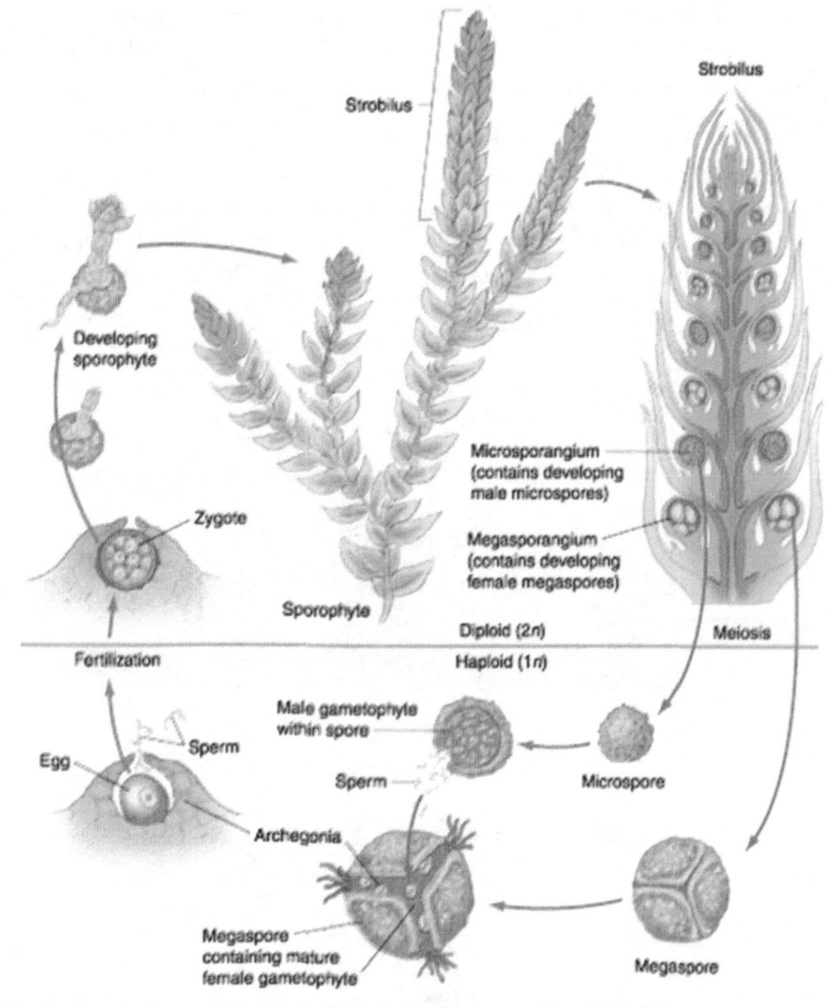

From Explorations in General Biology by Betty A. Rosenblatt and Sarah Warrington. Copyright © 2008 by Kendall Hunt Publishing Company. Reprinted by permission.

PHYLUM PTEROPHYTA

© Tatjana Romanova/Shutterstock.com

© Stephen Orsillo/Shutterstock.com

More than 12,000 species of ferns inhabit the earth in most habitats from the tropical rainforest to the dry desert. Ferns, more closely related to seed plants than to the lycophytes, possess branching *fronds* or *megaphylls*, *rhizomes* with adventitious *roots*. Most ferns have a definitive alternation of generations with the *sporophyte* being the dominant generation. The gametophytes of ferns are small in size, yet independent. Water is essential for fertilization. Most ferns are *homosporous*; that is, one type of spore is produced in the sporangia and each one can produce an *independent gametophyte* that will produce both archegonia/eggs and antheridia/sperm on the same gametophyte. The gametophyte is, therefore, *monoecious* or bisexual.[16]

Scientists refer to clusters of sporangia on the fern fronds as *sori* (singular - *sorus*). The sori often appear as brown or black spots on the fern frond and can be covered with a thin *indusium* (see below). When a fern *spore* lands in a suitable location that is warm and moist soil or the surface of a rock along a stream bank, the spore germinates and grows into an independent heart-shaped gametophyte called a *prothallus* (see below). Fertilization results in the formation of a zygote that divides mitotically to produce a young sporophyte.[16]

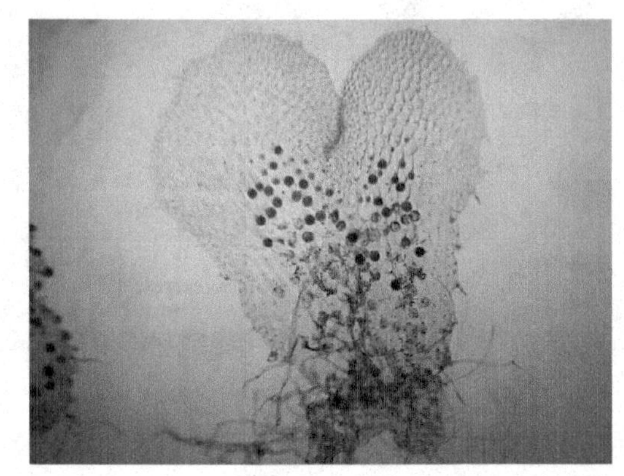

Prothallus 40x (Photos courtesy of Holly J. Morris and David T. Moat)

The life cycle of a fern is depicted below.

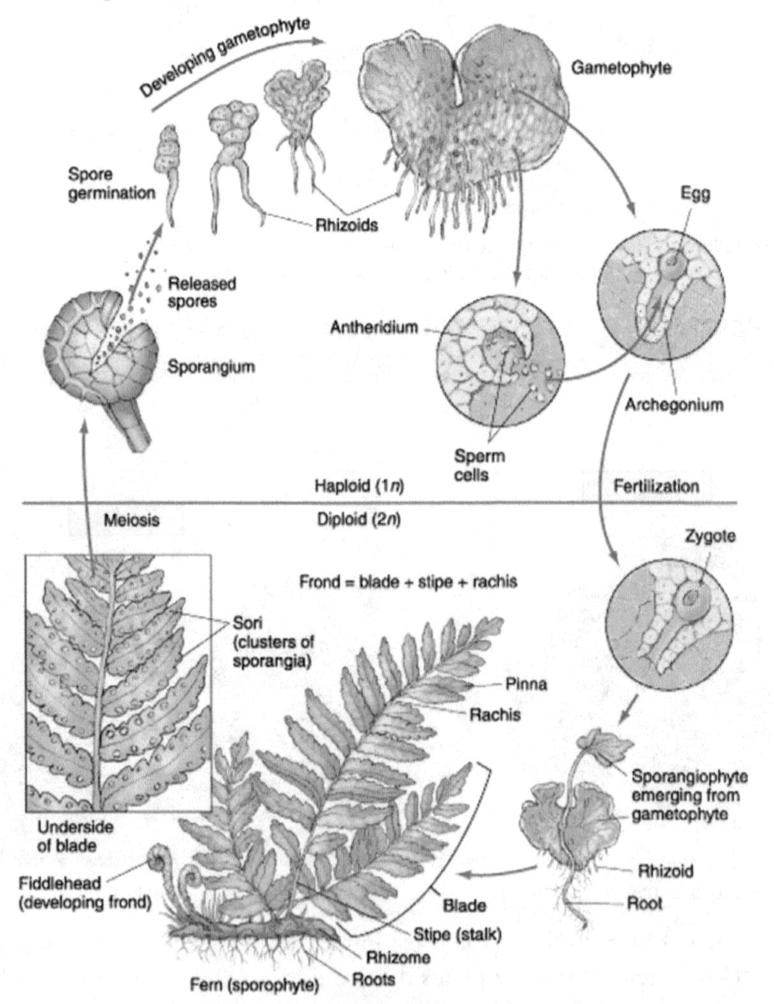

PHYLUM PSILOTOPHYTA

Whisk ferns (*Psilotum*) are extremely primitive in appearance with their lack of leaves. The dark green stems are dichotomously branched. If you gathered a bunch and wrapped string around the bunch, it would resemble a small whisk broom, hence its name, "whisk fern." On the naked aerial stems are the *tri-lobed sporangia*. There are only 12 species of whisk ferns. In the United States, one could find *Psilotum nudum* growing at the bases of buttressed cypress trees in the Louisiana swamps. *Psilotum* is *homosporous*, and the gametophyte is very tiny and rarely seen.[16]

PHYLUM SPHENOPHYTA

Horsetails (*Equisetum*) were important during the Carboniferous Period and attained great heights. Today, however, horsetails are reduced in size. Only 15 species are extant in the terrestrial environment. Most species exist along stream banks where they can exist underwater for periods of time. The leaves are barely noticeable minute scales. The stems are hollow and possess silica. Underground rhizomes allow for aggressive spreading. Campers know that you can make a ball of horsetail and clean pots and pans, hence the name, "scouring rush".[16]

EXERCISE 1 Observing Nonvascular Plants

Nonvascular plants do not have true leaves, stems, and roots. Although some nonvascular plants may have specialized tissue for water transport, these tissues are not considered to be true vascular tissue. Nonvascular plants are generally restricted to moist areas because the sperm require water to enable them to swim to the egg to fertilize it. The sporophyte is dependent on the gametophyte in nonvascular plants.

Purpose

To become familiar with the structure, function, and life cycle of nonvascular plants including Bryophytes and Hepatophytes.

Materials

- Compound light microscope
- Various Bryophytes and Hepatophytes on display
- Prepared *Marchantia* slides
- Magnifying glass

Methods

- Examine the various Bryophytes and Hepatophytes on display and record your observations in the space provided.
- Examine the *Marchantia* antheridia and archegonia prepared slides and record your observations in the space provided.

EXERCISE 2 Observing Seedless Vascular Plants

Unlike nonvascular plants, seedless vascular plants:

- Contain specialized tissue for transporting water and other nutrients.
- Contain true roots, stems, and leaves.
- Have independent gametophyte and sporophyte stages.

Similar to nonvascular plants, seedless vascular plants:

- Do not produce flowers or seeds.
- Reproduce with the use of spores.
- Are restricted to moist areas because the sperm need water to be able to swim to the egg to fertilize it.

Purpose

To become familiar with the structure, function, and life cycle of seedless vascular plants including Lycophytes, Pterophytes, Psilotophytes, and Sphenophytes.

Materials

- Compound light microscope
- Various Lycophytes, Pterophytes, Psilotophytes, and Sphenophytes on display
- Fern prothallia on display
- Various prepared *Lycopodium* and *Selaginella* prepared slides

- Various prepared *Pterophyta, Psilotophyta*, and *Sphenophyta* prepared slides
- Magnifying glass
- Disposable pipettes
- Slides
- Cover slips

Methods

- Examine the various Lycophytes on display and record your observations in the space provided.
- Examine *Lycopodium* and *Selaginella* prepared slides and record your observations in the space provided.
- Examine the various ferns, Psilotophytes, and Sphenophytes on display and record your observations in the space provided.
- Examine the Psilotophyte and Sphenophyte prepared slides and record your observations in the space provided.
- Make a wet mount of a fern prothallus. Examine the prothallus with a compound light and record your observations in the space provided.
- Examine the fern prepared slides and record your observations in the space provided.

Name:

Exercise 1 Results

1. What feature of the life cycle of nonvascular plants is different when compared with all other forms of land plants?

2. Why are nonvascular plants:
   - Restricted to moist environments?
   - Always small?

3. Make a **detailed drawing** of one of the *Bryophytes* on display.

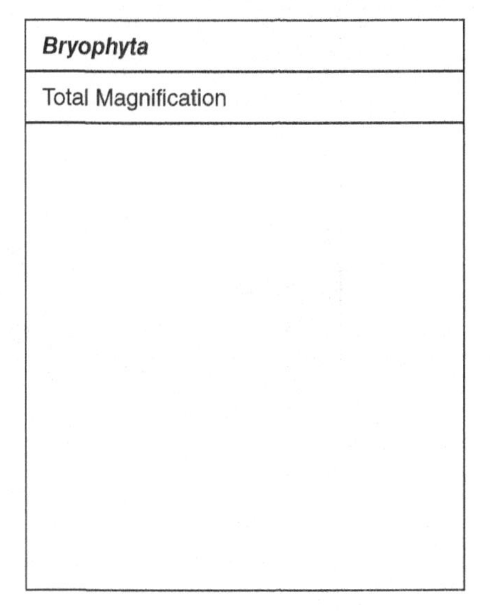

4. The spores produced by the moss sporophyte are:
   - Diploid or haploid?
   - Produced by meiosis or mitosis?
   - Belong to the gametophyte or sporophyte generation?

5. The gametes produced by the moss gametophyte are:
   - Diploid or haploid?
   - Produced by meiosis or mitosis?

6. Which is the dominant generation for the mosses?

- Gametophyte or sporophyte?

7. Make a **detailed drawing** of one of the Hepatophytes on display.

Hepatophyta
Total Magnification

8. Make **detailed drawings** of longitudinal views of *Marchantia* archegonium and antheridium slides.

***Marchantia* archegonium**	***Marchantia* antheridia**
Total Magnification	Total Magnification

9. Which is the dominant generation for the liverworts?

- Gametophyte or sporophyte?

10. Gemmae are responsible for:

- Asexual or sexual reproduction?

Exercise 2 Results

1. Make a **detailed drawing** of one of the *Lycopodium* and *Selaginella* members on display.

Lycopodium	*Selaginella*
Total Magnification	Total Magnification

2. Make a **detailed drawing** of the longitudinal view of a *Lycopodium* strobilus prepared slide and the transverse view of a *Lycopodium* stem prepared slide.

Lycopodium **strobilus**	*Lycopodium* **stem**
Total Magnification	Total Magnification

3. Make a **detailed drawing** of the longitudinal view of a *Selaginella* strobilus prepared slide and the transverse view of a *Selaginella* stem prepared slide.

***Selaginella* strobilus**	***Selaginella* stem**
Total Magnification	Total Magnification

4. The leafy plants are part of the:
   - Gametophyte or sporophyte generation?
5. Megaspores and microspores are produced by:
   - Meiosis or mitosis?
6. Make a **detailed drawing** of a *Psilotum* member on display, the longitudinal section of a Psilotum sporangia, and the transverse section of a *Psilotum* stem prepared slide.

Psilotum	***Psilotum* sporangia**
Total Magnification	Total Magnification

Psilotum stem
Total Magnification

7. Make a **detailed drawing** of an *Equisetum* member on display and the transverse section of an *Equisetum* stem prepared slide.

Equisetum	*Equisetum* stem
Total Magnification	Total Magnification

8. Make a **detailed drawing** of one of the *Pterophytes* on display and the wet mount of a *Pterophyta* (fern) prothallus.

Pterophyta	Fern prothallus
Total Magnification	Total Magnification

9. Make a **detailed drawing** of the longitudinal view of the indusium of a *Pterophyta* (fern) prepared slide and of a *Pterophyta* (fern) antheridia and archegonia prepared slide.

Fern indusium	**Fern antheridia and archegonia**
Total Magnification	Total Magnification

10. Make a **detailed drawing** of the transverse section of an *Osmunda* rhizome prepared slide.

***Osmunda* rhizome**
Total Magnification

11. The dominant generation for the fern is:
    - Gametophyte or sporophyte?
12. The spores produced by the fern sporophyte:
    - Are formed by meiosis or mitosis?
    - Belong to the gametophyte or sporophyte generation?
13. Once dispersed, these fern spores will produce the:
    - Gametophyte or sporophyte generation?
14. The gametes are produced by:
    - Meiosis or mitosis?

CHAPTER 25

Gymnosperms

Photo courtesy of Holly J. Morris

OBJECTIVES

After completing this exercise, you will be able to:

- Understand the characteristic features of the gymnosperm phyla.
- Understand the evolutionary advances of the gymnosperms.
- Describe the life cycle of a gymnosperm.

INTRODUCTION

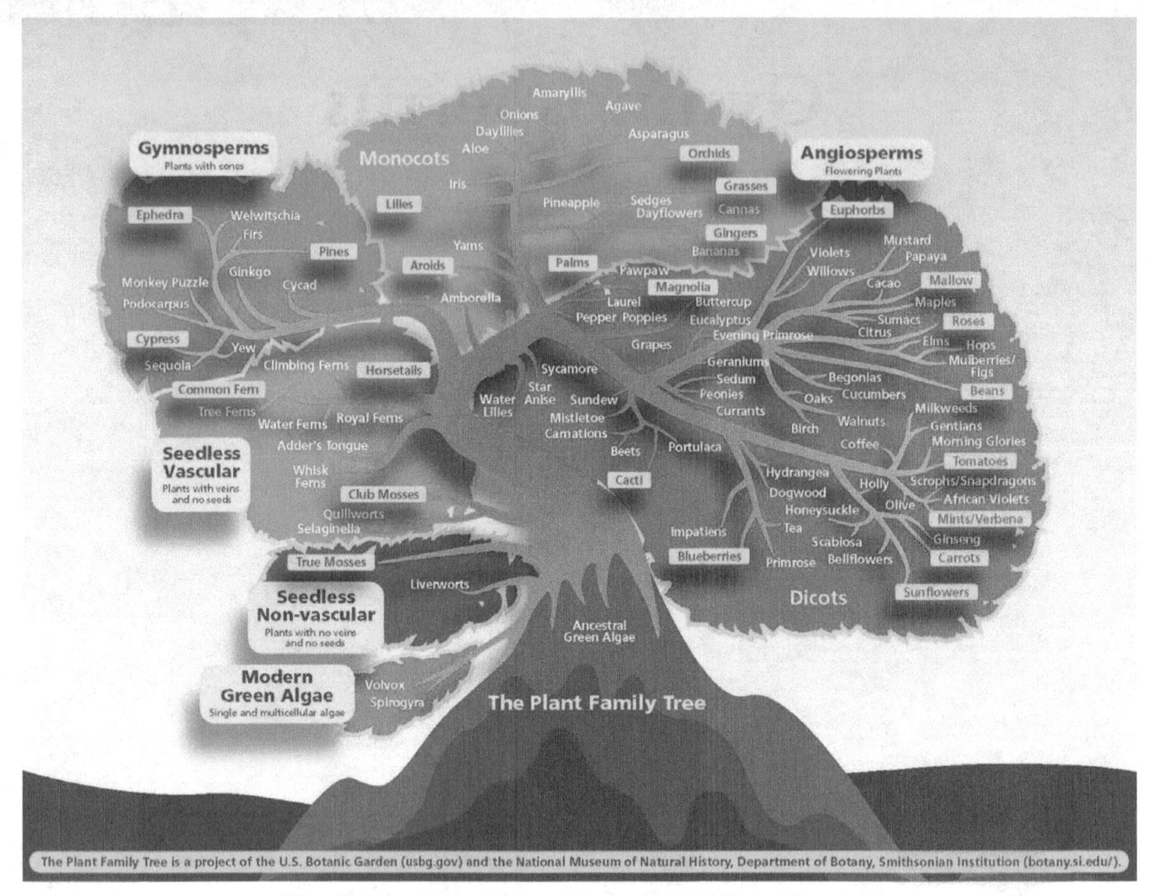

(https://www.usbg.gov/classroom-resources-plant-based-learning)

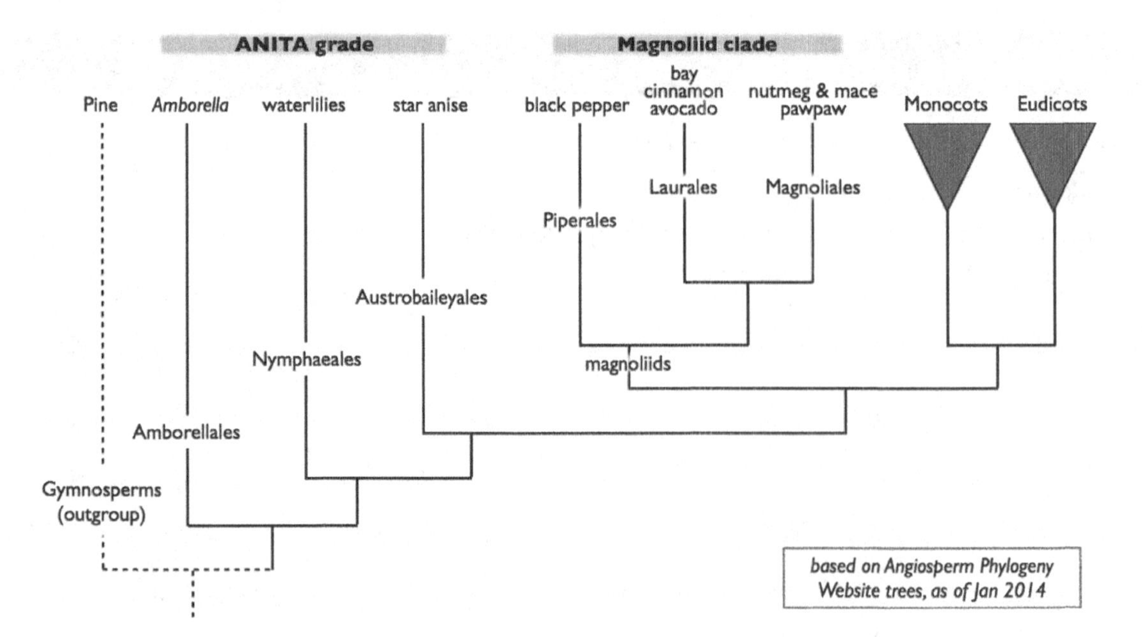

Vascular seed plants include the *gymnosperms* and the *angiosperms*. Gymnosperms include the giant sequoia and the sugar pine. The sugar pine cone is actually a megastrobilus with seeds borne on the exposed scales. A *seed* is an advance over the spore as a mode of reproduction because a seed has more protection from adverse environmental conditions. A spore is a single cell with its nuclear material. A seed is multicellular with an embryonic sporophyte, a food supply, and a protective outer coat. The seed is the main reason that the seed plants became dominant during the late Mesozoic and Cenozoic eras. Because a seed has stored nutritive tissue, the embryo can remain dormant until environmental conditions are favorable for germination. A spore, on the other hand, does not have that option. When the spore disperses, it must land in a suitable location for successful germination into the gametophyte generation.

Gymnosperms (*conifers*) had their heyday during the age of the dinosaurs, but angiosperms (*flowering plants*) became dominant since the end of the Mesozoic era or approximately 65 million years ago. The giant sequoia is one of the most massive of all known living organisms. Scientists believe that the bristlecone pine of the Western United States is one of the oldest living specimens on the planet Earth today. Some are older than 4000 years. Only approximately 800 species of gymnosperms exist today while there are over 300,000 species of angiosperms. We will study angiosperms in the next exercise.[16]

Giant Sequoia (© Felix Lipov/Shutterstock.com)

Bristlecone Pine (© Steve Cukrov/Shutterstock.com)

Seed plants have reduced gametophytes that are dependent upon the sporophyte for a food supply. Seed plants must be heterosporous, producing both megaspores and microspores, in order to produce seeds. In all seed plants, the female or megagametophyte that develops from the megaspore remains within the diploid sporophyte plant structure called the ovule. Pollen grains are another reason for success in the terrestrial environment. A pollen grain contains the male or microgametophyte with sperm nuclei. Wind or animals aid transfer of the sperm to the egg nucleus within an ovule. They do not require water for fertilization; in other words, the sperm of seed plants typically do not swim to an egg as we observed in the bryophytes and seedless vascular plants.

Gymnosperms are divided into four phyla: *Ginkgophyta, Cycadophyta, Gnetophyta,* and *Coniferophyta.* Gymnosperms produce *naked seeds.* The seeds are typically borne on scales of cones/strobili. In your study of gymnosperms, you will note that the sporophytes of gymnosperms are mostly the cone-bearing, evergreen trees. Pine, spruce, fir, redwood, sequoia, cedar, and hemlock are examples of evergreen gymnosperm conifers. There are some deciduous types, such as bald cypress. The leaves of the gymnosperm are often needle-like or scale-like. Before we get to the conifers, let us first study a few other types of gymnosperms.[16]

PHYLUM GINKGOPHYTA

Ginkgo biloba (Photo courtesy of Holly J. Morris)

Ginkgo biloba Fruit and Leaves (© Brzostowska/Shutterstock.com)

Biologists consider the ginkgo or maidenhair tree a living fossil because it dates back to the Mesozoic. Strangely, however, *Ginkgo biloba* seems to tolerate the modern day air pollution of metropolitan areas. It is unique with its fan-shaped leaf and dichotomously branched veins. The species is *dioecious*. The male trees produce male strobili with pollen grains. The female trees produce the fleshy ovules that ripen into smelly although edible seeds during the fall season. The branches on the tree are also unique.[16]

PHYLUM GNETOPHYTA

Ephedra (Mormon Tea) (© Jody/Shutterstock.com)

Welwitschia (© Anton_Ivanov/Shutterstock.com)

On a hike in Zion National Park, Utah, you will probably see a strange-looking plant that has dark green jointed stems. A closer look might reveal yellow "flowerlike" pollen cones as seen in the photo. The plant is *Ephedra*, commonly known as Mormon tea or joint fir. Researchers believe that the early Mormon settlers of Utah survived harsh winters by seeping the tea from this plant. The plant is a source of ephedrine, a medicinal substance, commonly used as a decongestant, a heart stimulant, and in a number of herbal products.[16] However, ephedrine is a potentially dangerous natural compound and as a result, has been banned from dietary supplements in the United States since 2004.

PHYLUM CYCADOPHYTA

© Peredniankina/Shutterstock.com

Cycads thrived during the days of the dinosaurs, and we often know them as the "Age of Cycads." Today there are approximately 150 or more species. Upon a first look at a cycad, the viewer might mistake it as a palm since the leaves are distinctly *compound* or fern-like or palm-like. The *dioecious* plants bear large fleshy cones. The fronds often appear to emerge from the trunk in a crown shaped fashion.[16]

PHYLUM CONIFEROPHYTA

The largest and best known members of the gymnosperms are the conifers. They are cone-bearing seed plants and have vascular tissue. Some examples include spruces, firs, cedars, and redwoods.

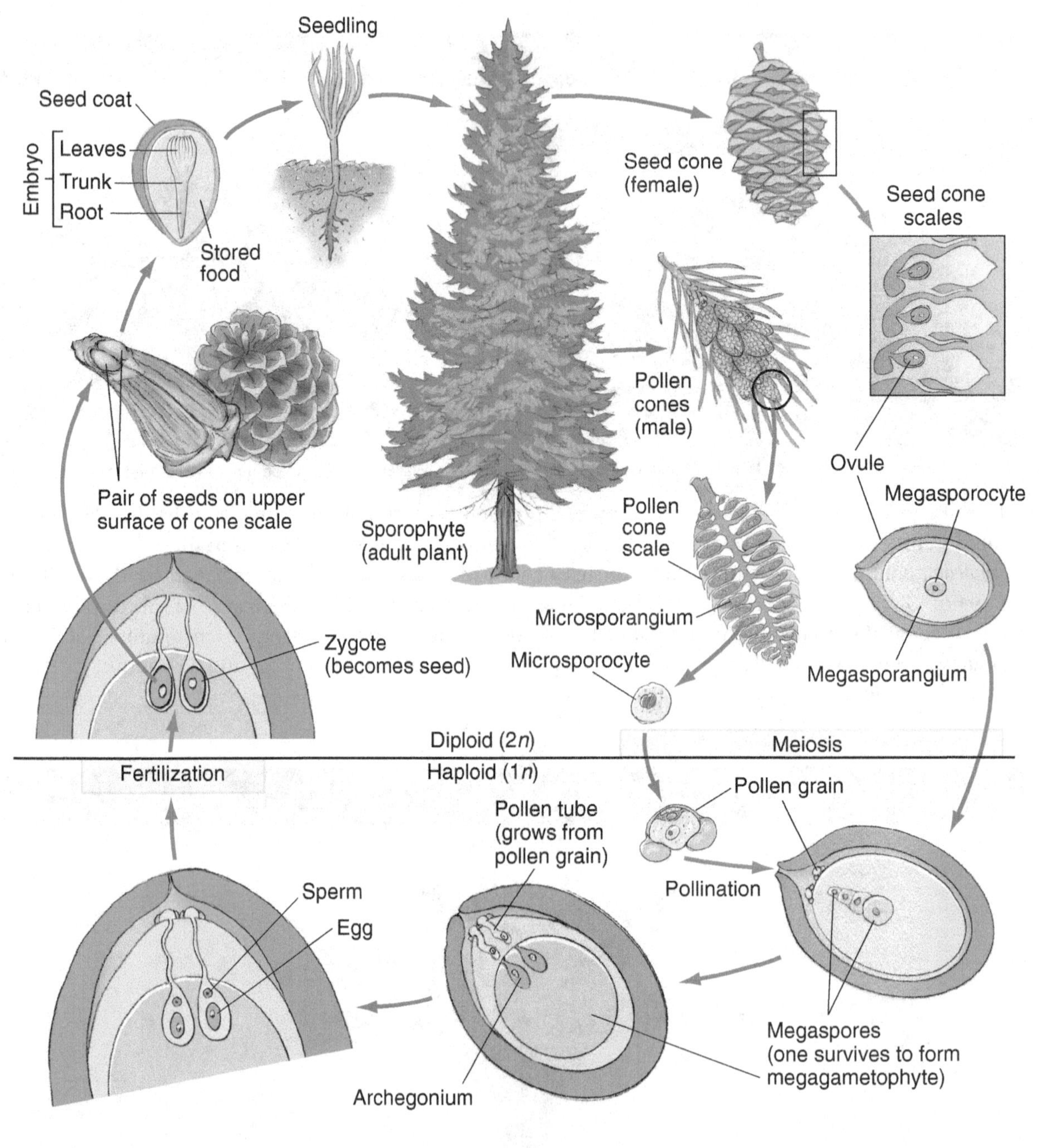

Conifer Life Cycle (© Kendall Hunt Publishing)

A mature pine tree is a diploid sporophyte. It produces two types of cones, male, and female.

Female pine cones are large and woody. In ovules on each scale of these cones, meiosis occurs and produces produce four haploid megaspores. Only one will remain viable and divide by mitosis to produce a single large haploid spore - the megaspore. The megaspore divides by mitosis, producing the multicellular haploid female gametophyte. This megagametophyte produces the egg.[16]

Female Pine Cone (© Trong Nguyen/Shutterstock.com)

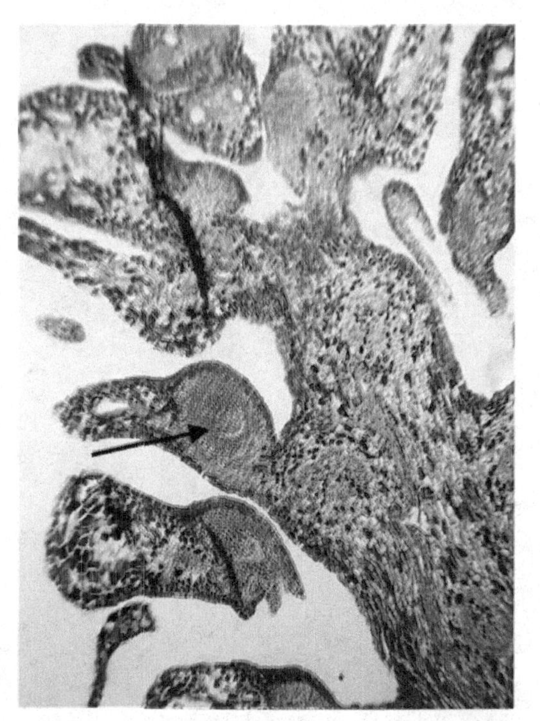

Female Pine Cone (longitudinal view) Arrow points to ovum (Photo courtesy of Holly J. Morris and David T. Moat)

Male cones are herbaceous and are much less conspicuous than female pine cones even at full maturity. These pine cones contain diploid microsporocyte cells that go through meiosis and produce 4 equally sized haploid microspores. Each microspore divides by mitosis and develops into a mature haploid male gametophyte (pollen grain) which eventually produces a tube cell and 2 haploid sperm cells by mitosis. After the pollen grain is released, the male pine cone begins to deteriorate.[16]

Male Pine Cone (longitudinal view)
(© rootstock/Shutterstock.com)

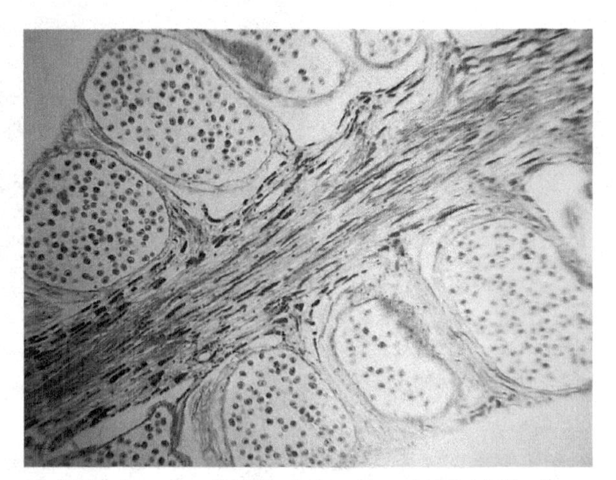

Male pine cone 40× (Photo courtesy of Holly J. Morris and David T. Moat)

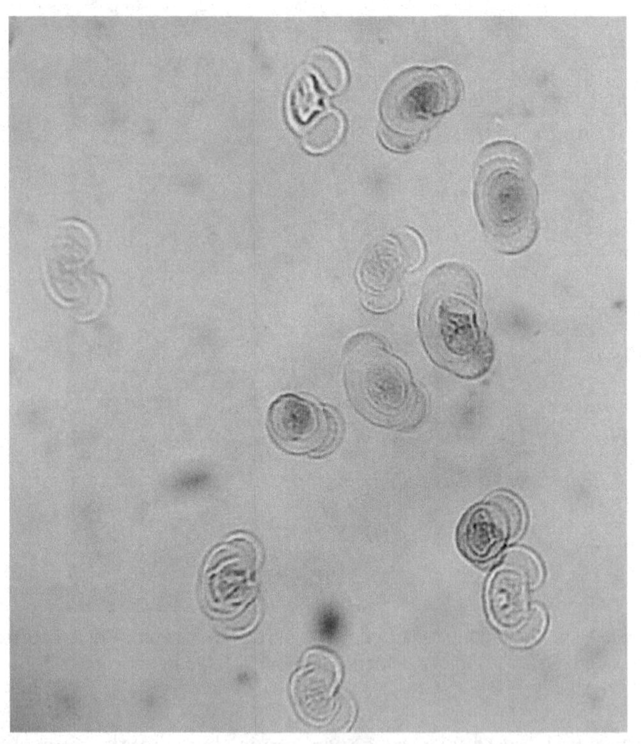

Pollen Grains 400× (Photo courtesy of Holly J. Morris and David T. Moat)

The mature gametophyte secretes a drop of fluid (pollen droplet). When pollen from the male cone contacts this droplet, it gets stuck on the droplet. The pollen tube cell then begins growing toward the ovule forming the pollen tube. The pollen tube approaches an opening in the ovule wall called the micropyle. The sperm cell is then transported to the ovule and fertilization occurs. After fertilization, a diploid zygote forms and results in the formation of a three layered seed containing the diploid embryo which is surrounded by haploid gametophyte tissue which is surrounded by a seed coat of sporophyte tissue. The seed can then be dispersed.

EXERCISE 1 Observing Gymnosperms

Gymnosperms are a group of seed bearing vascular plants that do not have flowers or fruits. Their seeds are considered to be "naked seeds" because they do not develop within an ovary. Some other features of the gymnosperms that have enabled them to be successful include:

- Size of the female and male gametophytes is reduced.
- Pollen is wind dispersed.
- Reproduction does not require water.
- Seeds contain nutritive materials to help the embryo survive.
- Seeds can be dispersed away from the parent plant.
- Advanced vascular tissues transport water, nutrients, and photosynthetic products.

Purpose

To become familiar with the structure, function, and life cycle of gymnosperms.

Materials

- Compound light microscope
- Various gymnosperms on display
- Various gymnosperms prepared slides

Methods

- Examine the various gymnosperm exhibits on display.
- Observe the various gymnosperm prepared slides and record your observations in the space provided.

Name:

Exercise 1 Results

1. Complete the table detailing the characteristics of the various gymnosperm phyla.

Gymnosperm Phyla		
Phyla	**Characteristics**	**Example**
Coniferophyta		
Ginkgophyta		
Cycadophyta		
Gnetophyta		

2. What are the key characteristics shared by all gymnosperms?

3. What is the ecological role of conifers in forest systems?

4. What economically important products are produced by conifers?

5. Make a **detailed drawing** of the longitudinal view of male pine cone and female pine cone prepared slides

Male Pine Cone (prepared slide)	Female Pine Cone (prepared slide)
Total Magnification	Total Magnification

6. Make a **detailed drawing** of a pine pollen grain prepared slide and a pine needle transverse section prepared slide.

Pollen Grain (prepared slide)	Pine Needle (prepared slide)
Total Magnification	Total Magnification

7. Make a **detailed drawing** of a one year pine stem transverse section prepared slide and an older pine stem transverse section prepared slide.

One Year Pine Stem (prepared slide)	Older Pine Stem (prepared slide)
Total Magnification	Total Magnification

8. What are the evolutionary advantages of the gymnosperms?

____

____

9. What is the value of wind dispersal of pollen?

____

____

10. What is the function of the wings on the pollen grain?

____

____

11. Megaspores and microspores are produced by:
    - Mitosis or meiosis?

Angiosperms

Photo courtesy of Holly J. Morris

OBJECTIVES

After completing this exercise, you will be able to:

- Identify the parts of a flower and cite the function of each part.
- Describe the life cycle of the flowering plants.
- Discuss the significance of double fertilization.
- Characterize the different types of fruits.
- Describe the various methods by which seeds are dispersed.
- Compare the differences between eudicots and monocots.

INTRODUCTION

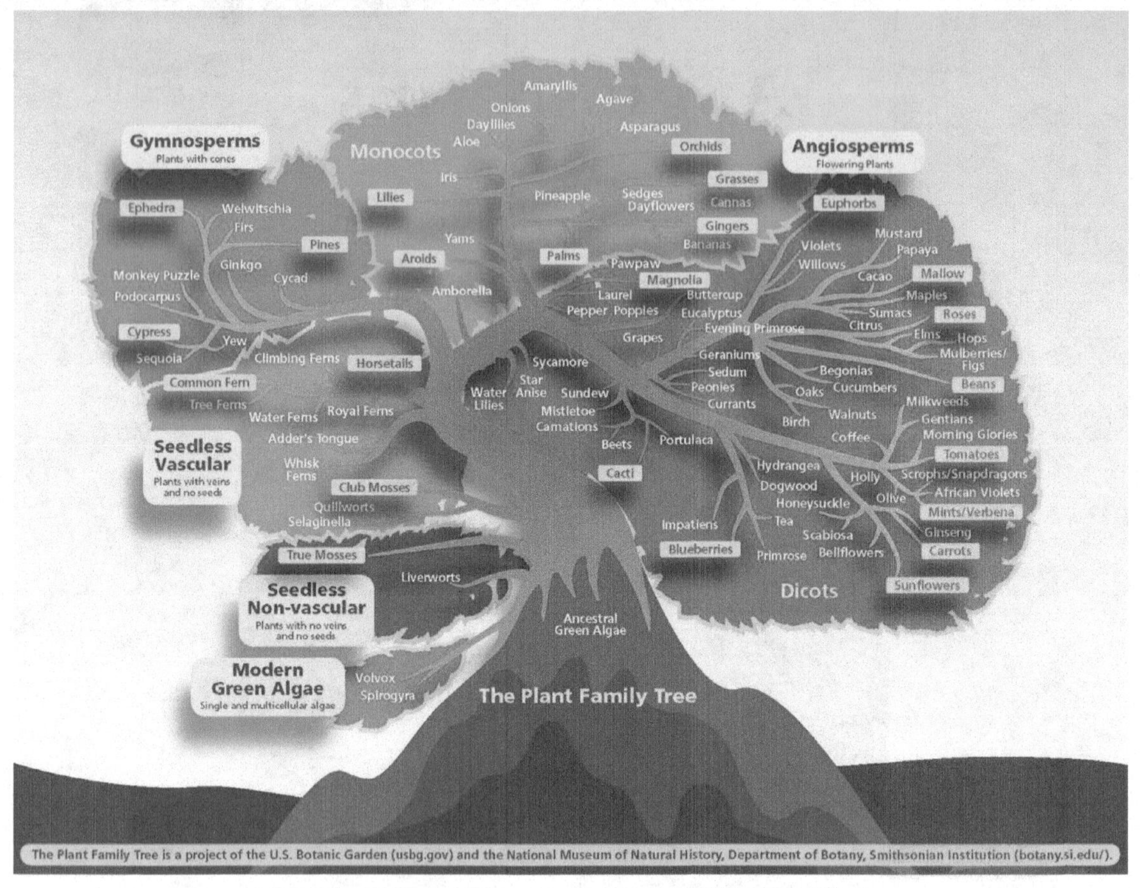

(https://www.usbg.gov/classroom-resources-plant-based-learning)

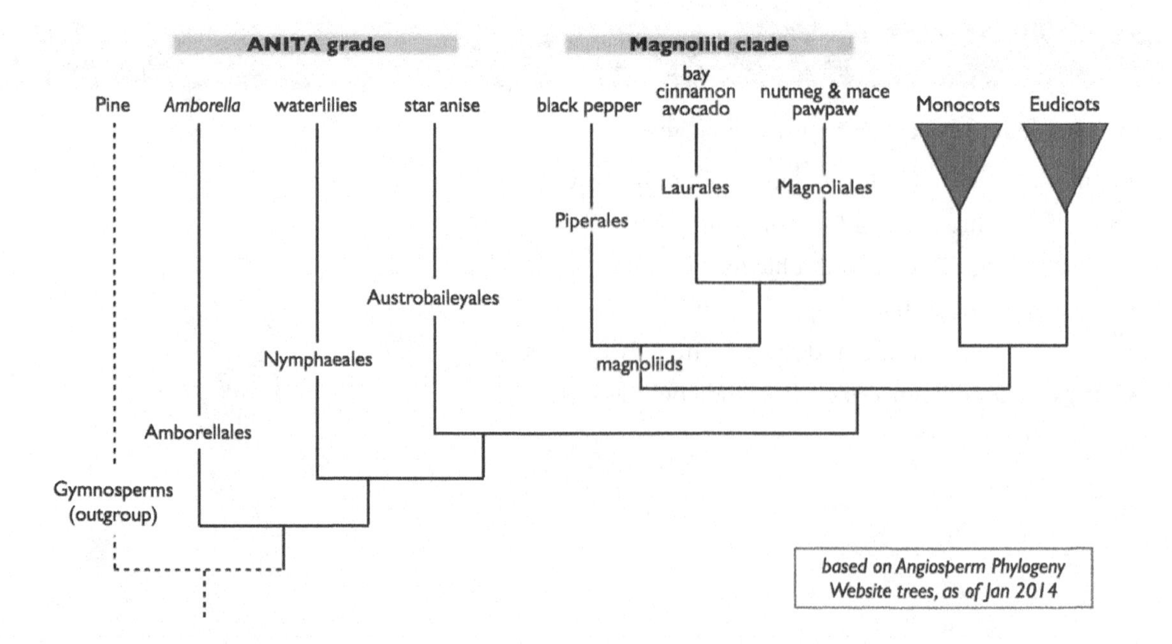

Angiosperms are flowering plants. There are about 300,000 species of angiosperms, making them the largest and most diverse group within the plant kingdom. The word "angiosperm" comes from the Greek words *angeion*, meaning vessel, and sperma, meaning seed. Like gymnosperms, angiosperms are vascular seeded plants. The difference between angiosperms and gymnosperms is that the seeds of angiosperms are covered with an integument, whereas gymnosperms have "naked" seeds.

FLOWER ANATOMY

The parts of a flower include:

Pedicel: The flower's stem attaching to a single flower.

Peduncle: A stem or branch from a main stem that holds a group of pedicels.

Receptacle: Expansion at the top of the pedicel to which the parts of the flower are attached

Sepal (calyx): Modified leaves that protect the bud

Petal (corolla): Colorful modified leaves that attract pollinators

Female Parts:

Carpel (pistil) The female part of the flower, containing the stigma, style, and ovary.

Stigma: The top part of the pistil, designed to "catch" pollen grains.

Style: The stalk that extends from the stigma to the ovary. A tube runs from the stigma, through the style, to the ovary.

Ovary: Holds the ovules, the immature seeds.

Ovule: The female sex cells. Develops into seed when fertilized.

Male Parts:

Stamen: The male reproductive structure of a flower, consisting of the anther and the filament.

Anther: Contains the microsporangia, or pollen grains.

Filament: Holds up the anther.

Two sets of terms that are used in relation to flower structures are *complete* or *incomplete*, and *perfect* or *imperfect*. **Complete flowers** have all four main structural components: sepals, petals, carpel, and stamens. **Incomplete flowers** are missing at least one of those components. **Perfect flowers** contain both male and female components, the carpels and stamens, whereas **imperfect flowers** contain only male or only female parts. Flowers that contain only male components are called staminate flowers. Flowers that contain only female parts are called carpellate flowers. All complete flowers are perfect; not all perfect flowers are complete.

Another set of terms that are used are *monecious* (single house), *dioecious* (two houses) and *bisexual* (hermaphroditic). **Dioecious** plants are easier to conceptualize. These are plants that are either male or female. Two examples of dioecious angiosperms are the ailanthus tree, commonly known as the tree of heaven, and holly. Dioecious flowers are imperfect. Only about 6-8% of angiosperms are dioecious.

Monecious plants have both male flowers and female flowers on the same plant but not on the same flower, therefore the flowers on monecious plants are also imperfect. Some examples of monecious plants are the flowers in papaya, watermelon, and corn.

A relatively new variation of terms is the idea of perfect flowers being called either **bisexual** or **hermaphroditic**, rather than monecious. Bisexual flowers have both male and female parts on the same flower. These are perfect flowers, and some examples are roses, lilies, and sweet peas.

Some texts combine monecious plants and bisexual plants and call both types of plants monecious. In this text we will make the distinction between plants that have separate male and female flowers, and are therefore imperfect flowers, and plants that have perfect flowers.

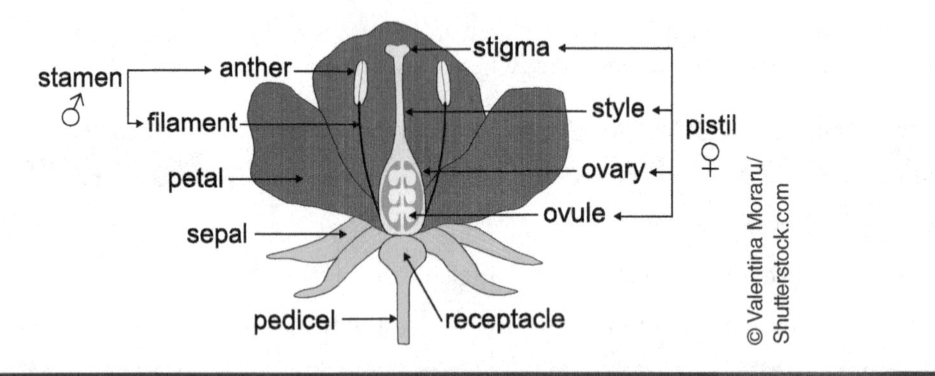

ANGIOSPERM LIFE CYCLE

In angiosperms, the diploid sporophyte, what we typically think of as the plant, is the dominant component of the life cycle. As seed plants evolved, the haploid gametophyte became reduced in size and dependent on the sporophyte. The angiosperm life cycle involves gametophyte development, followed by pollination, double fertilization, and seed development. We will examine each component starting with the gametophytes.

The sporophyte, which is diploid, produces spores by meiosis, thus the spores are haploid. In the development of male gametophytes, a diploid microsporocyte undergoes meiosis which results in four haploid microspores. Each microspore undergoes mitosis, producing two cells, a *generative cell* and a *tube cell*. These two cells, plus a spore wall, make up the pollen grain.

The development of the female gametophytes is more complex. In the ovule, a megasporocyte enlarges and undergoes meiosis, resulting in four megaspores. Three of the megaspores degenerate, leaving one haploid megaspore. The nucleus of the megaspore undergoes three rounds of mitosis, without cytokinesis, resulting in a large cell with eight haploid nuclei. This multinucleate cell forms the embryo sac, as cell membranes develop and separate the nuclei, forming seven cells, three antipodal cells, one central cell with two polar nuclei, two synergid cells, and the egg cell.

When a pollen grain lands on a receptive stigma, a pollen tube grows down the style. As the pollen tube elongates, the nucleus of the generative cell undergoes mitosis, producing two sperm. When the pollen tube reaches the synergid cells, it releases the two sperm into the embryo sac, which is going to result in *double fertilization*. One sperm fuses with the egg cell, forming a diploid zygote. The other sperm fuses with the central cell (which contains two polar nuclei), resulting in a triploid cell. This cell will develop into the endosperm, which is the nutritive tissue for the seed. Following double fertilization, each ovule develops into a seed.

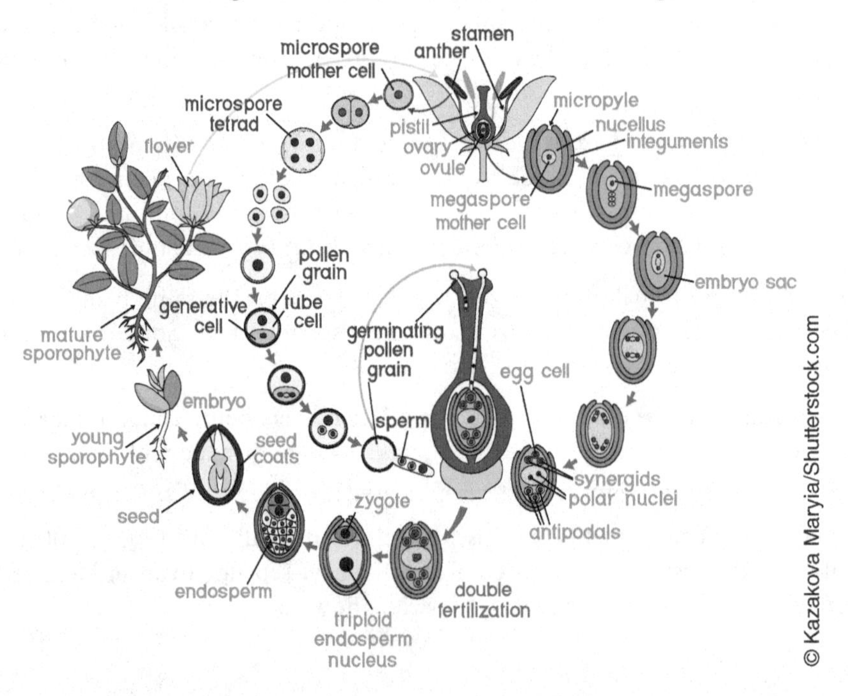

ANGIOSPERMS

Bromeliad

Red Clover

Hibiscus

Daisy

Grasses

Iris

Anthurium

Petunias

Photos courtesy of Holly J. Morris

Angiosperms are divided into two classes: *monocots* and *eudicots*. **Monocots** have flower parts in threes or multiples of three while **eudicot** flowers have flower parts in fours or fives, or multiples of four or five. If you are looking at a flower with three petals, you can identify it as a monocot based solely on the number of petals. If you are looking at a flower with 15 petals, you cannot identify, strictly by the number of petals, whether the flower is a monocot or a dicot. You need to look at other features.

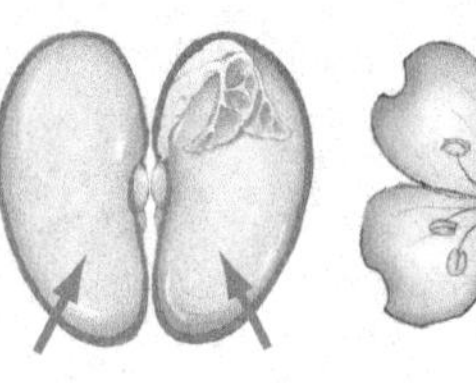

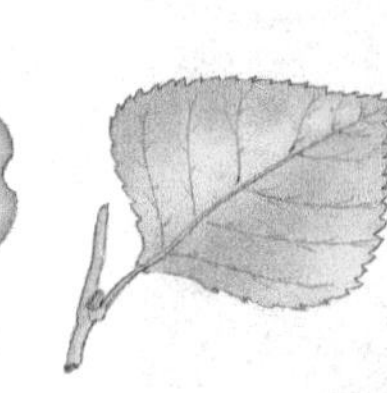

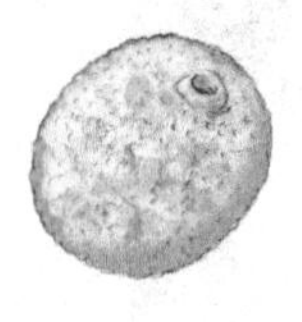

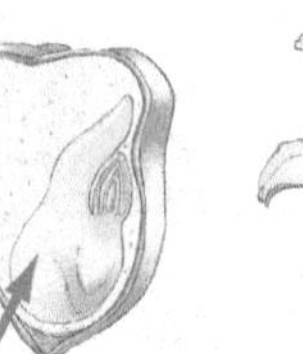

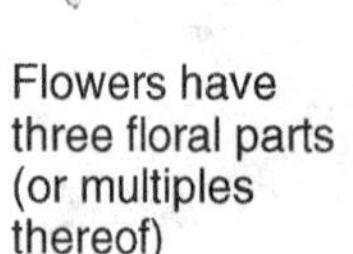

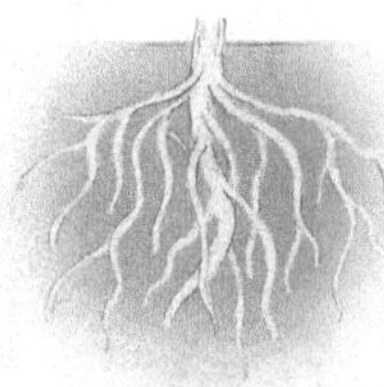

Monocot versus Eudicot Plants (© Kendall Hunt)

FRUITS AND SEEDS

The biological definition of a fruit is that it is the seed-bearing portion of a flowering plant that develops from the ovary. The fruit is designed to protect the seeds as they develop. Seeds are embryonic plants that, in angiosperms, have a protective cover. Seeds in angiosperms have three genetically distinct components, the embryo, the endosperm, and the seed coat.

The embryo is formed from the zygote (the fertilized egg) and is diploid. The endosperm is usually triploid and provides nutrition, in the form of starch, lipids, or proteins, to the developing embryo. The seed coat, or integument, develops from maternal tissue derived from the ovule.

Seeds are dispersed in a variety of ways. Some fruits, such as tomatoes or strawberries, are designed to entice animals to disperse the seeds by eating the fruit. The seeds have mechanisms to protect them from digestion in the animal's gut. They are then dispersed in a location away from the parent plant, along with feces acting as fertilizer. Other seeds, like those from dandelions or milk weed, are very light and can be dispersed by the wind.

The botanical definition of a fruit is that it is a seed-bearing structure that protects the seeds as they develop and helps in their dispersal. This definition is different than the everyday usage of the word "fruit." "Vegetables" such as tomatoes and cucumbers are actually fruits just as pineapples, oranges, and apples.

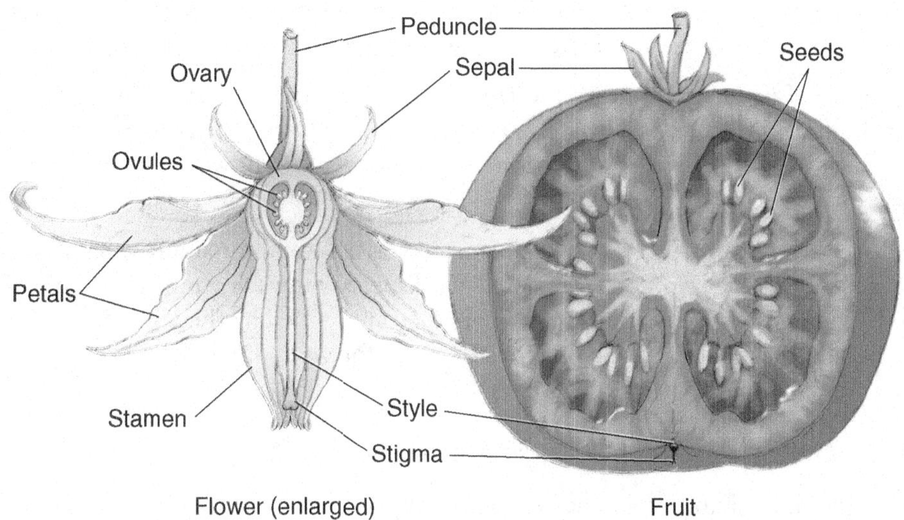

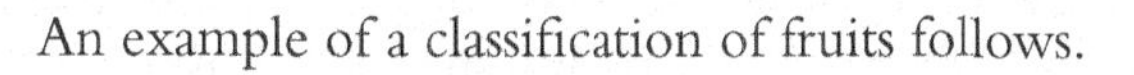

An example of a classification of fruits follows.

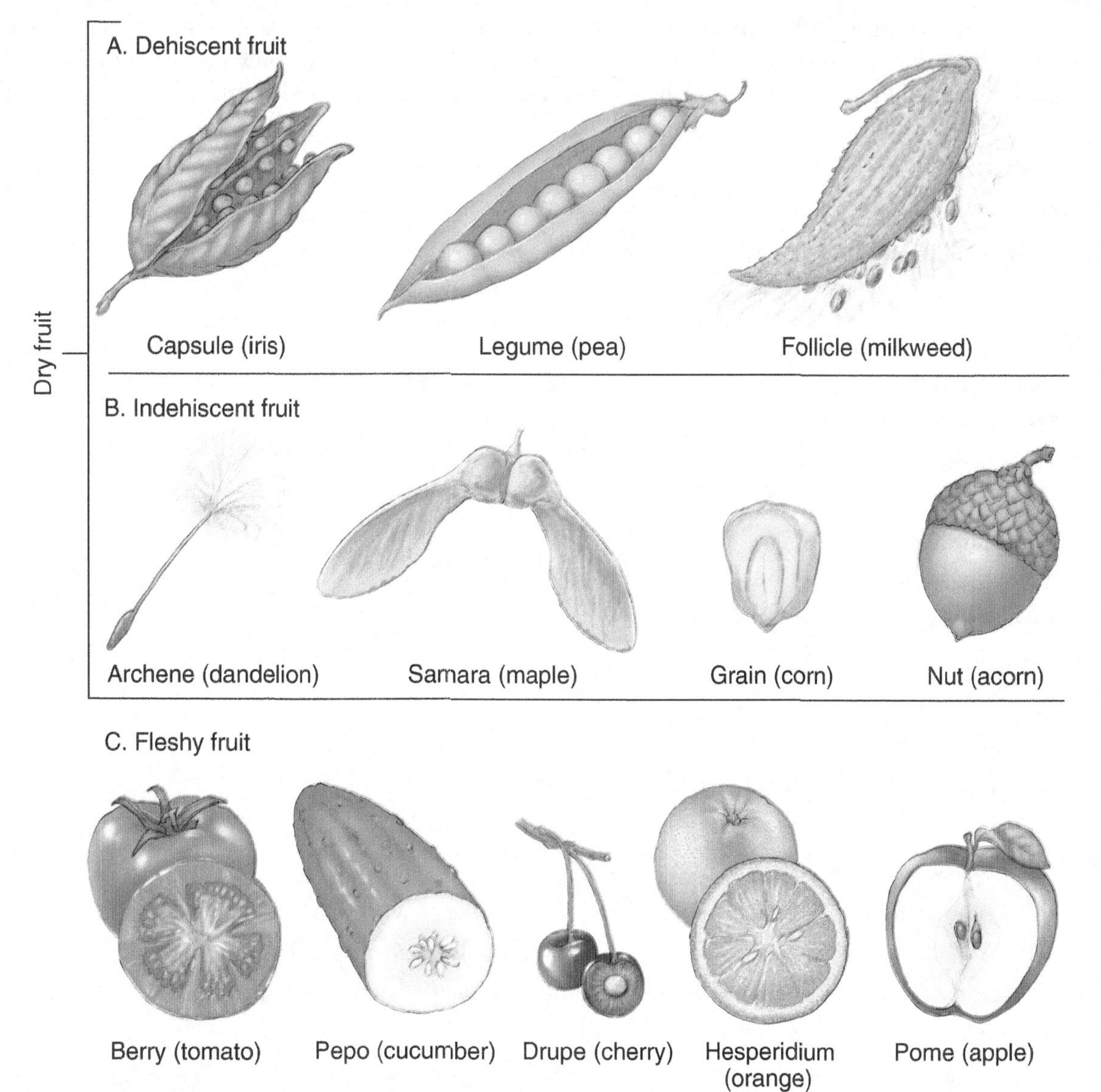

Fruit Types (© Kendall Hunt Publishing)

There are many ways to classify fruits. One way is whether the fruit was formed from a single ovary (simple) or from several ovaries of a single flower (aggregate), or from the fusion of the ovaries from several flowers (multiple). Simple fruits may be fleshy or dry. Dry fruits can be divided into dehiscent, those that split open, such as milkweed and indehiscent, those that don't split open, such as an acorn.

Another way to classify fruits is whether the fruit is or is not a true fruit. True fruits are formed by the ovary and the bulk of the fruit is usually derived from the ovary. In accessory fruits, the bulk of the fruit is not derived from the ovary but from other parts of the flowers. Pomes (apples and pears) are an example of accessory fruits.

Fruits can be dispersed in many ways. Fleshy edible fruits serve as food for animals that in turn spread the seeds. Some fruits, such as the coconut, are adapted for dispersal by water. Some dry dehiscent fruits may split explosively sending their seeds into the air where they are carried by the wind. Some fruits may have spines for attachment to animal fur while others have wings for wind dispersal.

EXERCISE 1 Understanding the Angiosperm Life Cycle

Angiosperms are the most abundant and diverse group of plants on Earth. Their life cycle, as with all vascular plants, is dominated by the sporophyte generation. Flowers and fruits, two angiosperm reproductive adaptations, have been largely responsible for their abundance. Flowers have highly specialized male and female reproductive organs. After fertilization, the diploid zygote develops into a seed which then develops into a fruit. The fruits are then dispersed by various means such as being windblown or eaten by animals. The seeds contained in the fruit will germinate and grow into a mature sporophyte starting a new cycle if the conditions are right.

Purpose

To become familiar with the life cycle of angiosperms.

Materials

- Compound light microscope
- Pollen growth medium
- Filter paper
- Depression slides
- Cover slips
- 37°C Incubator
- Petri plates
- Flowers
- Artist's brush
- Magnifying glass
- Prepared slides (Lily anthers, Lily ovaries, *Tilia* stem)

Methods

- Place filter paper in a Petri plate and moisten the filter paper with water.
- Obtain a clean depression slide and put a drop of the pollen growth medium on the slide.
- Using a magnifying glass, gently remove some of the pollen from the anthers of a flower with an artist's brush.

- Gently tap the artist's brush so that the pollen falls onto the pollen tube growth medium. Put a cover slip on the slide.
- Place the slide in the Petri plate, replace the Petri plate top and then place the Petri plate in the 37°C incubator for 20–30 minutes.
- Remove the slide from the Petri plate and examine the slide with the 40x objective.
- Make a detailed drawing of the growing pollen tubes in the space provided.
- Observe a transverse *Tilia* prepared slide and make a detailed drawing in the space provided.
- Observe transverse Lily anther and ovary prepared slides and make detailed drawings in the spaces provided.

EXERCISE 2 Observing Flowers

A flower is the reproductive part of a plant. Flowers enable plants to reproduce, and their colors and shapes facilitate pollination, seed growth and seed dispersal. The male part of the flower produces pollen that fertilizes the eggs in the female part of a flower. In a process called pollination, the pollen is transferred from flower to flower by various kinds of organisms attracted to the flowers' nectar. After fertilization, the seeds are produced by the flower.

In this exercise, you will examine several flowers and their morphologies and then use this information and the *Key to Pollination* to determine what kind of organism pollinates the flowers.

Key to Pollination

I. Sepals and petals reduced or inconspicuous; leathery or relatively large stigma; flower with no color — WIND

II. Sepals and/or petals large and easily identified; non-leathery stigma; flowers with or without color

a. Sepals and petals white or subdued (greenish or burgundy); distinct odor

i. Odor strong, heavy, or sweet — MOTH

ii. Odor strong, fermenting or fruit-like; flower parts and pedicel strong — BAT

iii. Odor of sweat, feces, or decaying meat — FLY

b. Sepals and/or petals colored; odor may or may not be present

i. Flower shape regular* or irregular** but not tubular

1. Flower shape irregular; sepals or petals are blue, yellow or orange; petal adapted to serve as a "landing platform"; may have dark lines on petals; fragrant odor — BEE

2. Flower shape regular; odor often fruity, spicy, sweet or carrion-like — BEETLE

ii. Flower shape tubular

1. Strong or sweet odor — BUTTERFLY

2. Little or no odor; flowers usually red — HUMMINGBIRD

* **Regular flower** – Radial symmetry (like a daisy or carnation) with similar parts (such as petals) having similar size and shape.

** **Irregular flower** – Bilateral symmetry

Purpose

To become familiar with flower morphology.

Materials

- Flowers
- Dissection kit

Methods

- Examine three flowers selected from those on display.
- Identify the parts of each flower.
- Dissect the flowers and record your observations in the **Flower Morphology and Pollinators** table.
- Use the **Key to Pollination** to determine the most likely pollinator of each flower and record your observations in the **Flower Morphology and Pollinators** table.

EXERCISE 3 Observing Fruits and Fruit Dispersal

A fruit is the seedbearing structure formed from the ovary after flowering. The seeds in a fruit are disseminated by various means. In this exercise, you will use the **Key to Fruits** and **Key to Seed Dispersal** dichotomous keys to identify fruits and then to determine how the seeds of that fruit are dispersed.

Key to Fruits

I. Simple fruits (one ovary)
- a. Dry fruits at maturity
  - i. Fruits with one seed
    - 1. Ovary wall and seed coat are not fused — ACHENE
    - 2. Ovary wall hard or woody but can be separated from the seed — NUT
  - ii. Fruits with 2 or more seeds
    - 1. Ovary with several cavities and several to many seeds — CAPSULE
    - 2. Ovary with 1 cavity
      - a. Mature ovary opens on both sides — LEGUME
      - b. Mature ovary opens on one side — FOLLICLE
- b. Fleshy fruits
  - i. Ovary with one seed which is surrounded by a very hard stone — DRUPE
  - ii. Ovary with many seeds which does not have a stone
    - 1. All of mature ovary is soft and fleshy; surrounding flower tissue does not develop into fruit — BERRY
    - 2. Fleshy fruit develops in part from surrounding tissue of the flower (base of sepals and petals); therefore ovary wall seen as "core" around seeds — POME

II. Compound fruits (more than 1 ovary)

- a. Fruit formed from ovaries of many flowers — MULTIPLE
- b. Fruit formed from several ovaries in 1 flower — AGGREGATE

Key to Seed Dispersal

I. Fruit or associated tissue is plump and juicy (perhaps sweet), seed is hard and indigestible — ANIMAL INGESTION

II. Fruit is dry

- a. Seed or fruit light weight, and has feathery or wing-like attachments — WIND BLOWN
- b. Seed or fruit has no attachments to catch a breeze
  - i. Seed or fruit has tiny hooks — HITCHHIKER
  - ii. Seed lacks sticky hooks
    1. Dry fruit splits open easily by itself
       - a. When fruit is very dry or mechanically stimulated, it bursts open; seeds are tiny and easily scattered — EXPLOSIVE
       - b. Dry fruit splits slowly rather than bursting; seed is too large to scatter very far
         - i. Hard shell formed around firm nutty edible seed — ANIMAL FOOD STOARGE
         - ii. Seed produced by plant living near water; has air pocket allowing it to float — WATER
    2. Dry fruit does not split by itself
       - a. Small seeds can be shaken out — SALT-SHAKER
       - b. Seeds larger; fruit has no holes for seed dispersal
         - i. Hard shell formed around firm nutty edible seed — ANIMAL FOOD STOARGE
         - ii. Seed produced by plant living near water; has air pocket allowing it to float — WATER

Purpose

To become familiar with the life cycle of angiosperms.

Materials

- Fruits on display

Methods

- Examine four fruits selected from those on display.
- Use the **Key to Fruits** to determine the fruit type and record your observations in the space provided.
- Use the **Key to Seed Dispersal** to determine how the seeds of the fruit are dispersed and record your observations in the space provided.

Notes

Name:

Exercise 1 Results

1. Compare and contrast monocots and eudicots.

____

____

2. Make a **detailed drawing** of a growing pollen tube.

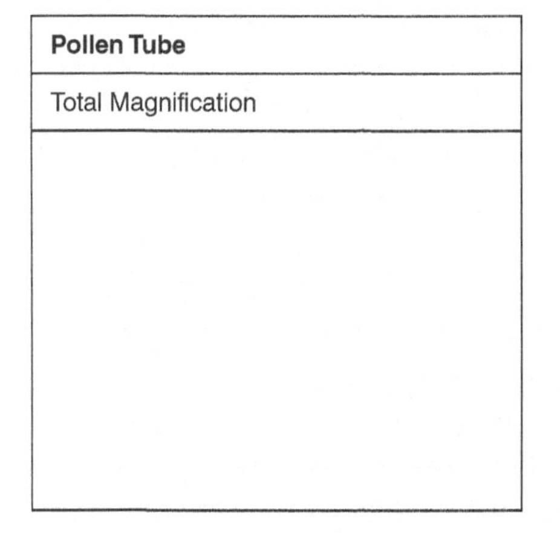

3. Make a **detailed drawing** and label the transverse section of a prepared Tilia stem (1 year or 3 years).

Tilia Stem (1 year or 3 years)
Total Magnification

4. What part of the life cycle is represented by the mature pollen grain?

____

5. How does the female gametophyte in angiosperms differ from the female gametophyte in gymnosperms?

____

____

6. Do you think that pollen germinates indiscriminately on all stigmas? Explain.

____

____

7. Make **detailed drawings** and label the transverse sections of prepared slides of Lily anthers and ovaries.

Lily Anther	**Lily Ovary**
Total Magnification	Total Magnification

Exercise 2 Results

1. How has the evolution of angiosperms transformed the face of the planet?

____

____

2. Complete the table below based on your flower dissections.

Flower Morphology and Pollinators			
	Plant Name		
Features			
Number of petals			
Number of sepals			
Parts absent (petals, etc.)			
Color			
Scent (yes/no)			

Nectar (yes/no)			
Flower shape (tubular, star, etc.)			
Special features			
Predicted pollinator			

Exercise 3 Results

1. What is the function of a fruit?

____

____

2. Complete the table below based on your examination of the fruits on display.

Fruit Types and Methods of Dispersal		
Fruit	**Fruit Type**	**Method of Dispersal**

Additional Exercise

Comparison of Land Plant Characteristics				
Features	***Moss***	**Fern**	**Conifer**	**Flowering Plant**
Gametophyte or sporophyte dominant?				
Water required for fertilization (yes/no)?				
Vascular tissue (yes/no)?				
Homosporous or heterosporous?				
Seed (yes/no)?				
Pollen grain (yes/no)?				
Fruit (yes/no)?				
Example				

Notes

CHAPTER 27

Embryology

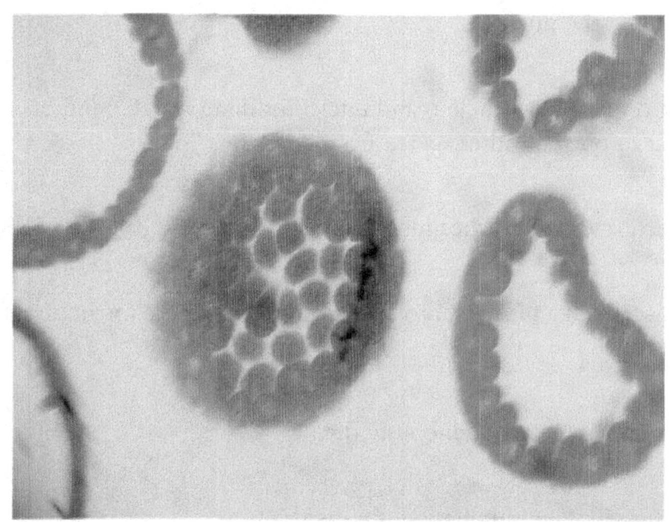

Photos courstesy of Holly J. Morris

OBJECTIVES

After completing this exercise, you will be able to:

- Identify the following stages of development:
  - Fertilization
  - Cleavage
  - Morula
  - Blastula
  - Gastrula.
- Understand the trend in cell number and cell size as development proceeds from fertilized egg to blastula.
- Know the three germ layers of the gastrula.
- Identify the major portions of the body which develop from each germ layer.

INTRODUCTION

The general pattern of reproduction in higher types of animals is the same. Sex cells (eggs and sperm) are produced by meiosis. Sperm and egg join in a process of fertilization forming a zygote. In some organisms, fertilization is external, usually occurring in water. In other organisms, fertilization occurs in the reproductive tract of the female. Following fertilization, the zygote (fertilized egg) begins a series of mitotic cell divisions. These early embryological divisions are called cleavage since the large egg is successively divided or cleaved into smaller cells. For a while, the cells divide so quickly that they do not have time to grow between divisions. Consequently, each new generation of cells becomes progressively smaller. Finally the speed of mitosis decreases allowing the subsequent generations of cells time to grow between divisions.

The early stages of development are:

1. **Unfertilized egg**

   The unfertilized egg is a large single cell with a jelly-like coat around it. There is a distinct nucleus and a single nucleolus inside the nucleus.

2. **Zygote**

   In the fertilized egg (zygote), the nucleus and nucleolus disappear. Chemicals released from the zygote harden the jelly coat to prevent other sperm from entering.

3. **Two-cell stage**

   The single-celled fertilized egg pinches into two cells. Each new cell is half the size of the egg cell.

4. **Four-cell stage**

   Each cell of the two-cell stage pinches into two, forming a four-cell structure. Each new cell is half the size of its parent cell.

5. **Eight-cell stage**

   Each cell of the four-cell stage simultaneously divides.

6. **Sixteen-cell stage**

   Each cell of the eight-cell stage simultaneously divides.

7. **Morula or thirty-two-cell stage**

   The morula is a solid ball of 32 cells looking like a raspberry. Cell division proceeds so quickly that the new cells do not have time to grow.

8. **Blastula**

   The blastula is a hollow ball of cells with a central cavity called the blastocoel.

9. **Gastrula**

   The gastrula stage begins when the cells at one spot on the hollow ball stage begin to migrate into the central cavity (blastocoel). This appears as a depression on the surface of the blastula. Imagine pushing your finger into a tennis ball. This is the way the gastrula will appear. The in-folding of the outer surface will continue until the blastocoel is eliminated. At this point, two of the three embryonic germ layers can be seen: the ectoderm and the endoderm. The different cell layers of the gastrula are called germ layers because they contain the initial cells, which will later develop into specific body structures. Soon the third germ layer, the mesoderm, will form between the first two layers.

The three embryonic germ layers are:

(a) **Ectoderm** or outer layer

(b) **Endoderm** or inner layer, which has just in-folded

(c) **Mesoderm** or middle layer, which forms between the ectoderm and endoderm.

The ectoderm will form the skin and nervous system. The endoderm will form primarily the digestive tract. The mesoderm will form primarily muscle and bone.[1]

The various embryological stages are shown below:

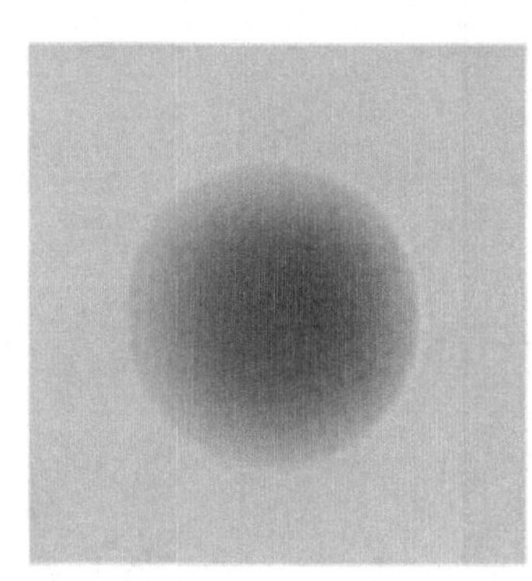

Fertilized Egg

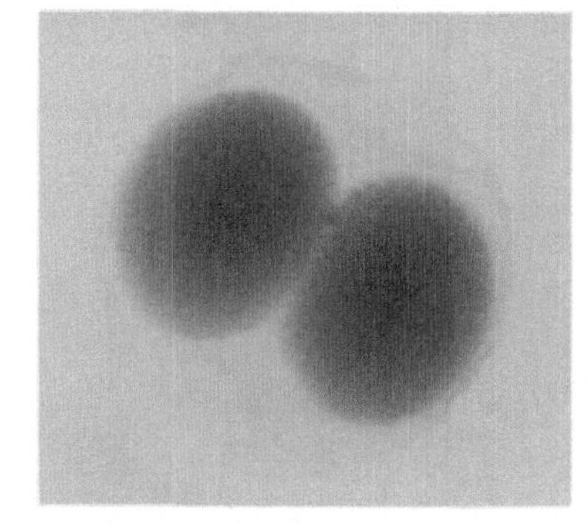

2-Cell Stage

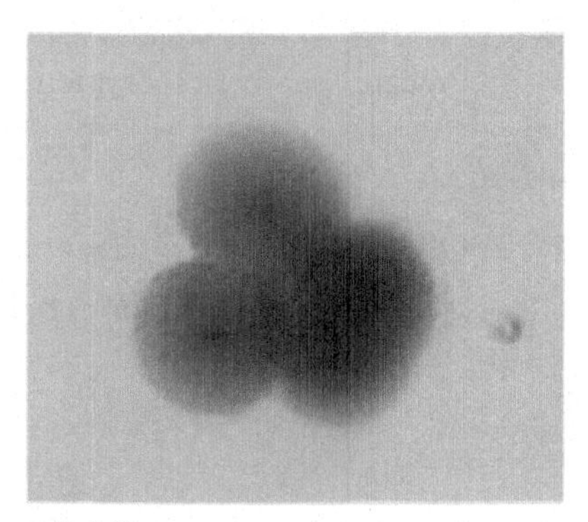

4-Cell Stage

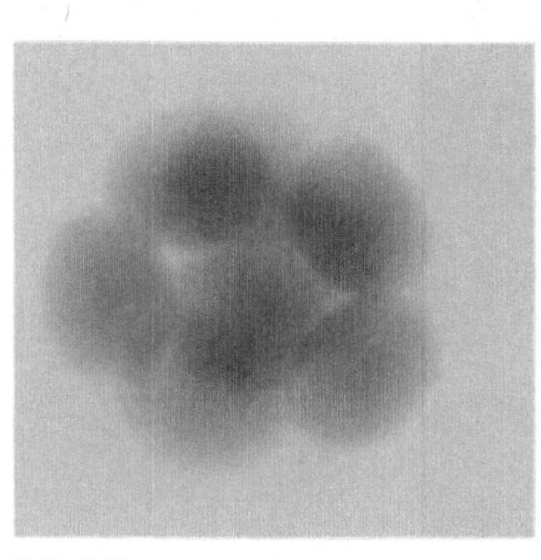

8-Cell Stage

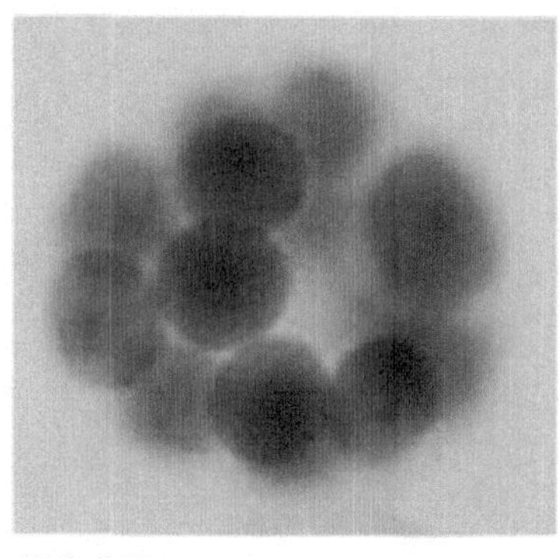

16-Cell Stage

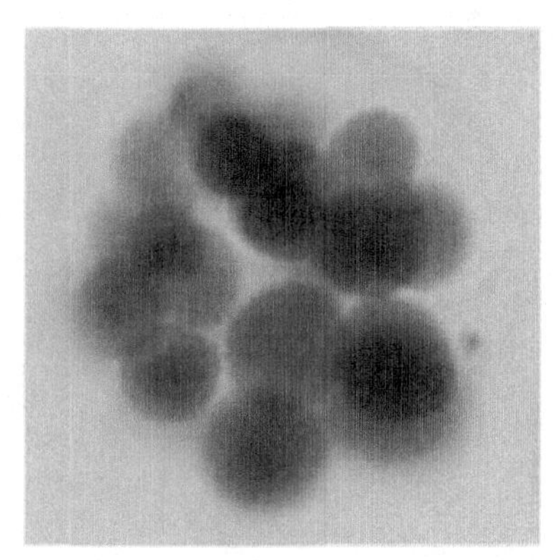

Morula

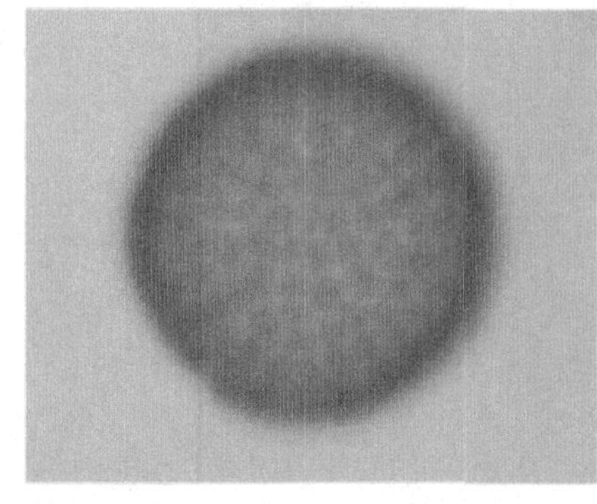

Blastula

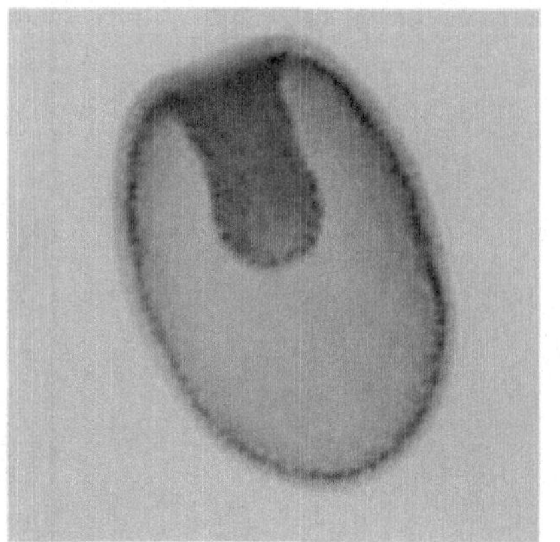

Gastrula (All Photos courtesy of Holly J. Morris and David T. Moat)

EXERCISE 1 Observing Embryological Development in Sea Stars

Embryology is the study of the early development of organisms after fertilization. Observing the embryological development of sea stars provides the opportunity to see these various early embryological stages.

Purpose

To become familiar with the various stages of embryological development by observing sea stars.

Materials

- Compound light microscope
- Prepared sea star embryology slides

Methods

- Observe the earliest stages of sea star development.
- Record your observations below.

Name:

Exercise 1 Results

1. Make **detailed drawings** of the identified stages of the embryological development of a sea star at 400×.

Unfertilized Egg	Fertilized Egg
4 cell stage	**Morula**
Blastula	**Gastrula**

2. What happens to the size of the embryo as development proceeds through all of the stages?

____

____

3. What happens to the amount of cytoplasm in each of the cells of an embryo as development proceeds through all of the stages?

____

____

4. Most "big" differences separating phyla have to do with embryological features. Why might a genetic change or mutation that occurs in an embryo have a greater impact on the organism than one that occurs at a later stage of development?

____

____

CHAPTER 28

Invertebrates

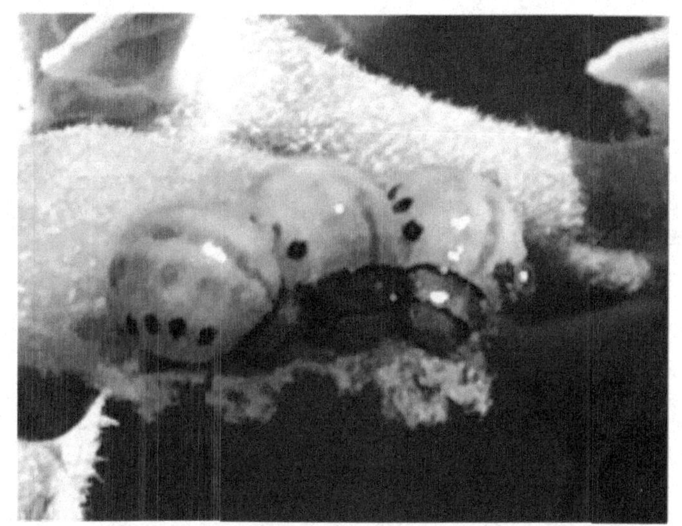

Photo courtesy of Holly J. Morris

OBJECTIVES

After completing this exercise, you will be able to:

- Understand that animals are classified based upon certain criteria: body symmetry, level of organization, number of germ/tissue layers, type of body cavity, segmentation, and digestive techniques.
- Identify representatives of each group of animals.
- Identify the distinguishing features of each group of animals
- Identify anatomical features of each group of animals.

INTRODUCTION

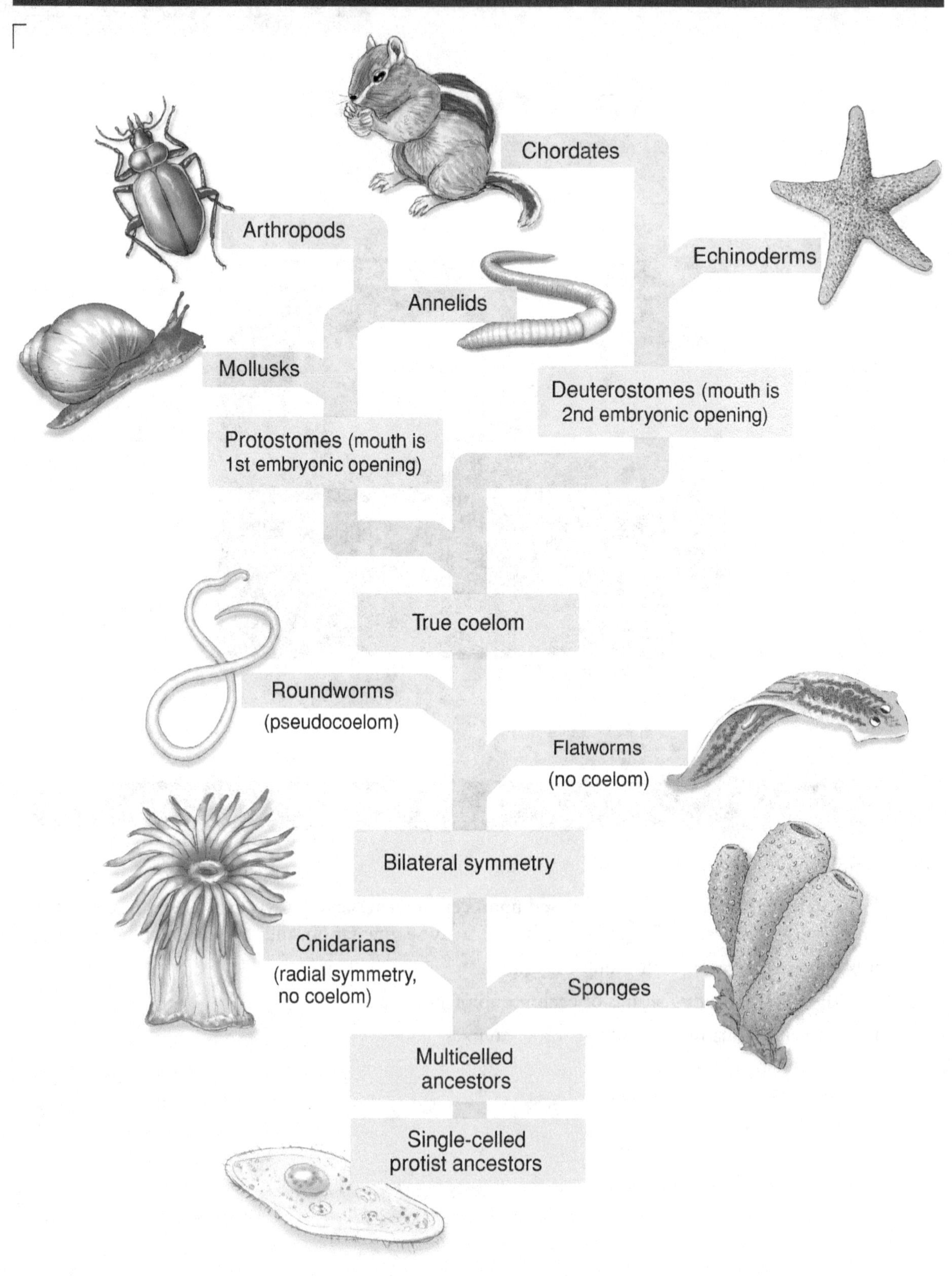

Based on all the evidence, the Animal Kingdom is divided into two major clades: the Parazoa and the Eumetazoa. The Parazoa consists of the sponge phylum. These organisms possess simple bodies and little or no symmetry. The Eumetazoa are all the animals with true tissues and diverged into groups with radial symmetry or bilateral symmetry.

With well over one million species of animals, it is easy to understand that animals are grouped in many different phyla; however, in spite of all this diversity, all animals have some basic features.

BASIC ANIMAL CHARACTERISTICS

Animals are *eukaryotic*, *multicellular*, and *heterotrophic* with varying *digestive* modes, and *locomotive* to some extent. Most animal *body plans possess tissues and organs* derived from germ layers (ectoderm, endoderm, mesoderm) early in the development of the embryo. The term for an animal that possesses only an outer ectoderm and inner endoderm is *diploblastic*. The term for an animal with a third mesoderm layer from which tissues and organs develop is *triploblastic*. Regarding reproduction, sexual reproducing animals are *diploid* with only the gametes being haploid.

A simple family tree of the animal kingdom on the previous page. Note that *body symmetry and body cavity* are part of the basis for placement on this tree. An asymmetric animal does not possess any type of body symmetry. An animal with radial symmetry can be divided into equal halves when cut by a plane going through the central axis. An animal with bilateral symmetry has a right and a left side when cut along a midline.[14]

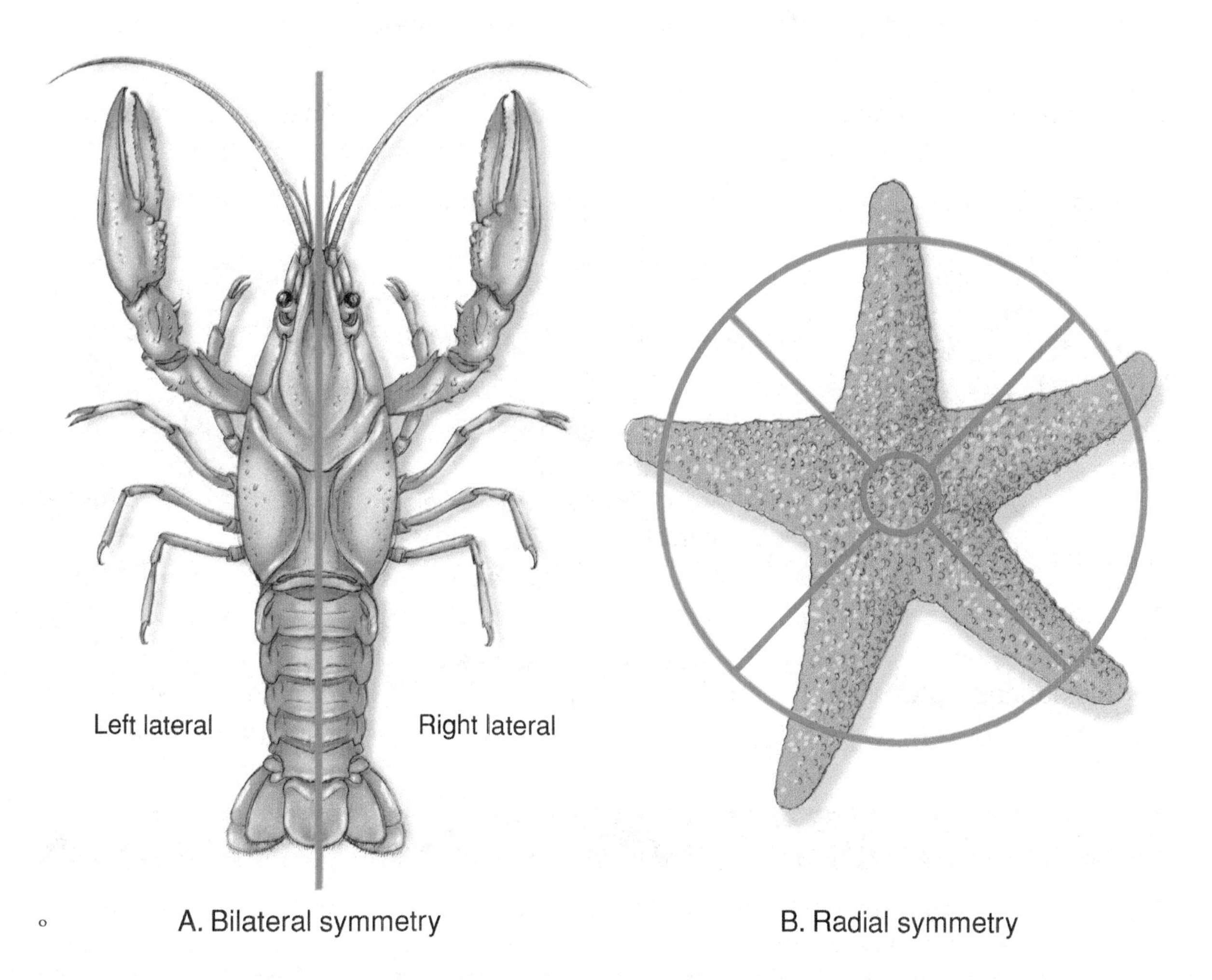

o A. Bilateral symmetry B. Radial symmetry

Certain traits are more primitive or more advanced. The higher branches of the family tree indicate more advanced phylogenies. Bilateral symmetry is an advance over radial symmetry. It will becomes obvious that the presence of a head (*cephalization*), a *complete digestive tract* (mouth to anus), a triploblastic body plan, segmentation, and the presence of a fluid-filled body cavity (*coelom*) are all advanced traits.[14]

Bilaterally symmetrical animals diverged into two major evolutionary lineages: the *protostomes* and the *deuterostomes*. Basic differences are shown in the following table.

Protostomes	Deuterostomes
First opening forms the mouth of the animal	First opening forms the anus of the animal
Spiral cleavage in embryo with upper cells spirally arranged between lower cells	Radial cleavage with early cells at right angles to a polar axis or stacked on top of each other
Determinate cleavage	Indeterminate cleavage
Mesoderm splits to form body cavity	Mesoderm "outpockets" from primitive gut

The protostomes, based upon molecular evidence, evolved along two distinct branches: the Lophotrochozoa and the Ecdysozoa. The Ecdysozoa branch of the protostomes includes two major animal phyla that exhibit the process of *ecdysis* (*molting*). This branch represents the largest and most diverse group of species: *Nematodes* and *Arthropods*.

The *body cavity type* of animal phyla is significant. If the animal does not have a *coelom* (body cavity) it is *acoelomate*. If an animal has mesoderm derived tissue that completely lines a true coelom, it is *coelomate*. If the coelom is incompletely lined with mesoderm, the animal is a *pseudocoelomate*.[14]

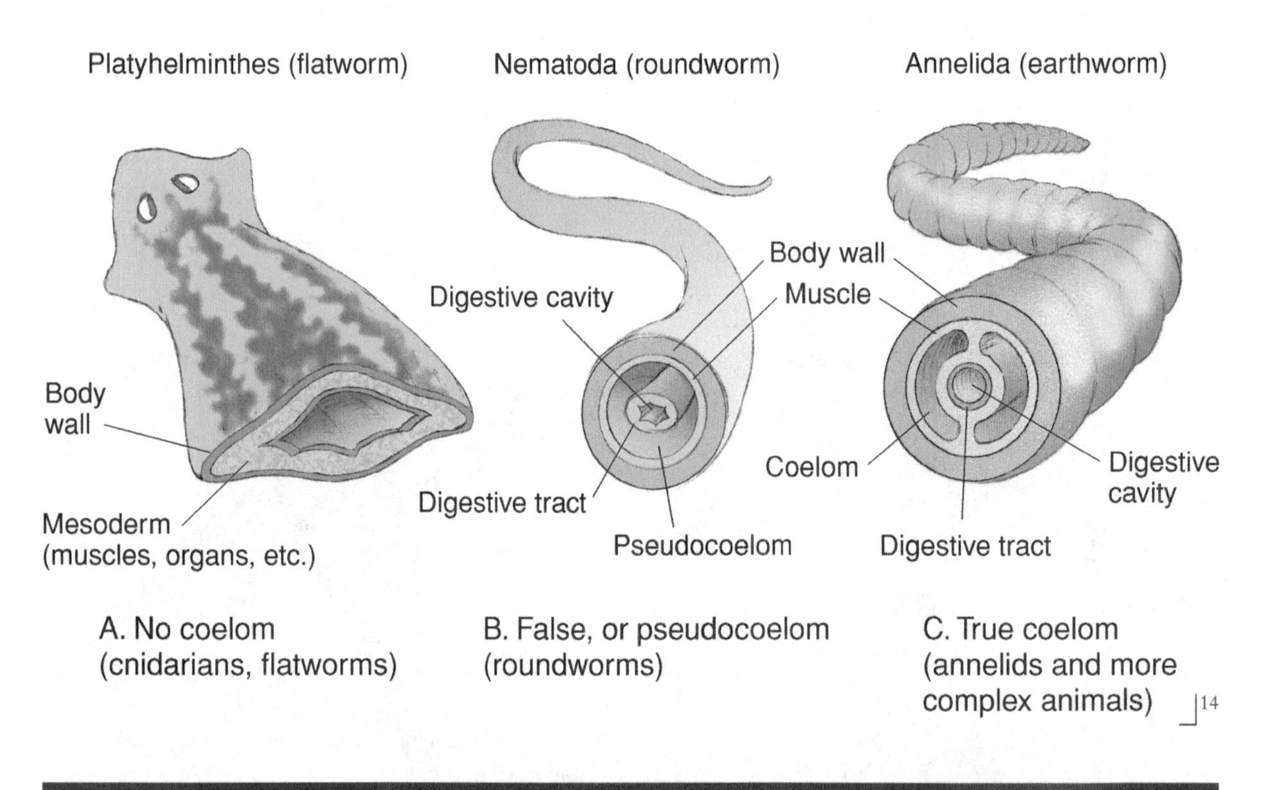

PHYLUM PORIFERA

Sponges do not possess defined tissues but are diploblastic. Biologists consider the two layers of cells intermediate between the cellular and the tissue level of organization since there is some division of labor among the cell types. Sponges have numerous *pores* and possibly canals through which the water flows. The simple body may be asymmetrical or radial. Sponges inhabit the warmer marine waters. The adult is *sessile*; the *ciliated planula larva* is the only stage that exhibits motility.

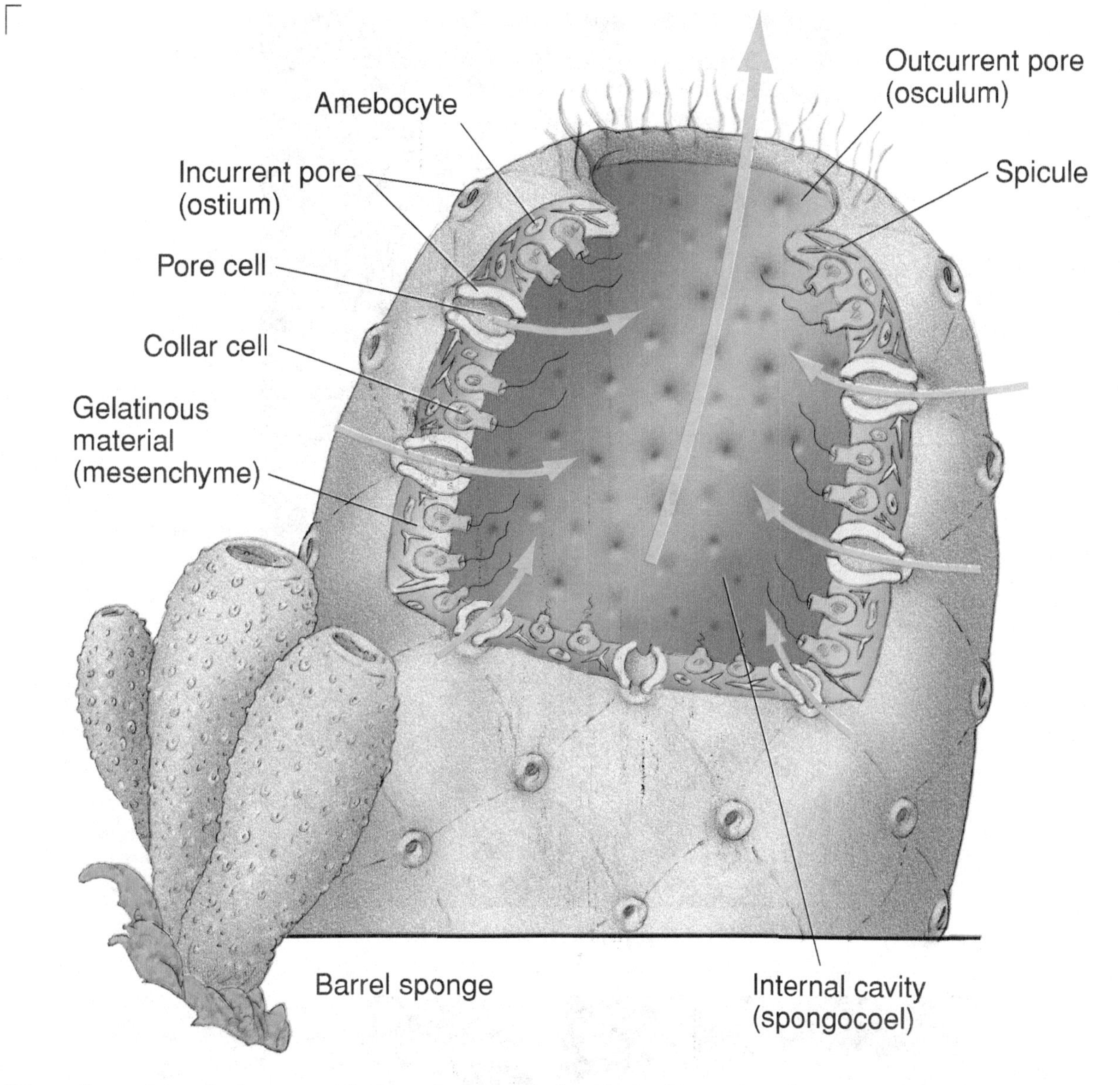

Water flows through the pore cells, into the *spongocoel*, and out the *osculum*. Sponges possess *choanocytes* or *collar cells* in the inner layer of cells. These flagellated cells take in particulate matter for food and digest these molecules in their food vacuoles. Other cells receive nourishment by diffusion or from the wandering *amoebocytes* located in the gelatinous mesenchyme. Amoebocytes also produce the supportive *spicules* in some sponges, collect waste, and can differentiate into gametes.

Scientists classify Porifera into three classes based upon the components of the skeleton:

- Chalky sponges with spicules of calcium carbonate (Grantia)
- Glass sponges with spicules of silica (Venus flower basket)
- Commercial bath sponges with proteinaceous fibers[14]

Chalky Sponge (© Angelo Giampiccolo/Shutterstock.com)

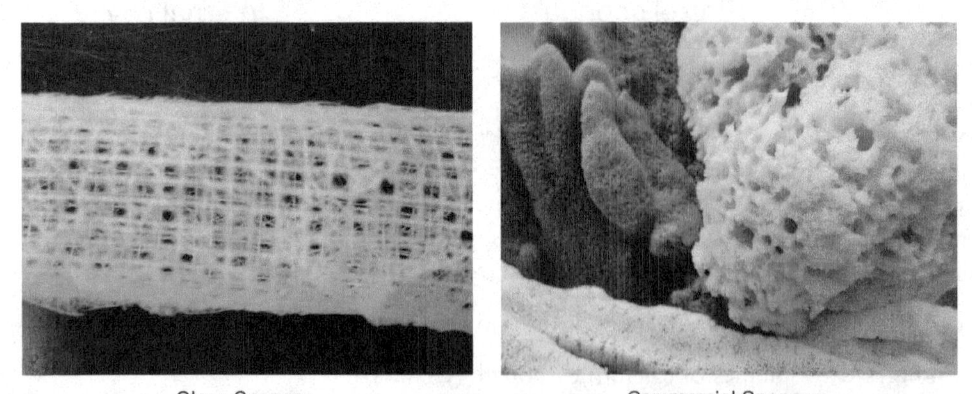

Glass Sponges Commercial Sponges

From Explorations in General Biology by Betty A. Rosenblatt and Sarah Warrington. Copyright © 2008 by Kendall Hunt Publishing Company. Reprinted by permission.

PHYLUM CNIDARIA

All animals other than the sponges are eumetazoans possessing true tissues. The *Cnidaria* (ni-dare'-ee-ah) are the first of the two phyla of radiate animals to study. Formerly, the cnidarians were called the *coelenterates.* Most of these species are marine. These animals are *diploblastic* with two distinct tissue layers, the outer *ectoderm* and the inner *endoderm*, separated by a gelatinous *mesoglea*. A prominent feature of the phylum is the stinging *nematocyst* coiled within a *cnidocyte* on the *tentacles*. They use the nematocysts in defense and in the capture of prey for food. The nematocyst is released in a fraction of a second like a harpoon, and its toxin flows from the tip into the prey. Immediately paralyzed, the prey is held, and tentacles move the food source into the mouth for digestion. Two distinct body forms are present in cnidarians: the sessile *polyp* and the swimming *medusa*. There are four classes: Hydrozoa, Scyphozoa, Cubozoa, and Anthozoa.[14]

1 Class Hydrozoa

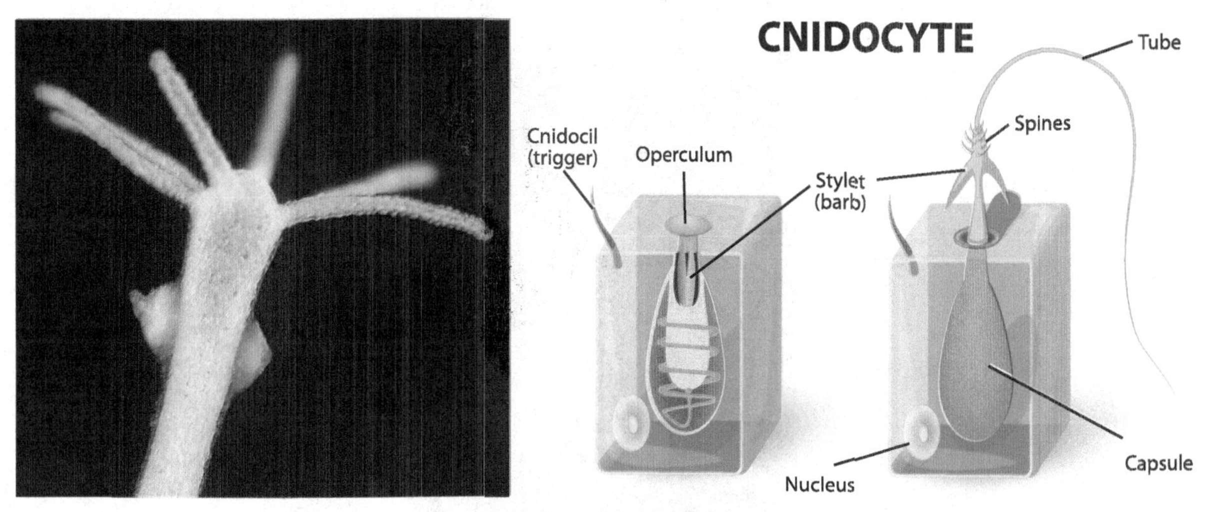

Hydra and Cnidocyte
(*Left:* © Lebendkulturen.de/Shutterstock.com; *Right:* © Designua/Shutterstock.com)

Some Hydrozoa, such as *Hydra*, possess only a polyp stage. Obelia, a colonial hydrozoan, alternates between the polyp and medusa stages. The medusae produce the sex cells. Egg and sperm unite to form a ciliated swimming larva stage that will settle, attach, and begin formation of a new colony.[14]

Obelia Colony, 40x (Photos courtesy of Holly J. Morris and David T. Moat)

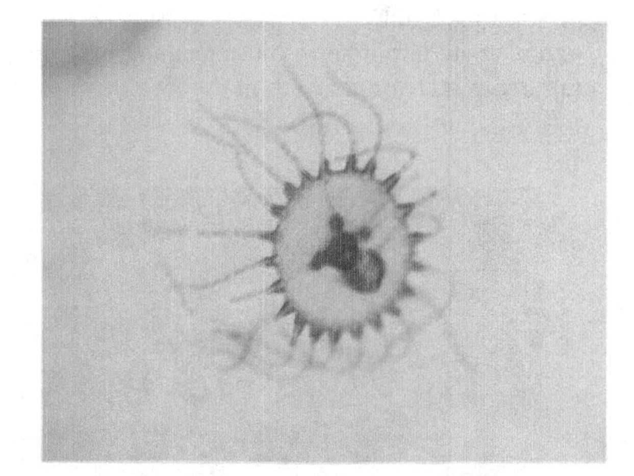

Obelia Medusa 40x (Photos courtesy of Holly J. Morris and David T. Moat)

2 Class Scyphozoa

The jelly-filled *medusa* is the dominant stage in this "true" jellyfish group. *Oral arms* are apparent. The polyp is reduced in size. Tentacles with nematocysts capture the prey. Some have bioluminescence.

© Vladimir Wrangel/Shutterstock.com

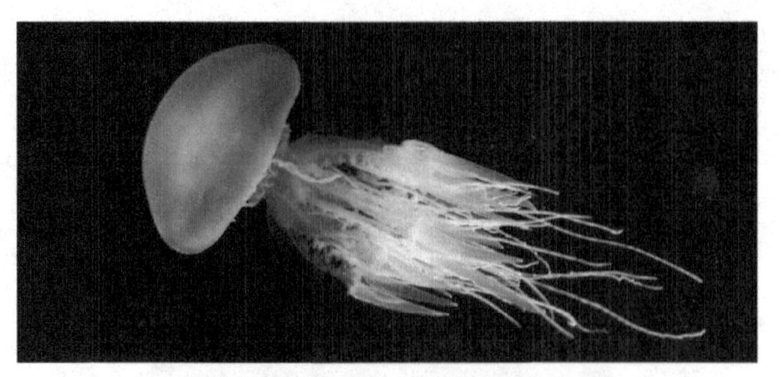

© Vladimir Wrangel/Shutterstock.com

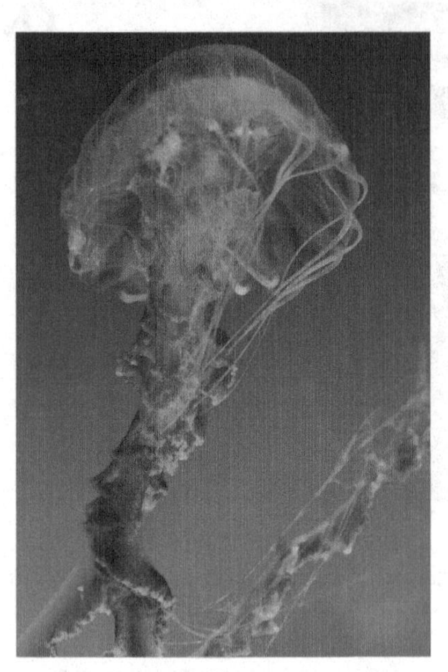

© Sann von Mai/Shutterstock.com

3 Class Anthozoa

Anthozoa have only polyp stages. These advanced cnidarians include corals, sea fans, and sea anemones. Coral reefs are important habitats teeming with diverse marine life forms. Millions of coral polyps secrete a calcareous skeleton that aids in the formation of reefs.

© Manamana/Shutterstock.com

© Annetje/Shutterstock.com

© Ingvars Birznieks/Shutterstock.com

© Dwight Smith/Shutterstock.com

PHYLUM CTENOPHORA

Comb jellies and sea gooseberries or sea walnuts belong to the phylum Ctenophora. These are usually abundant in warm seas. Each one can be as small as a pea or as large as a tomato. The diploblastic, translucent body is jelly-filled and possesses eight rows of fused cilia (ciliary combs). In place of stinging nematocysts, there are adhesive glue cells that release sticky threads to snare small prey. Comb jellies are bioluminescent and glow in the dark.[14]

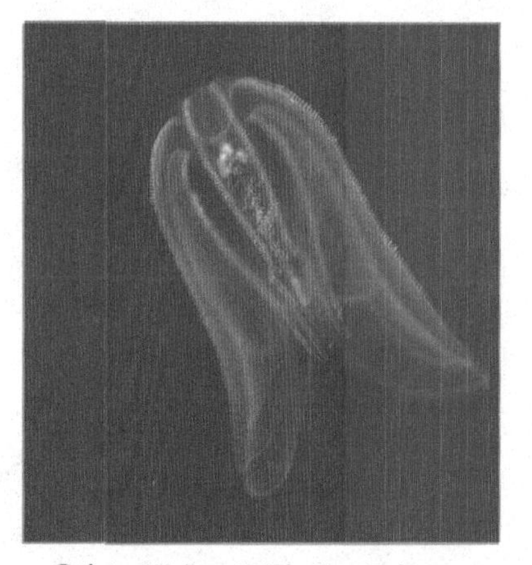
© John Wollwerth/Shutterstock.com

PHYLUM PLATYHELMINTHES

Flatworms are primitive, dorsoventrally flattened, triploblastic, worms with, bilateral symmetry. These worms are *acoelomate* having a digestive tract is incomplete or lacking. The mesoderm is present and developed into muscles. The nervous system development is an advance over the parazoa. These worms do have incredible abilities to regenerate when fragmented. Most species are hermaphroditic.

Flatworms are subdivided into three classes:

- Turbellaria
- Trematoda
- Cestoda

1 Class Turbellaria

Turbellarians are mostly free-living, marine flatworms. We will study the ciliated freshwater flatworm, Planaria. Planaria lives in shallow, clear streams and springs. There are no organs for circulation and gas exchange. These processes rely mostly on osmotic balance. The presence of *protonephridia* or *flame cells* assists in excretion of liquid. Flatworms possess both male and female sexual organs. These worms mate and cross-fertilize each other. Regeneration is an asexual mode of reproduction in this group.

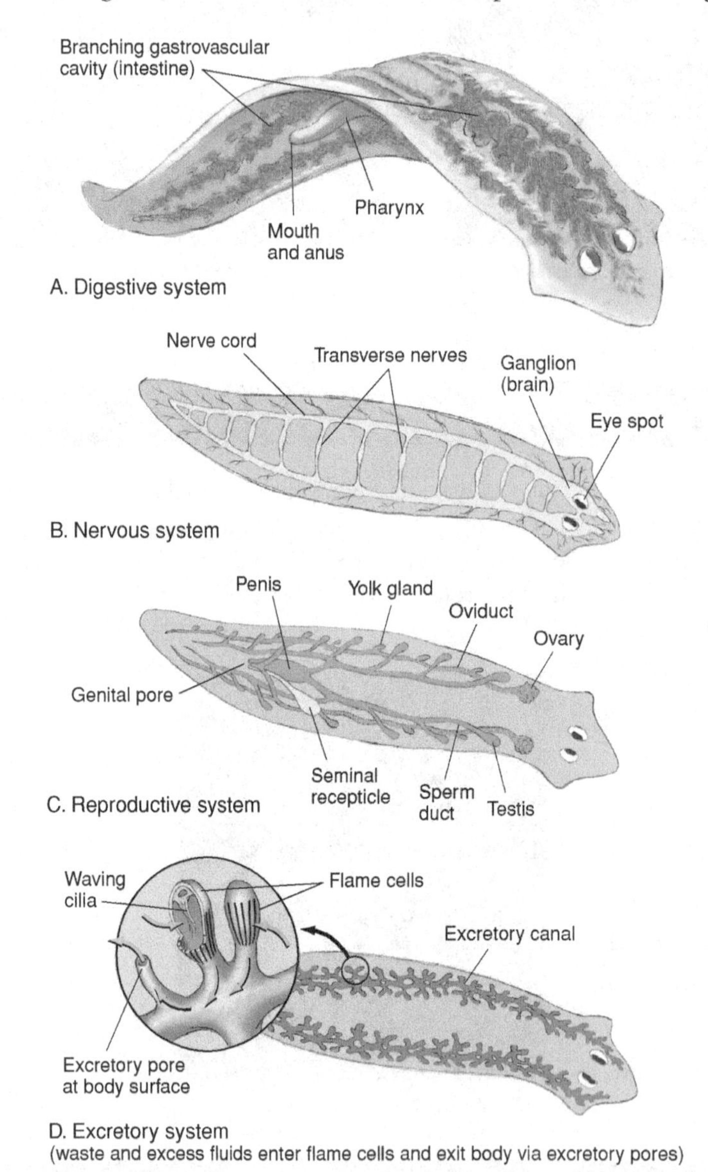

2 Class Trematoda

Trematoda are parasitic flatworms of vertebrates. Commonly known as flukes, the worm of this class has a thick cuticle to resist host enzymatic digestion and suckers for attachment to the host. The hermaphroditic fluke releases fertilized eggs that enter complex life cycles involving alternate hosts such as snails and fish before entry into the final vertebrate host for the adult stage.

Members of this class include the human liver fluke (*Fasciola*) and the human blood fluke (*Schistosoma*). *Fasciola* infections occur through eating fluke infested, fresh-water raw or undercooked fish. Adult flukes attach to the human liver in their life cycle. *Schistosoma* infections occur when the human host picks up the larval stage through capillaries in the skin when wading in tropical waters contaminated with human feces. The adult worms copulate in human blood vessels, and the eggs leave the host in the feces.⌋[14]

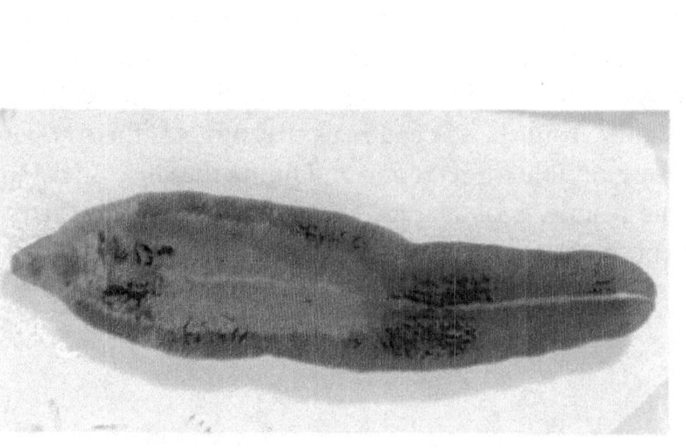

Helminth, *Fasciola hepatica,* Liver Fluke (Photos courtesy of Holly J. Morris and David T. Moat)

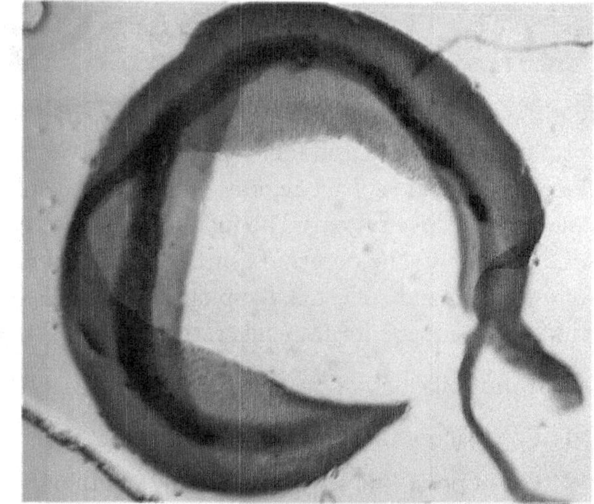

Schistsoma 40× (Photos courtesy of Holly J. Morris and David T. Moat)

3 Class Cestoda

Another class of parasites is called tapeworms. These end parasites inhabit the intestines of vertebrates and absorb nutrients from the host; therefore, the tapeworm has no digestive system. The worm is ribbon-shaped with a distinct head, the scale, and segments, the proglottids. These segments bud asexually from the neck region. The proglottids on the far end are the mature ones in which gravid eggs are located. These mature segments break off and pass with the feces.[14]

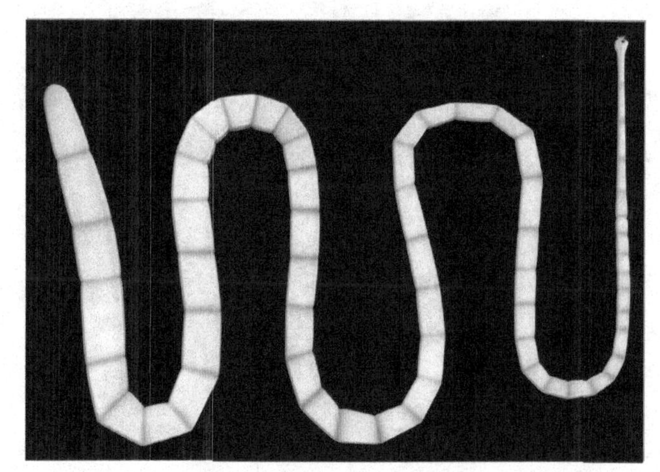

© sciencepics/Shutterstock.com

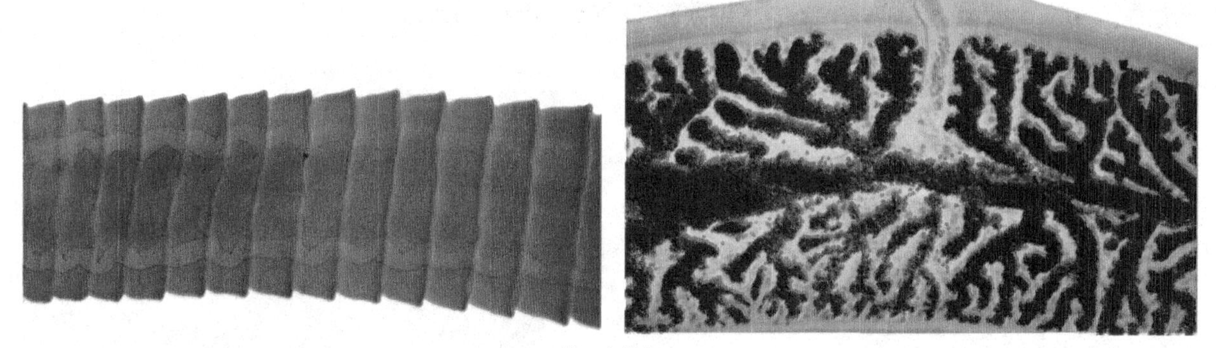

Proglottids

(*Left*: © D. Kucharski K. Kucharska/Shutterstock.com; *Right*: © D. Kucharski K. Kucharska/Shutterstock.com)

PHYLUM MOLLUSCA

Chitons, snails, clams, scallops, oysters, slugs, squids, and octopuses are all members of Mollusca . The body plan consists of the head foot region and the visceral mass region. The head foot region contains the feeding, locomotive, and cephalic regions while the visceral mass region contains the digestive, respiratory, circulatory, and reproductive systems. The mantle is a sheath extending from the visceral mass that protects the soft part of the mollusk. The outermost surface of the mantle is responsible for secreting and lining the shell. Shells are often prized and collected. Respiration is by gills or the mantle. A larval stage, the trochophore, is characteristic of the aquatic forms, while terrestrial mollusks deposit eggs that hatch as miniature adults.

Most mollusks belong to four classes:

1. Polyplacophora: chitons
2. Gastropoda (univalves): snails, slugs, nudibranchs
3. Bivalvia (Pelycypoda): clams, oysters, mussels, scallops
4. Cephalopoda: squid, octopus, chambered nautilus, cuttlefish

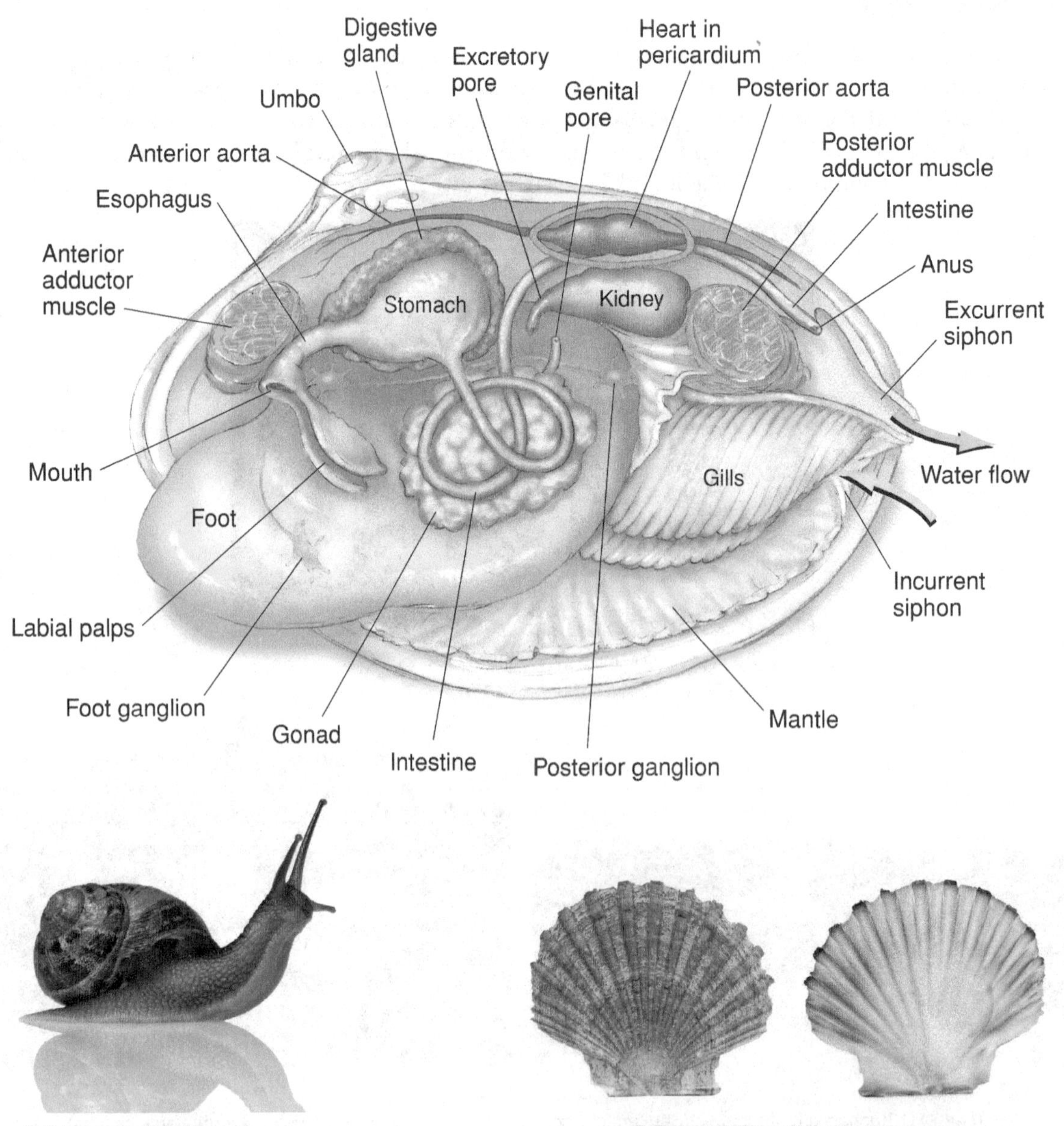

PHYLUM ANNELIDA

An earthworm is a good representative *annelid* to observe. An earthworm has well-developed muscles for movement. Its body is divided into numerous repeating *segments* (*somites*). A *septum* or partition separates each somite. The body plan is a tube-within-a-tube. The outer epidermis surrounds a fluid-filled coelom that contains a complete digestive tract and other internal organs. Scientists often refer to earthworms as the "plowers of the soil" since they burrow through the soil. This aerates and also enriches the soil environment as the worm's excrement is like a natural fertilizer.

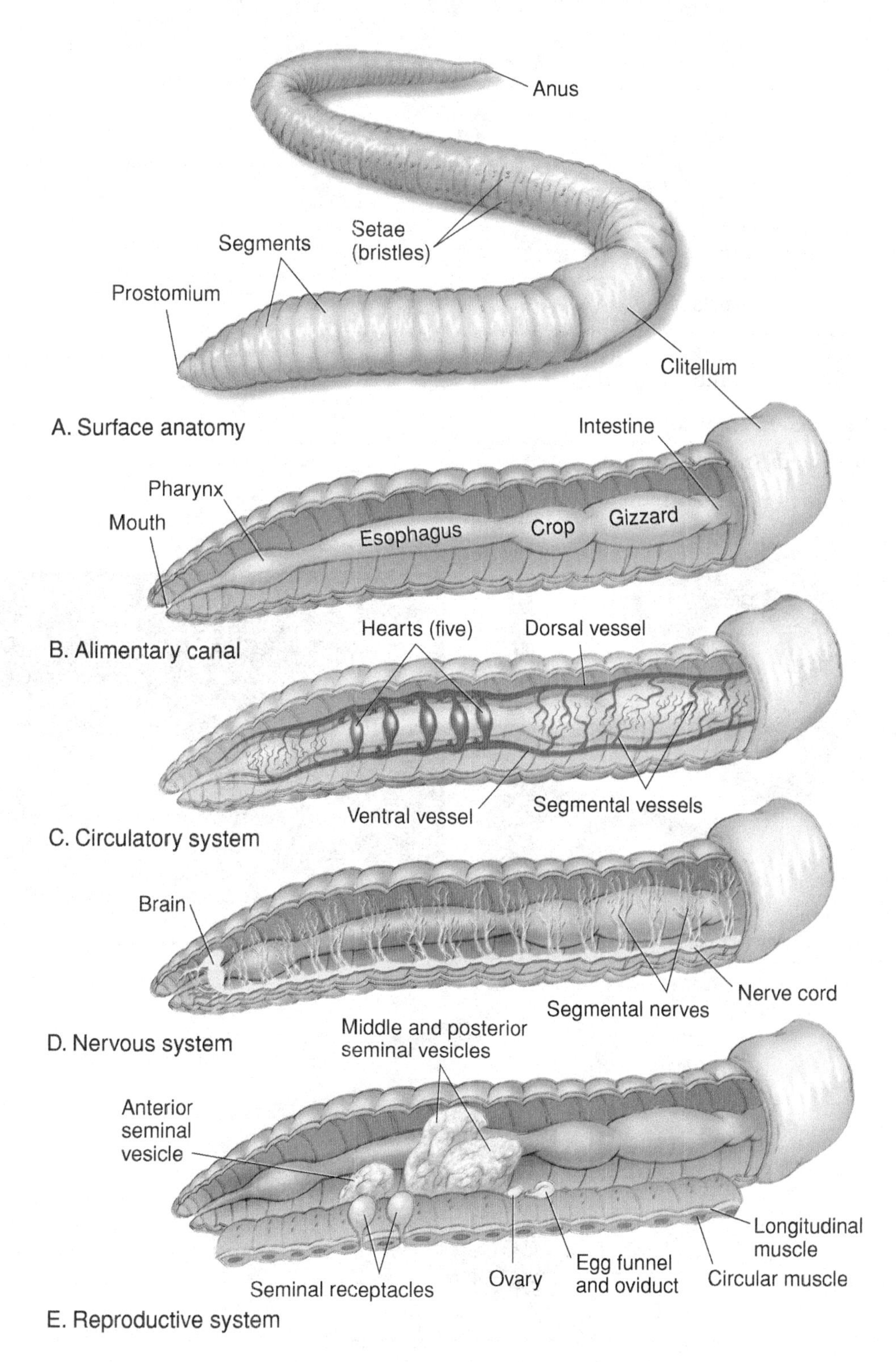

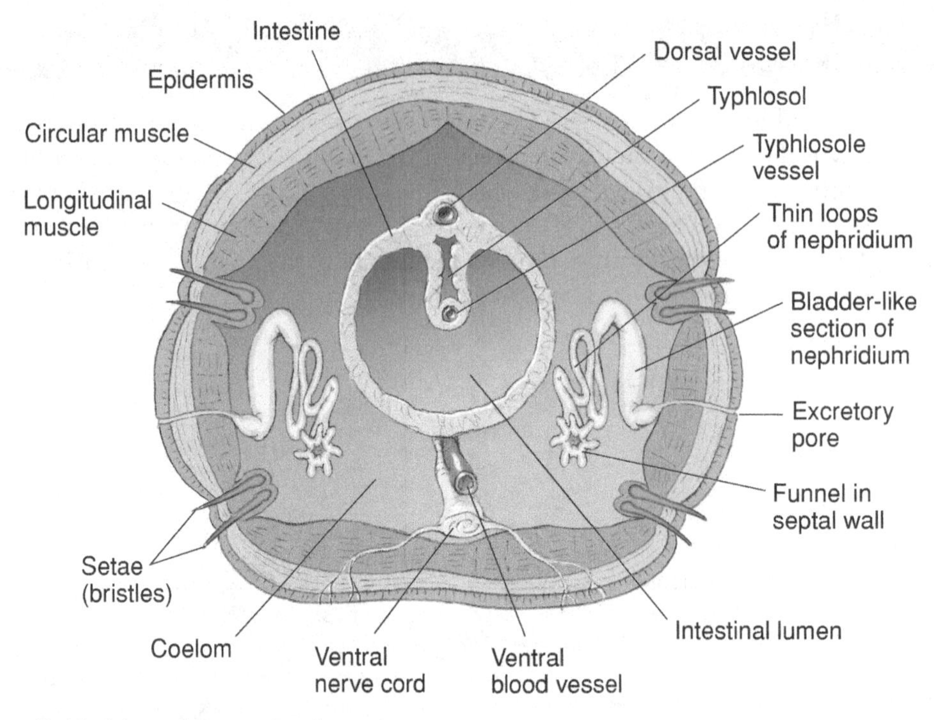

Annelids are divided into three main classes:

- **Class Oligochaeta** (earthworms).
- **Class Polychaeta** (marine worms).
- **Class Hirudinea** (leeches - most are blood-suckers).

© Dusty Cline/Shutterstock.com

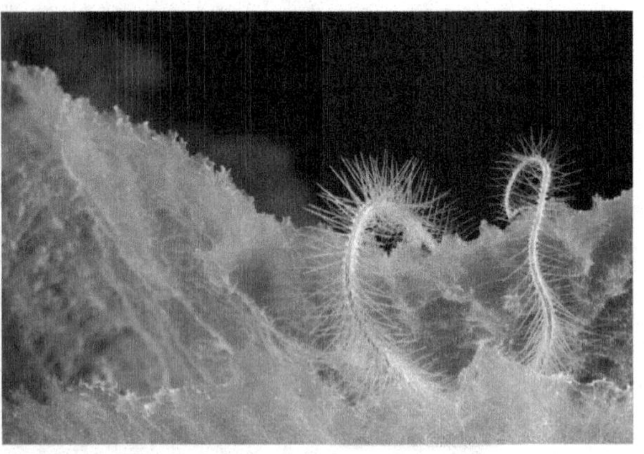

© Joe Barbarite/Shutterstock.com

© aaltair/Shutterstock.com

PHYLUM NEMATODA

Members of the phylum Nematoda are *roundworms* which are widespread in virtually every habitat from aquatic to moist soils and inside the tissues of plants and animals. Some are parasitic in humans, but most are free-living and significant ecological species in that many species are decomposers of soil microbes.

Roundworms are bilaterally symmetrical, triploblastic and *unsegmented*. The worms range in size from microscopic to a meter in length. The tough outer *cuticle* molts or sheds as the worm grows. Roundworms have a unique characteristic number of cells in the body. Most animals grow by increasing the number of cells due to mitotic divisions. Roundworms grow beyond the embryonic stage by cell growth, not cell multiplication. Sexes are separate. Sexual organs and the digestive tract largely pack the inside of the roundworm. There is no circulatory system. The nervous system consists of a "brain" and dorsal, ventral, and lateral nerves. Roundworms possess a tube-within-a-tube body plan with a complete mouth to anus digestive system. All are *pseudocoelomate* with a fluid-filled hydrostatic skeleton. The nematode has only longitudinal muscles. The motion is a whipping or thrashing unlike the burrowing motion of the annelid worm, which has circular muscles in addition to the longitudinal ones.[14]

Representative Nematodes include:

- *Ascaris*

Ascaris is a parasite of humans and pigs. These adult worms thrive inside the intestines of the host by taking food from the host's gut. If left untreated, these roundworms can multiply, becoming so crowded that they seek other locations in the human body for residence. The female worm can generate one to two hundred thousand eggs daily. These eggs pass with the feces, and if accidentally ingested by a vertebrate host will complete the life cycle from larva to adult. The larvae burrow into the blood vessels of the intestine and circulate to the lungs where these larvae burrow into air passages. The larvae migrate through the air passageways to the throat of the host and are then swallowed. In the intestine the young worms mature.[14]

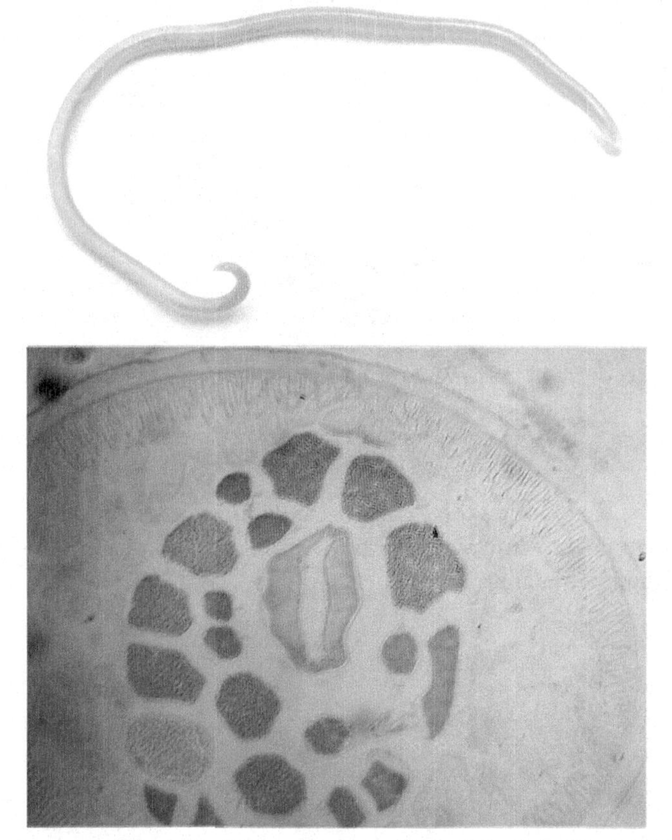

Ascaris (cross section) 40× (Photos courtesy of Holly J. Morris and David T. Moat)

- *Trichinella*

Trichinella is an encysted roundworm in vertebrate muscle tissue. When a human eats undercooked pork or bear meat infested with the trichina worms, the human will have these parasites encysted in his/her muscles. Trichinosis is the potentially fatal condition that brings serious muscle and cardiac complications to the host.[14]

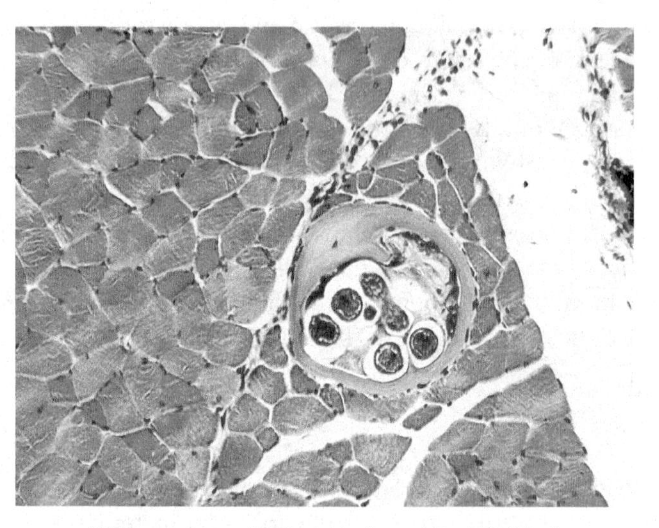

(Photo courtesy of Holly J. Morris and David T. Moat)

- *Enterobius* (Pinworms)

Pinworms are common in children, usually of preschool age. The female worms crawl around the anal area at night causing disruptive sleep patterns and even nightmares. Intense itching in the anal area is symptomatic of pinworm infestation. These roundworms are easily transferred to other family members or others who share public toilets with an infested individual. Medications can rid the host of these parasites.[14]

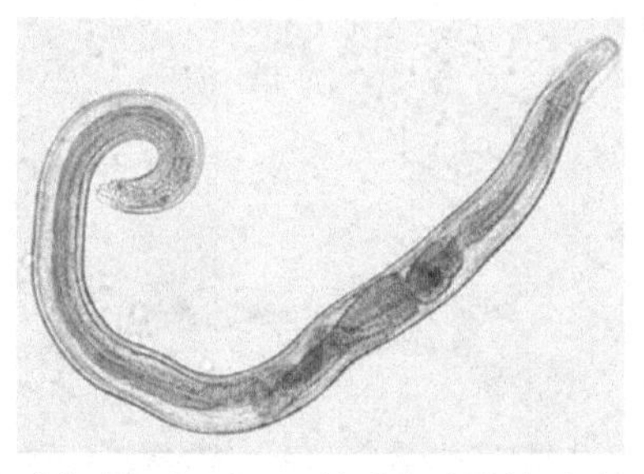

(http://www.cdc.gov/dpdx/enterobiasis/)

- *Hookworms*

Hookworms feed on intestinal blood in the capillaries and blood vessels of the host. Hookworms typically gain entry into the human host by burrowing into the capillaries of a barefooted individual walking into a barnyard or field with infested manure.[14]

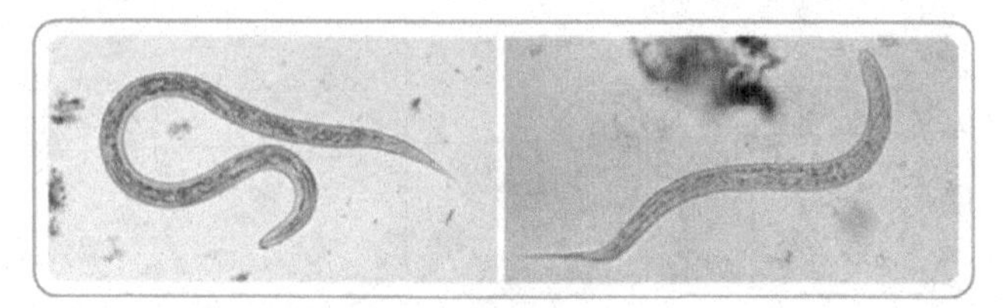

(http://www.cdc.gov/parasites/hookworm/)

PHYLUM ARTHROPODA

The list of Arthropoda goes on and on since this is the most diverse and most successful group of animals. Scientists estimate there are over one million arthropods in this biosphere and most of these are insects highly adapted to terrestrial life.

Characteristics of the phylum are:

1. Paired, jointed appendages
2. Hard outer exoskeleton of chitin
3. Segmented body with specialization of parts
4. Open circulation
5. Keen sensory organs for touch and smell
6. Presence of compound eyes
7. Molting

Current molecular evidence supports four major groups:

Subphylum	# Body Divisions	Appendage Type	Antennae	Mouthparts	Legs
Myriapoda (centipedes, millipedes)	Head/Segmented Body	Unbranched (uniramous)	One pair	Mandibles Maxillae	One or two pair per segment
Chelicerata (horseshoe crabs, arachnids)	Fused head and thorax (cephalothorax) and abdomen	Uniramous	None	Chelicerae, pedipalps	Horseshoe crabs: 5 pairs Arachnids: 4 pairs
Crustacea (lobsters, crabs, barnacles, copepods)	Head, Thorax, and Abdomen	Branched with two joints at base (Biramous)	Two pairs	Mandibles, Two pair of maxillae	One pair per segment as a rule
Hexapoda (insects)	Head, Thorax, and Abdomen	Uniramous	One pair	Mandibles and maxillae	Three pairs attached to thorax

A brief look at some members of these groups follows:

- Subphylum Myriapoda
  1. Class Chilopoda

     Centipedes or "hundred-legged" worms possess one pair of legs per segment. Centipedes are predators and can inflict a serious sting.

© Audrey Snider-Bell/Shutterstock.com

2. Class Diplopoda

Millipedes or "thousand-legged" worms possess two pairs of legs per segments. Millipedes are herbivores and do not sting; however, they can secrete an irritating chemical.

16

© Dr. Morley Read/Shutterstock.com

- Subphylum Chelicerata
  1. Class Merostoma

  Horseshoe crab dates back 350 million years. They can be found along the seaboard when migrating to reproduce. They feed on smaller invertebrates.

© mtd/Shutterstock.com

  2. Class Arachnida

  Arachnids include spiders, ticks, mites, scorpions, and daddy longlegs. Most are predators seeking other arthropods for food. They use the chelicerae to inject a toxin into the prey. Pedipalps are for food handling. The head and thorax fuse into the cephalothorax. The larger abdomen is the more evident part of the body.

© Shutterstock.com

© J.Y. Loke/Shutterstock.com

- Subphylum Crustacea

There are over 65,000 species in this group. They inhabit mostly marine habitats but some are fresh-water inhabitants and some are terrestrial inhabitants. These animals use their mouthparts and antennae for scavenging. There is a hard, protective *carapace* over the cephalothorax. The first walking leg is the obvious *chelipeds* or pincer4s. On the abdominal segments are the *swimmerets* used for swimming and for holding eggs in the female. The *telson* and *uropod* at the posterior end are used for swimming backwards. The heart, located under the carapace, is in a pericardial sinus. Blood enters the heart and leaves through openings into arterial branches. The blood is forced onto the tissues directly. Sinuses collect the fluid, which moves into the ventral sinus in the thorax, then to the gills, and finally back into the pericardial sinus. This is called *open circulation*.

© mtd/Shutterstock.com

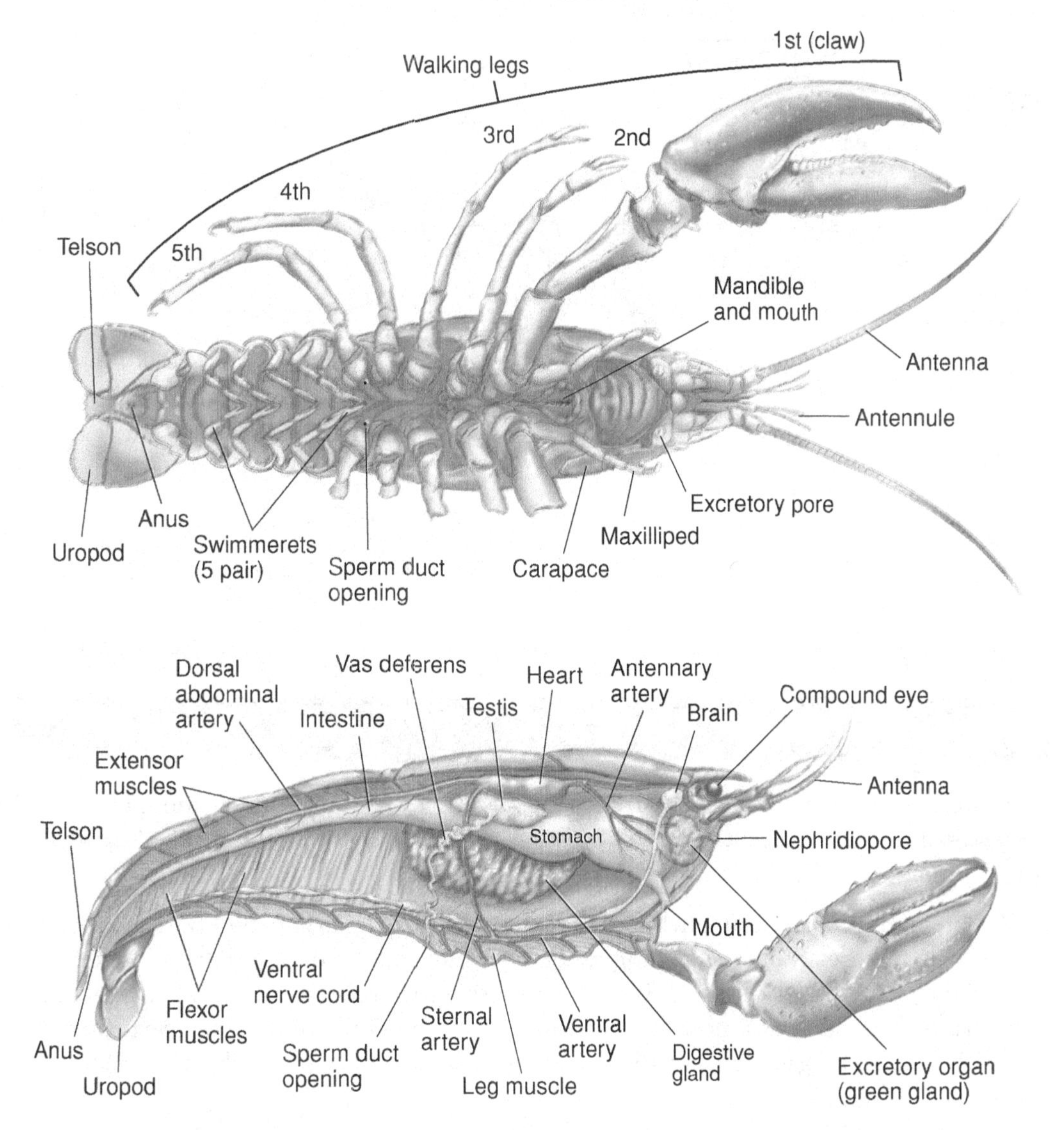

- Subphylum Hexapoda

Class Insecta

Insects are the largest group of animals in the world with over one million species. Most insects fly, but there are some species that are wingless. The insect body plan includes a *head*, a *thorax*, and an *abdomen*. There are typically *three pairs of walking legs, one pair of antennae, and two pairs of wings.* Both *simple eyes (ocelli) and compound eyes* are present.

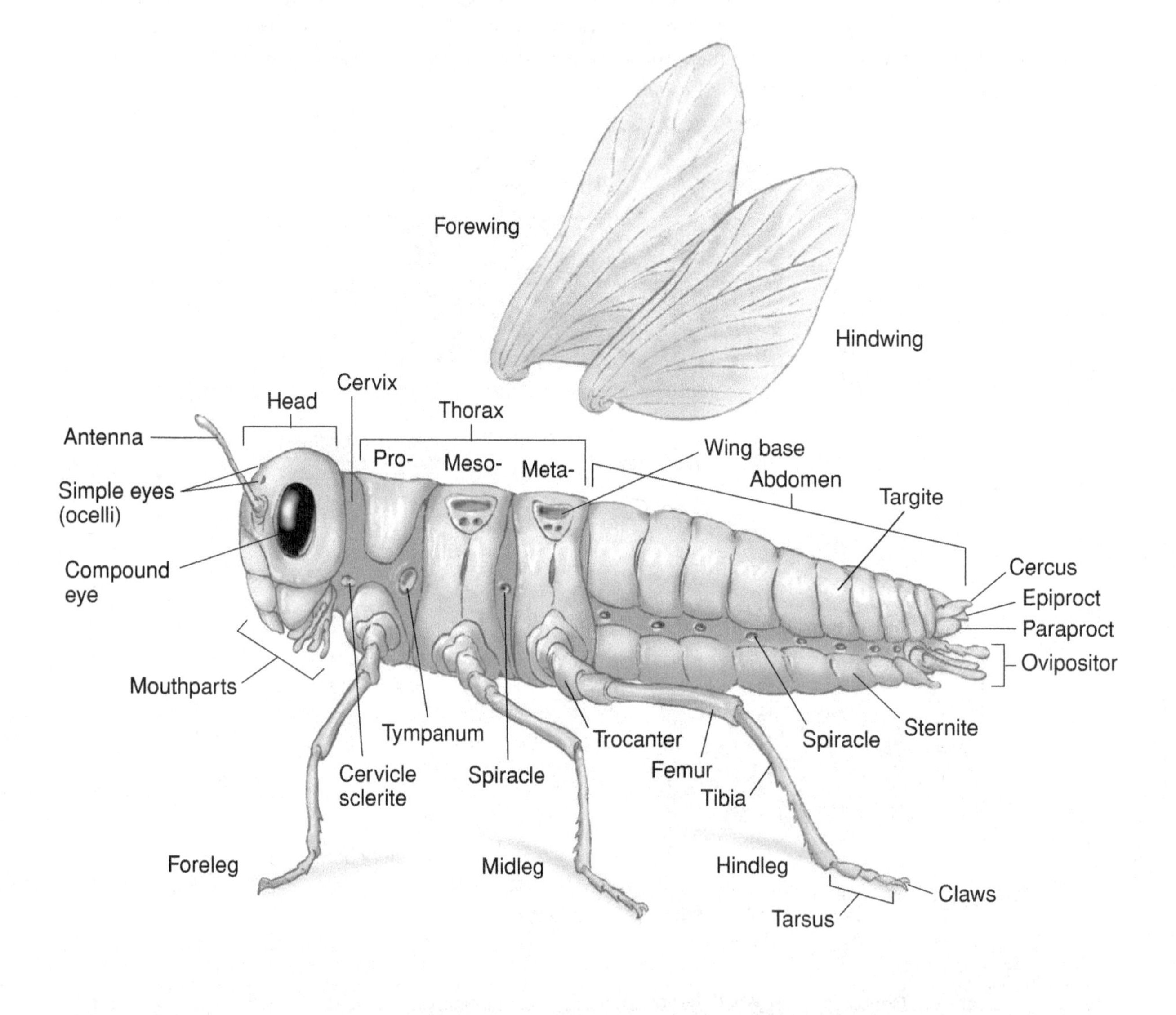

PHYLUM ECHINODERMATA

Echinoderms are spiny skinned moderately sized *coelomate* animals without segmentation. The adult has *five radially symmetrical parts* about a central axis. The *larval stage is bilaterally symmetrical and free-swimming.* A limy *endoskeleton* of movable or fixed limy plates and spines is present.

A unique feature of the echinoderm phylum is the *water vascular system* for locomotion and food handling. Through this system of canals and *tube feet*, salt water flows into and out of the animal. Gas exchange is also a function of the water vascular system. Skin gills or respiratory trees accomplish respiration. Modified spines called *pedicellariae* serve as pincers on the skin to keep the skin free of debris. Sexual reproduction occurs with the release of gametes into the sea water. The echinoderms exhibit remarkable powers of regeneration.

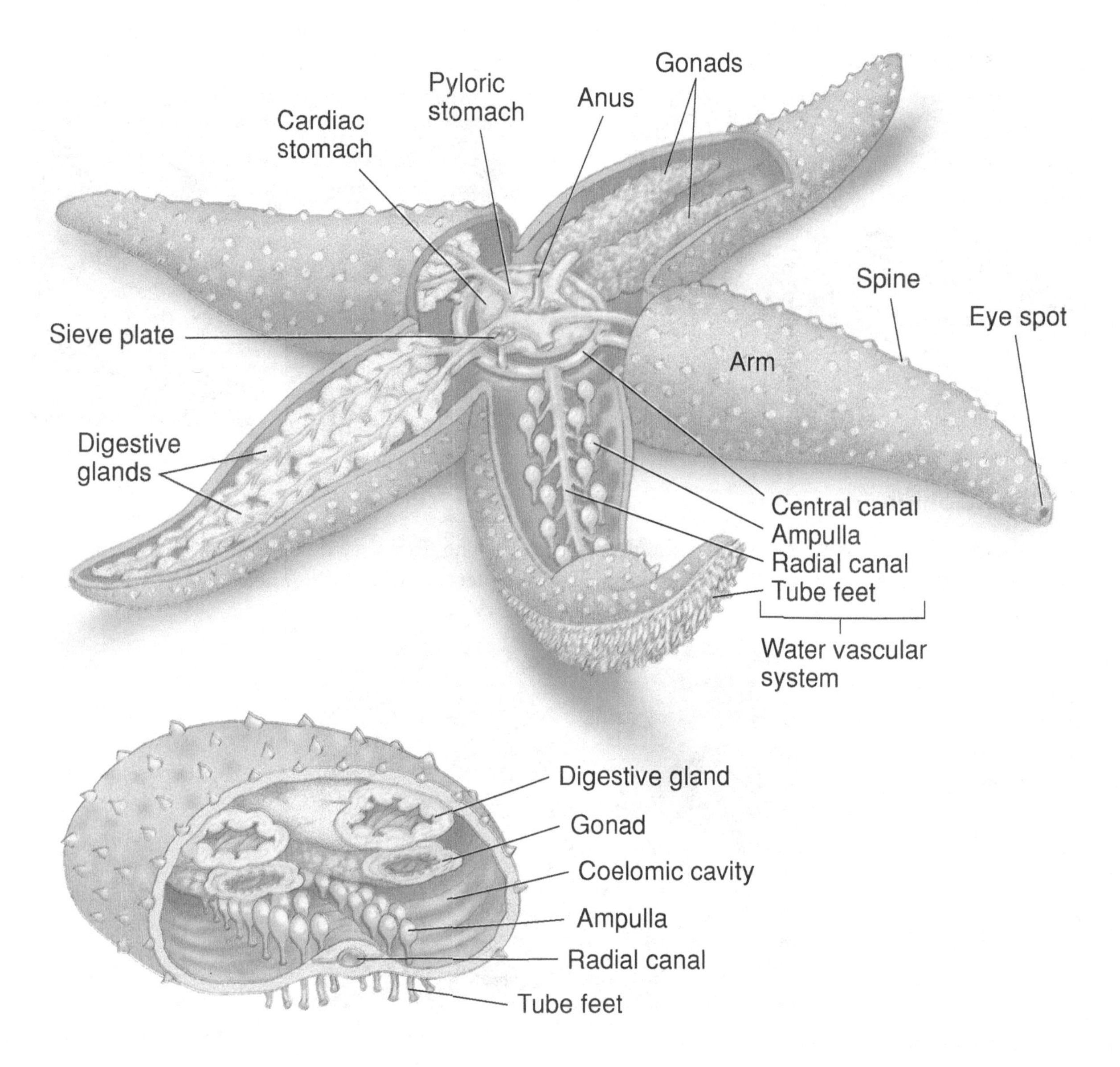

1 Class Crinoidea

Crinoids are living relics. The ancestors of present day Crinoids lived in the Paleozoic seas. Common names of members of this class are sea lilies and feather stars. Sea lilies are attached to a substrate in the ocean and appear more like flowers than animals. They filter food from the surrounding water. Feather stars are mobile.

© Richard Williamson/Shutterstock.com

© Richard Williamson/Shutterstock.com

2 Class Asteroidea

© Ricardo A. Alves/Shutterstock.com

© Khoroshunova Olga/Shutterstock.com

3 Class Ophiuroidea

Basket stars and brittle stars possess solid, fragile arms or rays. These are very flexible allowing these animals to move about at a faster pace than the asteroids. The tube feet have a tactile function rather than being involved in locomotion.

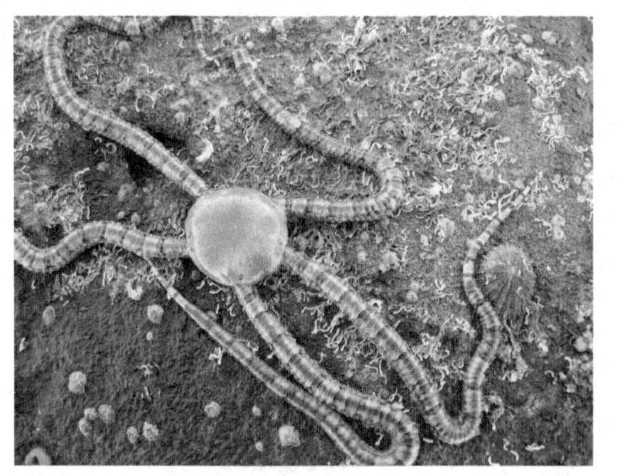

© Robyn Butler/Shutterstock.com

4 Class Echinoidea

Spiny sea urchins and sand dollars have a rounded body encased in the endoskeleton (test) with fused limy plates. Sea urchins with spines, which can be short or long, can sometimes inflict pain. Sea urchins are mostly herbivorous. Marine animals, including sea stars and fish, prey upon them. Sand dollars burrow in the sand and feed on tiny particulate matter.

© Cico/Shutterstock.com

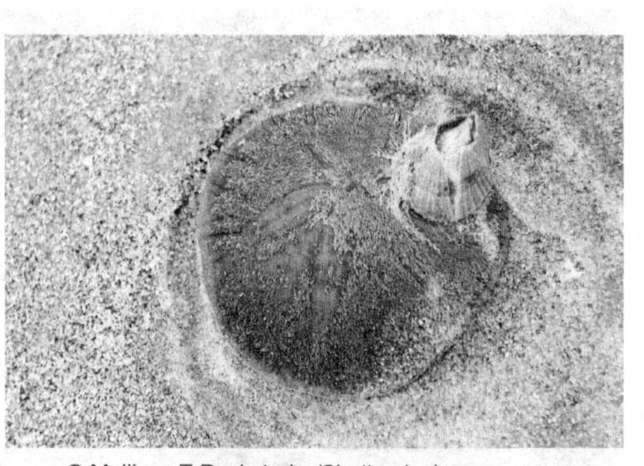

© Mellissa E Dockstader/Shutterstock.com

5 Class Holothuroidea

The sea cucumber is cucumber-shaped with tentacles that exhibit tube feet. These animals are quite slow in movement, but they have tactics to allow for escape when a predator attacks. They eject sticky threads from the anal region that confuse the predator. Cucumbers like other echinoderms have regeneration capability.

© Nikita Tiunor/Shutterstock.com

PHYLUM CHORDATA

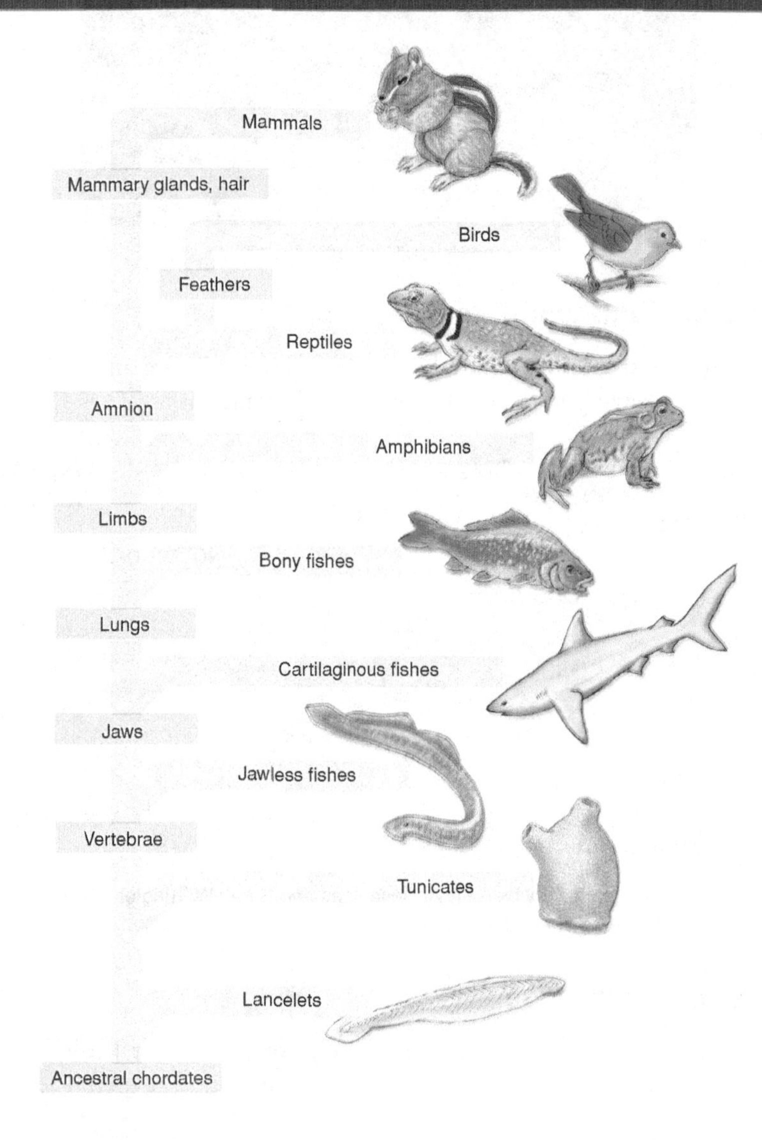

Chordates are divided into two invertebrate phyla, the Urochordata and the Cephalochordata, and one vertebrate phylum. At some stage of all invertebrate and all vertebrate chordate life cycles, there are four shared derived characteristics that classify the organism as a chordate:

1. The presence of a *dorsal hollow nerve cord*
2. The presence of a dorsal supportive *notochord*
3. The presence of *pharyngeal gill slits or pharyngeal pouches*
4. A *postanal tail*

A. Phylum Chordata, Subphylum Urochordata

Tunicates, or the more common sea squirts, are typically sessile marine animals that are filter feeders. Some do retain their swimming capabilities. Only the gill slits distinguish these animals as chordates in the adult stage. However, the larval stage does possess a notochord and dorsal hollow nerve chord.[16]

B. Phylum Chordata, Subphylum Cephalochordata

Lancelets are small marine animals also commonly called amphioxus. These animals resemble a tiny fish in shape. You can find them burrowing in the sand along coastal tidal regions.[16]

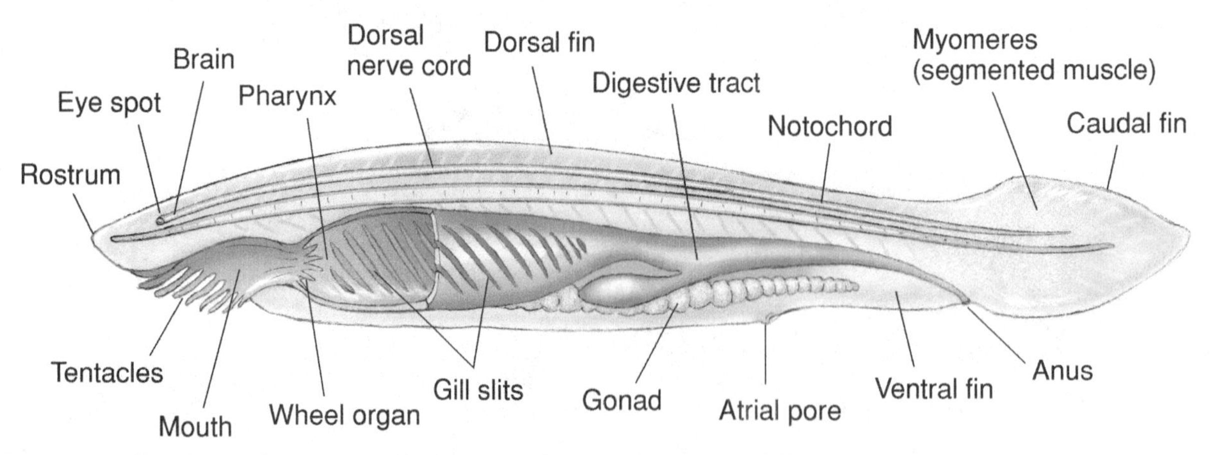

From Explorations in General Biology by Betty A. Rosenblatt and Sarah Warrington. Copyright © 2008 by Kendall Hunt Publishing Company. Reprinted by permission.

EXERCISE 1 Observing Invertebrates

An invertebrate is an animal that does not have or does not develop a vertebral column. Examples of invertebrates include insects; crabs, lobsters, snails, clams, octopuses, sea stars, sea-urchins, and worms.

Purpose

To become familiar with the animal family tree, the representatives of each group, the distinguishing features of each group, and the anatomical features of each group.

Materials

- Compound light microscope
- Magnifying glass
- Various invertebrate exhibits on display
- Various invertebrate cultures.
- Various invertebrate prepared slides
- Pipettes
- Slides
- Cover slips

Methods

- Examine the various invertebrate exhibits on display.
- Make wet mounts of the various invertebrate cultures and examine them with a compound light microscope.
- Examine the various invertebrate prepared slides with a compound light microscope.
- Record your observations in the space provided.

Name:

Exercise 1 Results

Porifera

1. Make **detailed drawings** of one of the *Porifera* on display and the prepared slide of the skeletal elements of *Porifera*.

Porifera on Display	*Porifera* Skeletal Elements
Magnification:	Magnification:

2. How does the sponge keep water flowing through its ostia and out through the osculum?

____

____

3. Describe the 3 classes of sponges.

____

____

4. How are sponges similar to plants?

____

____

5. How are sponges different than plants?

____

____

Cnidaria

1. Make **detailed drawings** of a *Hydra* wet mount, an *Obelia* medusa prepared slide, and an *Obelia* colony prepared slide.

***Hydra* Wet Mount**
Magnification:

***Obelia* Medusa**	***Obelia* Colony**
Magnification:	Magnification:

2. What are the major classes of cnidarians?

____

____

3. How is *Obelia* similar to a plant?

____

____

4. How is *Obelia* different than a plant?

____

____

Platyhelminthes

1. Make **detailed drawings** of a wet mount of a *Planaria* and a prepared transverse section of a *Planaria*.

***Planaria* (Wet Mount)**	***Planaria* (Transverse Section)**
Magnification:	Magnification:

2. Compare and contrast turbellarians, cestodes, and trematodes.

3. *Planaria* moves with its head first. Why is it an advantage for an organism to have sensory structures in its anterior rather than its posterior?

4. Few parasitic worms that live inside their hosts have eyespots or pigmentation. However, many in the free living stages (outside of the host) in the life cycle do. Why might this be an advantage for the organism?

5. What is a proglottid? List the various types of proglottids.

6. Make **detailed drawings** of a prepared slide of *Fasciola hepatica* and *Schistosoma*.

Fasciola hepatica	***Schistosoma***
Magnification:	Magnification:

Mollusca

1. Make **detailed drawings** of the external anatomy of a univalve and bivalve shell.

Univalve	**Bivalve**
Magnification:	Magnification:

2. Describe the 2 major body regions of a typical mollusk.

____

____

Annelida

1. Make **detailed drawings** of the external anatomy of an earthworm on display and an earthworm prepared transverse section slide.

Earthworm (External Anatomy)	Earthworm (Transverse Section)
Magnification:	Magnification:

2. Describe the 3 major classes of phylum *Annelida*.

____

____

____

3. Annelids exhibit body segmentation. Why is this a major evolutionary innovation?

____

____

Nematodes

1. Make **detailed drawings** of the external anatomy of an *Ascaris* and a prepared transverse section slide of an *Ascaris*.

Ascaris (External Anatomy)	*Ascaris* (Transverse Section)
Magnification:	Magnification:

2. Make a **detailed drawing** of a wet mount of *Turbatrix aceti* (vinegar eels).

Turbatrix
Magnification:

Arthropods

1. Compare and contrast centipedes and millipedes.

____

____

2. List the major characteristics of phylum *Arthropoda*

____

____

____

____

3. Why are insects so successful?

____

____

CHAPTER 29

Vertebrates

Photos courtesy of Holly J. Morris

OBJECTIVES

After completing this exercise, you will be able to:

- Identify the vertebrate classes.
- Cite the distinguishing features of each vertebrate class.

INTRODUCTION

Vertebrates belong to the Subphylum Vertebrata of the Phylum Chordata. Vertebrates are coelomate deuterostomes with bilateral symmetry. Vertebrates possess segmentation with highly specialized anatomical features. Vertebrates have a vertebral column composed of cartilaginous or bony segments called vertebrae that encases and protects a dorsal hollow nerve cord. Vertebrates have an endoskeleton and two pairs of appendages. A cranium/skull encases the brain of the vertebrate animal. The organ systems are well developed with higher degrees of specialization in the advanced classes as compared to the more primitive classes. Recent molecular evidence has redistributed the vertebrates into 10 extant classes. Six of these classes are *fishes*; four are *tetrapods* (four appendages).

I. Class Myxini (Hagfish)

Hagfishes are jawless, marine scavengers. These are very primitive. While there is a cranium, there is no vertebral column. The notochord remains throughout the life of the hagfish. While referred to as a fish, there is little similarity to a fish. The hagfish resembles an eel with a smooth skin.

II. Class Cephalaspidomorphi (Lampreys)

Lampreys are agnathans that, like the hagfishes, have a notochord and do not possess jaws or paired fins. These animals are predators in the marine world but migrate into freshwater habitats for reproduction. Lampreys are often parasitic upon fish. They attach to the side of a fish and use their rasping tongue and circular rows of teeth to suck blood and tissues from the host. The larval stage, ammocoete, resembles a lancelet.

III. Class Chronrichthyes (Cartilaginous Fishes)

Sharks, rays, and skates are marine fishes that possess a cartilaginous skeleton. To maintain an osmotic balance with the salt water, they retain urea in the blood. Other characteristics include five pairs of gill openings, jaws, paired pectoral and pelvic fins, a body covered with placoid (tooth-like) scales, no lungs, and no swim bladder. Oil concentrated in the liver adds buoyancy, but these fish will sink unless they move continuously. A lateral line organ along the side of the fish allows it to sense its prey's motion. The tail fin is unsymmetrical. Some members lay eggs (oviparous); some retain the eggs inside the female where they hatch (ovoviviparous); and some are viviparous (embryos develop in uterus and receive nourishment).

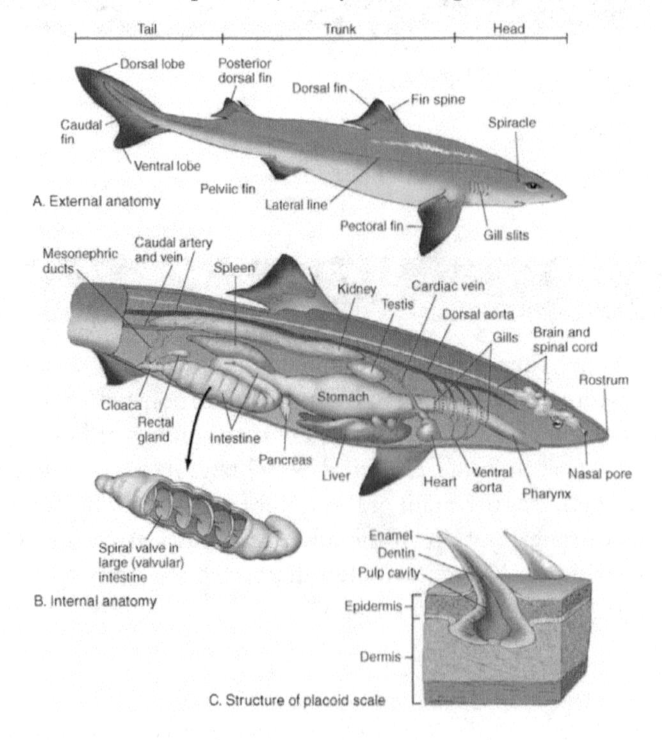

IV. Class Actinopterygii (Ray Finned Fishes)

This is a diverse group of fishes inhabiting both marine and freshwater environments. An operculum covers the gills. The skeleton is bony; jaws are present. Fins are paired. The tail fin is symmetrical unlike that of a shark. A gas-filled swim bladder enables the bony fish to alter buoyancy with little muscular effort. The scales are larger than the placoid type. Most fish are oviparous.

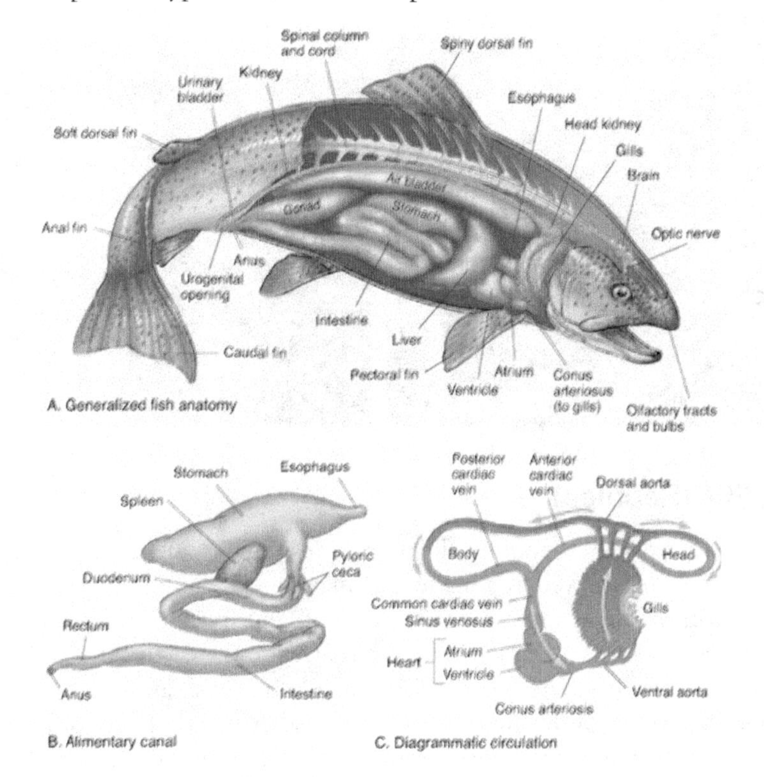

V. Class Actinistia (Lobe Finned Fishes)

Lobe-finned fishes were present during the Devonian Period at the same time the ray-finned fishes were evolving. Lobe-fins differ from ray-fins in that their pectoral and pelvic fins possess bones and muscle. The lobe-fin had lungs and gulped air. Living today is a member of the ancient lobe-finned fishes known as a coelacanth, *Latimeria*. This fish is approximately six feet in length. Scientists thought that coelacanths became extinct over 75 million years ago; but fishermen caught a few in the twentieth century, and now scientists believe that over 200 coelacanths inhabit the Pacific waters near southeastern Africa.

VI. Class Dipnoi (Lungfishes)

Researchers consider lungfishes ancestral to the tetrapod or the ancient amphibians. Ancient members of this group possessed lobe-fins and lungs. When conditions during the Devonian and the Carboniferous Periods were adverse in the aquatic environment, the lungfishes could "walk" from pool to pool in search of more favorable conditions. Extant lungfishes exist in the tropical rivers in the southern hemisphere.

We will next focus our study on the vertebrate classes of *tetrapods* (two pairs of appendages). Amphibians, Reptiles, Birds, and Mammals are tetrapods. We also refer to the reptiles, birds, and mammals as *amniotes*.

VII. Class Amphibia (Amphibians)

Amphibians include frogs, toads, salamanders, newts, and caecilians. They are tetrapods. While the amphibians are transitional between aquatic and terrestrial environments and considered the first land vertebrates, they must rely on a watery medium for sexual reproduction. Eggs are without shells or protective coverings and laid into the water. The male releases the sperm into the water onto the eggs. Amphibians are ectothermic or cold-blooded; activity is restricted when the temperature drops. Capillaries for gas exchange richly supply the skin. Lungs and gills are also present in various amphibian species for gas exchange. All organ systems of the amphibian are advanced over the members of the Superclass Pisces.

VIII. Class Reptilia (Reptiles)

Turtles, snakes, lizards, tortoises, crocodiles, alligators, and tuataras are all extant reptiles. They inhabit most terrestrial environments as well as aquatic and marine environments. The leathery amniote egg and internal fertilization largely account for this group of land dwellers becoming highly successful in the Mesozoic Era. The dinosaurs were king during that time on earth. The reptilian skin is dry and scaly. There are claws on the toes. The heart is four chambered with an incomplete septum. The lungs are better developed than those of the amphibian. Reptiles have teeth, fangs, or hooked beaks. Reptiles are carnivorous. The most common modern reptiles are the members of the Order Squamata, the snakes and lizards.

IX. Class Aves (Birds)

Scientists may eventually classify birds within the reptilian class as they continue to collect evidences. An ancient fossil *Archeopteryx* shows reptilian traits and feathers. Birds have adaptations for flight. Birds are endothermic (warm-blooded) and maintain a constant body temperature. Birds have a completely separated four-chambered heart and an advanced nervous system featuring keen vision and acute hearing.

X. Class Mammalia (Mammals)

Mammals possess hair and glandular skin. The cerebral cortex is highly advanced. A diaphragm separates the thorax from the abdomen and assists in moving air into and out of the respiratory system. Teeth with diversified functions are the incisors, canines, premolars, and molars. Reproduction involves internal fertilization (except for monotremes) and a placenta. The females nourish young with milk.[17]

CHAPTER 30

Vertebrate Tissues

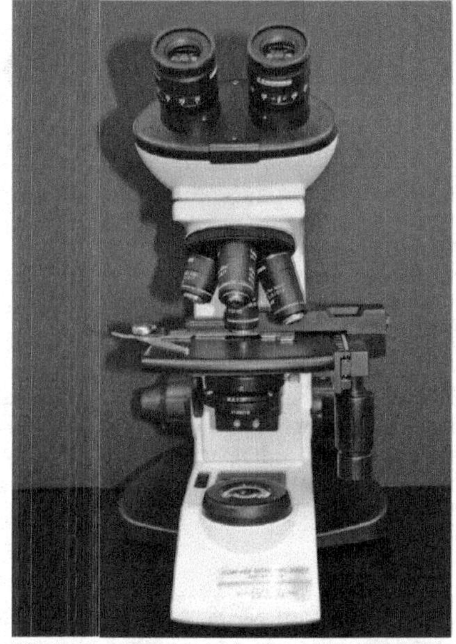

Photos courtesy of Holly J. Morris

OBJECTIVES

After completing this exercise, you will be able to:

- Understand the characteristics of the four basic tissue categories:
  - Epithelial
  - Muscle
  - Connective
  - Neural
- Identify the major subtypes of cells of the four basic tissue categories.
- Understand the structure, function, and location of each of these cell subtypes.

INTRODUCTION

There are over 75 trillion cells in the human body. Anatomists have categorized the cells to make it easier to study and understand their functions. All of the cells of the human body have been placed into groups of tissue categories. There are four main types of tissues that make up the human body.

	Tissue Type	Characteristic
1	Epithelial tissue	The cells that belong in this category make up the inside or outside lining of an organ.
2	Muscular tissue	The cells that belong in this category have the ability to contract and relax.
3	Neural tissue	The cells that belong in this category have the ability to transmit an impulse or are involved in providing protection for the cells that transmit impulses.
4	Connective tissue	The cells that belong in this category have a matrix of some sort. The cells are embedded in the matrix. This tissue type serves to connect, bind, surround, or enclose various structures. Some of the cells in this tissue category do not actually connect anything.

EPITHELIAL TISSUE

Epithelial Tissue	
Squamous cells	In deeper tissue, these cells appear round. As you approach the surface of the tissue, the cells become flattened. These cells provide physical protection. Cells can be found making up the inside lining of the mouth and the outside lining of the skin.
Cuboid cells	These cells appear to be cube-shaped. These cells secrete and absorb material. Cells can be found making up the lining of the urinary tubes and some glands.
Columnar cells	These cells appear to be longer than they are wide. These cells secrete and absorb material. Cells can be found making up the lining of the trachea and the small intestine.

Epithelial Cell Types ( ) = location	
Simple epithelial tissue	Simple squamous. Single layer of squamous cells. (alveoli and capillary walls) Simple cuboidal. Single layer of cuboid cells. (glands and kidney tubes) Simple columnar. Single layer of columnar cells. The nuclei are pretty much in a row along the basement membrane. (uterine tubes and respiratory tract)
Stratified epithelial tissue	Stratified squamous. Several layers of squamous cells. The apical cells are flat shaped. (lining of mouth and esophagus) Stratified cuboidal. Several layers of cuboidal cells. The apical cells are cube- shaped. (rare – found in parts of the male urethra) Stratified columnar. Several layers of columnar cells. The apical cells are columnar-shaped. (rare – found in parts of the male urethra)
Pseudostratified epithelial tissue	Pseudostratified columnar. One layer of cells but of varying heights. The nuclei are also at different heights within each columnar cell. These cells can be ciliated or non-ciliated. (trachea – ciliated) (ductus deferens – non ciliated)
Transitional epithelial tissue	Tissue consists of cells that have a variety of shapes. This tissue is extremely stretchable – (lines the urinary bladder)[18]

18

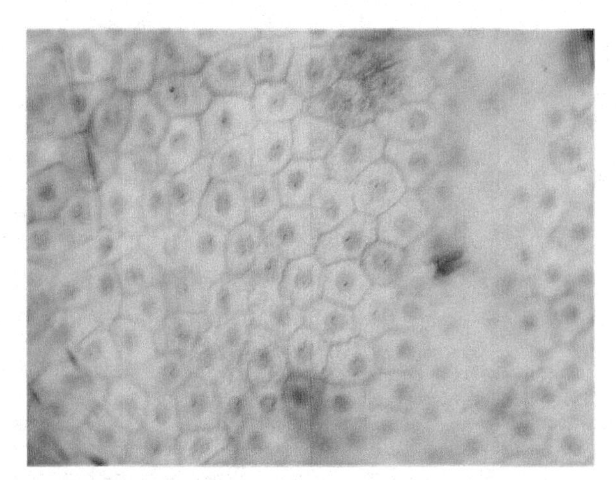

Simple Squamous 400x
(Photo courtesy of Holly J. Morris and David T. Moat)

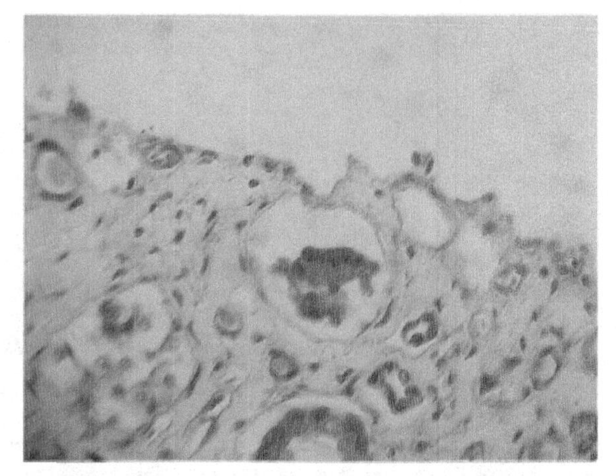

Simple Squamous edge view 400x
(Photo courtesy of Holly J. Morris and David T. Moat)

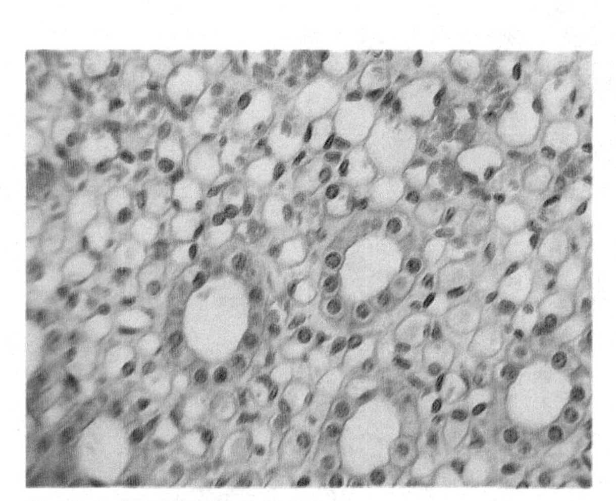

Simple Cuboidal Epithelium 400x
(Photo courtesy of Holly J. Morris and David T. Moat)

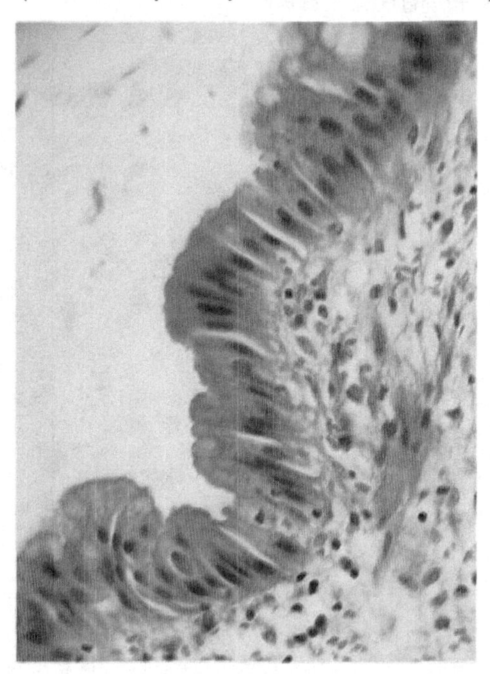

Simple Columnar Epithelium 400x
(Photo courtesy of Holly J. Morris and David T. Moat)

MUSCLE TISSUE

Muscular Tissue	
Skeletal muscle cells	When viewed under a microscope, these cells appear to be striped (striated). These cells provide voluntary contraction. Cells can be found making up the arm and leg muscles.
Smooth muscle cells	These cells appear to have pointed ends (spindle-shaped). The nucleus appears flattened with pointed ends as well. These cells provide involuntary contraction. Cells can be found making up sphincter muscles, myometrium, and the blood vessels.
Cardiac muscle cells	These cells are associated with intercalated discs. These cells provide rhythmic or pulsating contractions. Cells are found only in the myocardium of the heart.

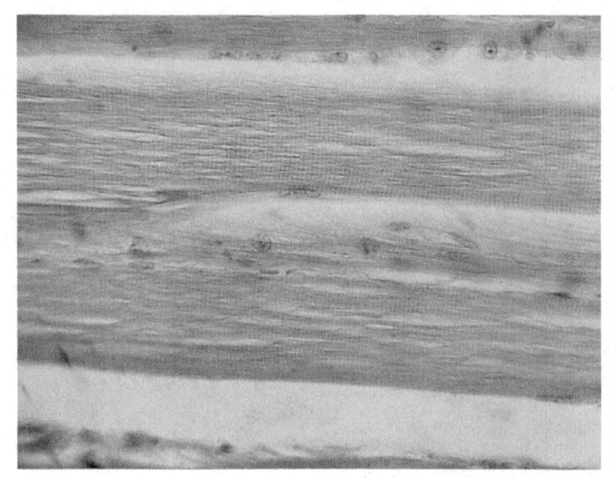
Skeletal Muscle 400× (Photo courtesy of Holly J. Morris and David T. Moat)

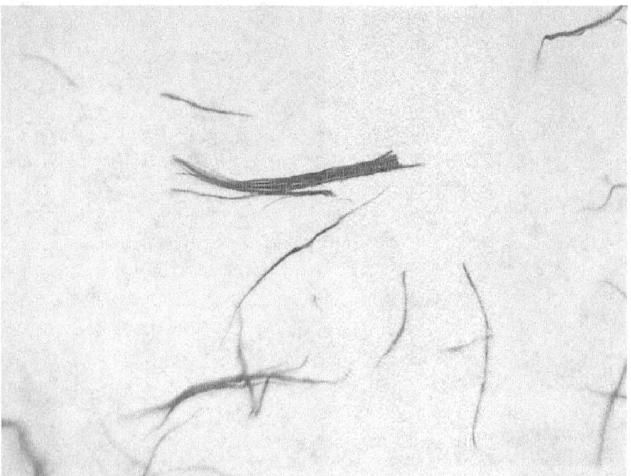
Smooth Muscle 400× (Photo courtesy of Holly J. Morris and David T. Moat)

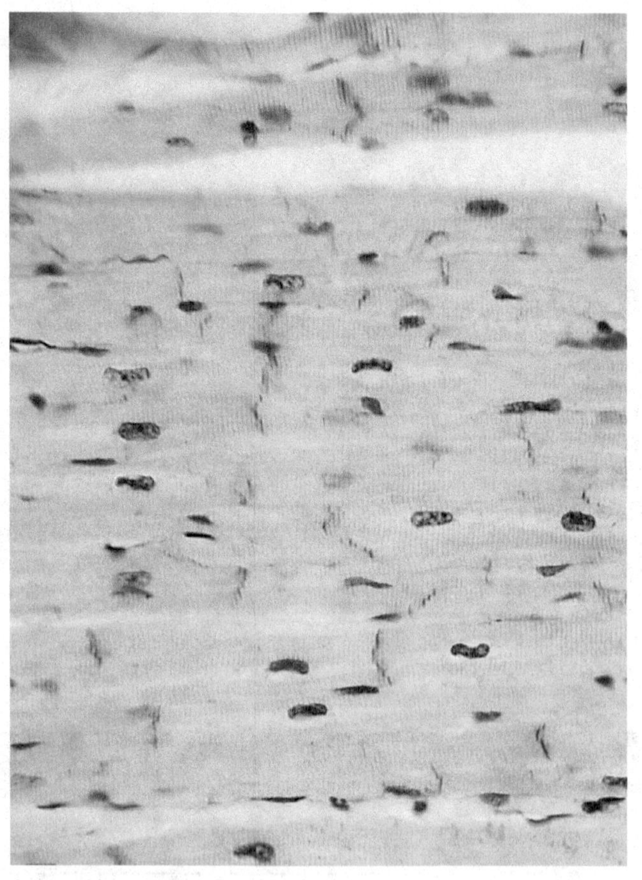
Cardiac Muscle 400× (Photo courtesy of Holly J. Morris and David T. Moat)

CONNECTIVE TISSUE

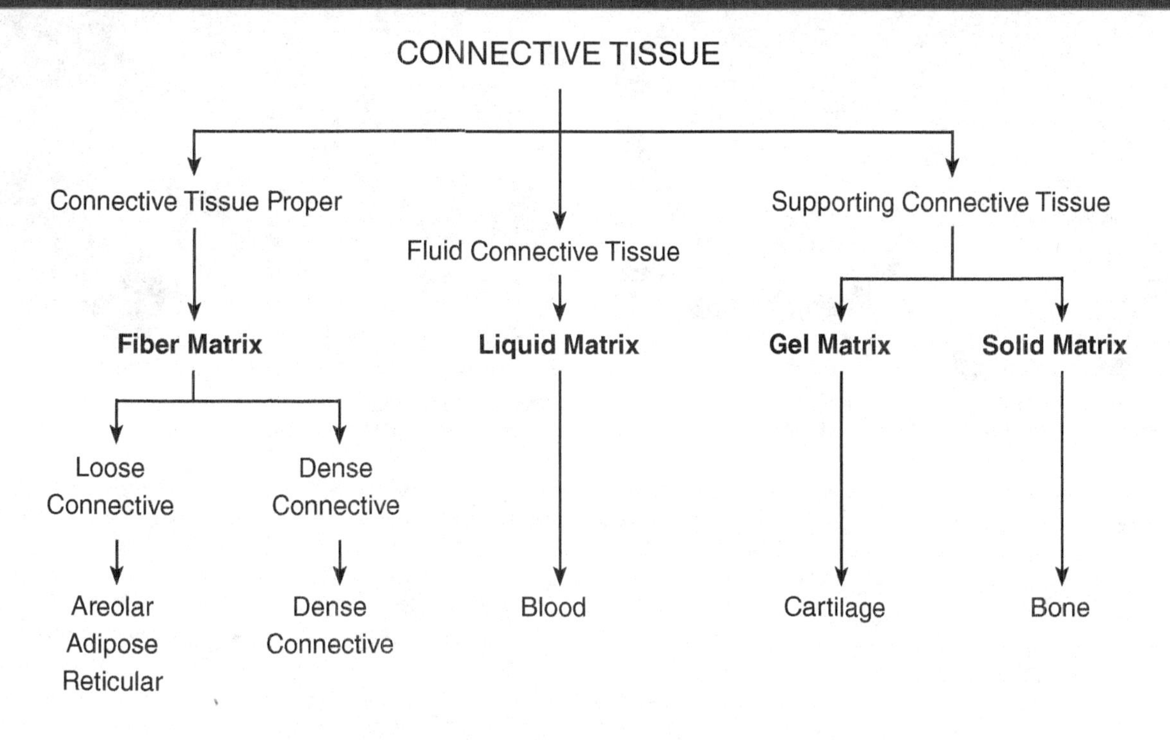

Connective Tissue Cell Types

Adipose tissue (adipocytes)	These cells generally appear round and have the appearance that they are empty. These cells provide: 1. insulation for organs of the body 2. protection and cushion 3. stored energy (li[ids) Cells can be found surrounding body organs.
Areolar tissue (areolar cells)	These cells (fibrocytes) are small and have numerous, thin hair-like fibers coursing between the cells. These fibers provide a physical attachment of the skin to muscle. Cells are located between the skin and muscle.
Reticular tissue	These are small cells that have short, thick fibers coursing between them. These cells make up the framework of many organs. Cells make up the liver, spleen, appendix, tonsils, and thymus gland (for example)
Dense tissue	These cells appear to be packed tightly together to form strong fibrous strands of material. These cells form strong connections. Cells are found associated with tendons, ligaments, and aponeuroses.
Blood tissue (blood cells)	These cells are small and have plasma flowing between them. Refer to the "Additional Information" section to see the functions. Cells are found in the circulatory system.
Bone tissue (osteocytes)	The osteocytes form concentric rings around a central canal. These cells provide strength. Cells are found associated with our skeleton.
Cartilage tissue (chondrocytes)	These cells appear to have a lot of white area around them. These cells provide flexibility. Cells are found within the joints and flexible parts of the helix of the ear and the ala of the nose.

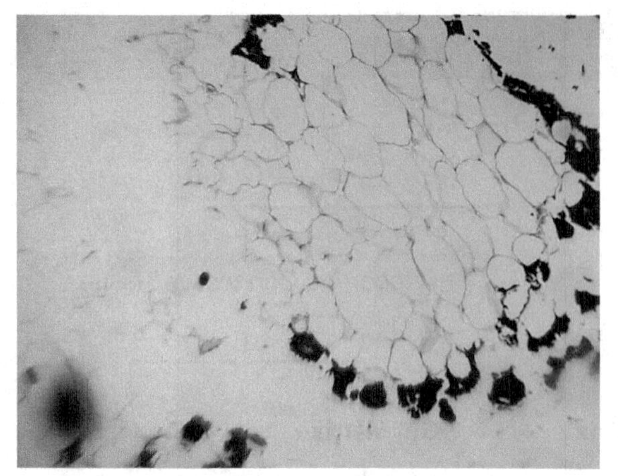
Adipose Tissue 100×
(Photo courtesy of Holly J. Morris and David T. Moat)

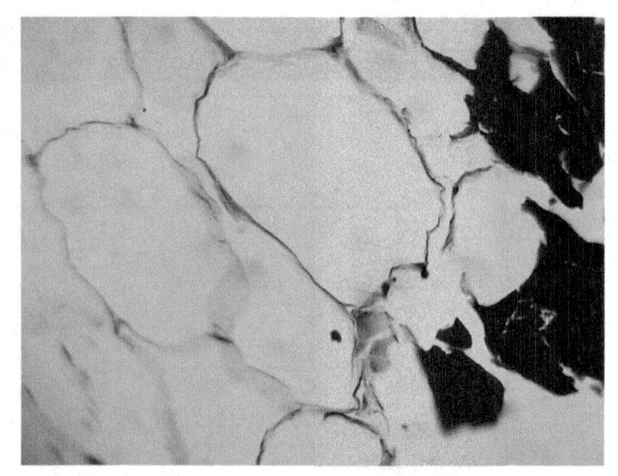
Adipose Tissue 400×
(Photo courtesy of Holly J. Morris and David T. Moat)

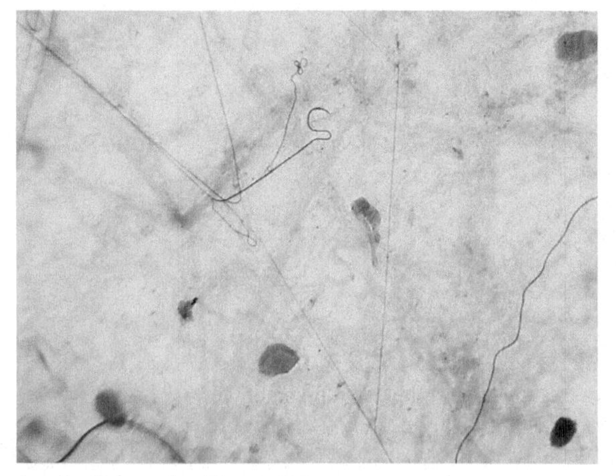
Areolar Connective Tissue 400×
(Photo courtesy of Holly J. Morris and David T. Moat)

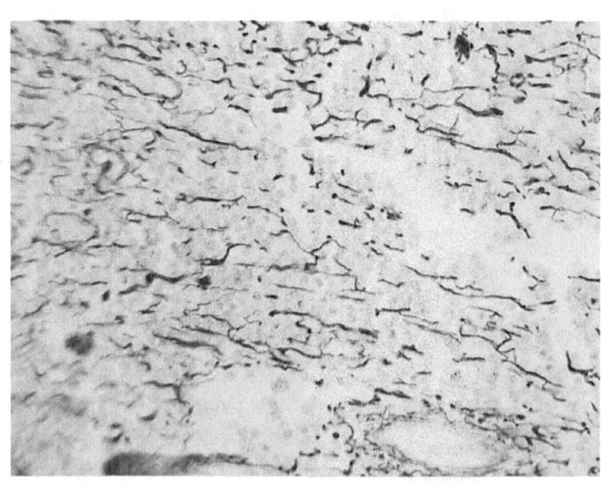
Reticular Connective Tissue 400×
(Photo courtesy of Holly J. Morris and David T. Moat)

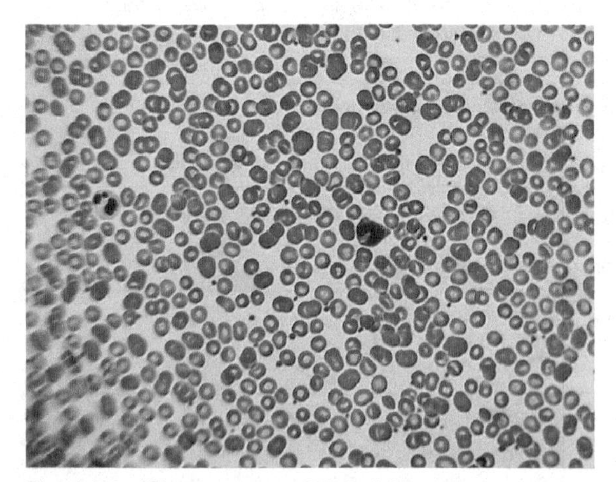
Blood 400× (Photo courtesy of Holly J. Morris and David T. Moat)

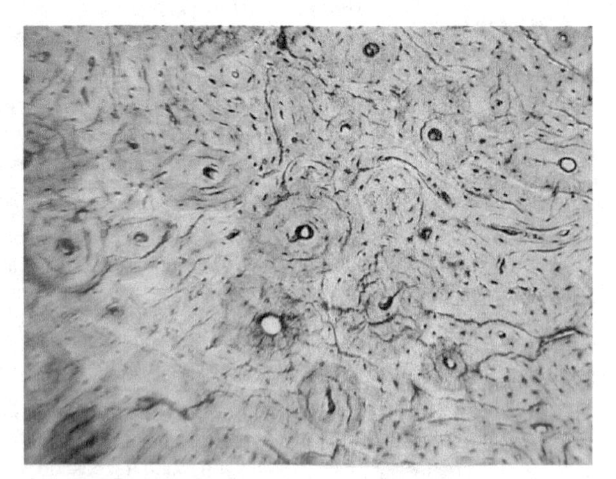
Compact Bone 400× (Photo courtesy of Holly J. Morris and David T. Moat)

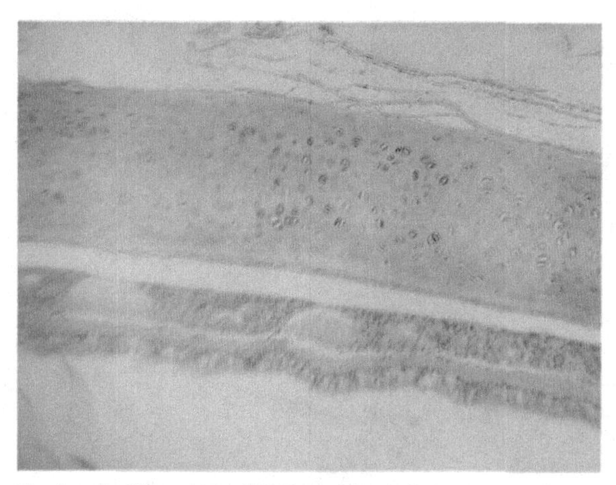

Hyaline Cartilage 100x (Photo courtesy of Holly J. Morris and David T. Moat)

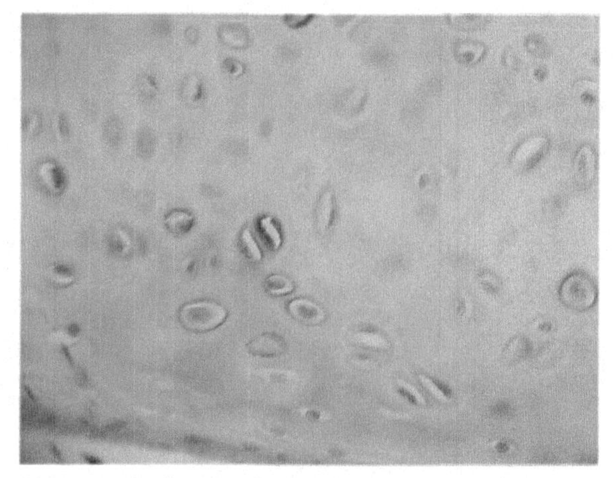

Hyaline Cartilage 400x (Photo courtesy of Holly J. Morris and David T. Moat)

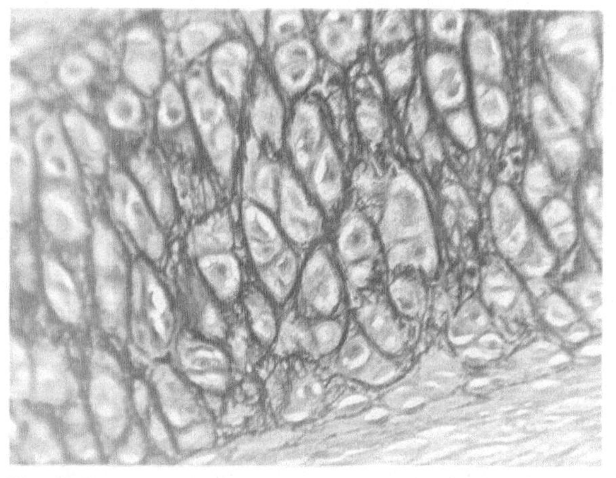

Elastic Cartilage 400x (Photo courtesy of Holly J. Morris)

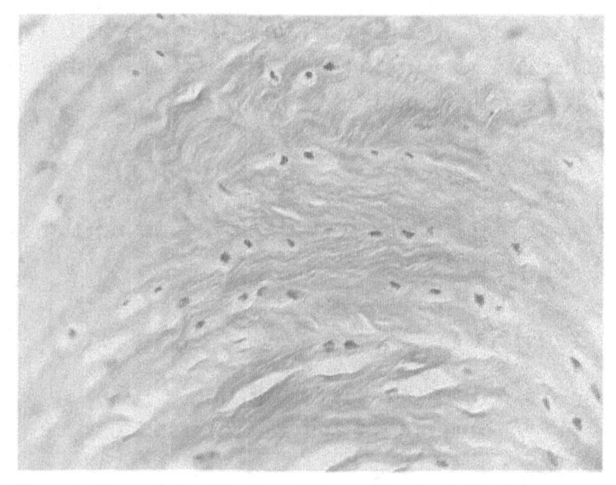

Fibrocartilage 400x (Photo courtesy of Holly J. Morris)

NEURAL TISSUE

Neural Tissue	
Neurons	These cells have extensions branching off the soma. These cells conduct impulses. Cells can be found making up the brain tissue.
Glial cells	These cells take on a variety of shapes. These cells provide protection for the neurons. Cells can be found surrounding the axon of the neuron or in close proximity to the neuron.

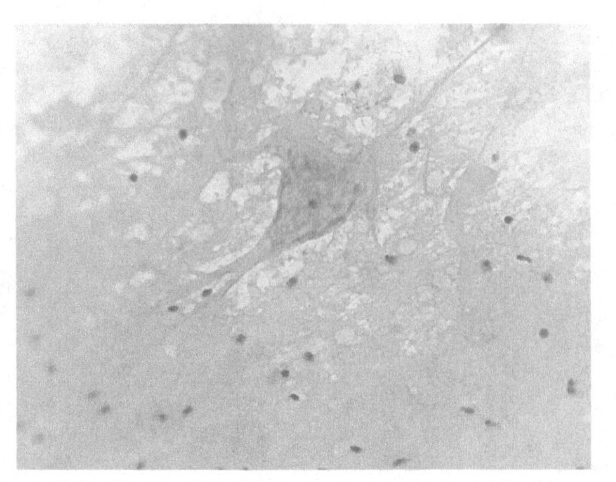

Motor Neuron 400x (Photo courtesy of Holly J. Morris)

EXERCISE 1 Observing Vertebrate Tissues

Vertebrate animals are composed of four basic types of tissues (epithelial, muscle, connective, and neural). These tissues vary in their composition and their function. A basic understanding of the role of each tissue makes understanding the specific functions easier.

Purpose

To become familiar with the structure, function, and location of the cell subtypes of the four basic tissue categories.

Materials

- Compound light microscope
- Various prepared slides of vertebrate tissues.

Methods

- Examine the various prepared slides including:
  - Cell's tissue type.
  - Appearance of the cell.
  - Function of the cell.
  - Location of the cell in the body.
- Record your observations in the space provided.

Name:

Exercise 1 Results

1. Make **detailed drawings** of each of the vertebrate tissues at medium (100×) or high (400×) total magnification.

Simple Squamous Epithelium	**Stratified Squamous Epithelium**
Total Magnification:	Total Magnification:

Simple Columnar Epithelium	**Pseudostratified Columnar Epithelium**
Total Magnification:	Total Magnification:

Simple Cuboidal Epithelium
Total Magnification:

Skeletal Muscle	Cardiac Muscle
Total Magnification:	Total Magnification:

Smooth Muscle
Total Magnification:

Hyaline Cartilage	Elastic Cartilage
Total Magnification:	Total Magnification:

Areolar Tissue	Adipose Tissue
Total Magnification:	Total Magnification:

Reticular Tissue
Total Magnification:

Fibrocartilage
Total Magnification:

Compact Bone	**Human Blood**
Total Magnification:	Total Magnification:

CHAPTER 31

Fetal Pig and Sheep Brain Dissection Instructions

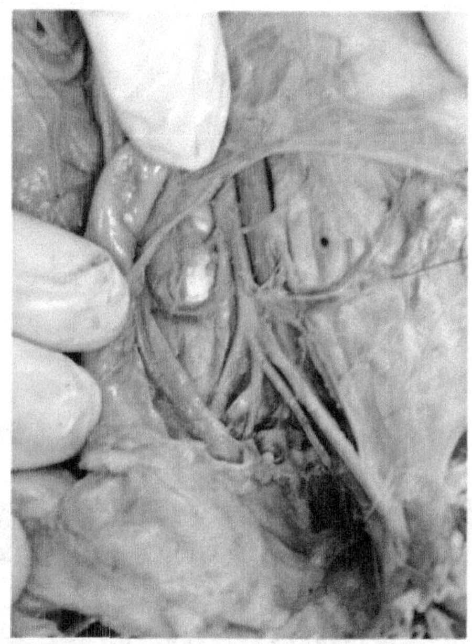
Photos courtesy of Holly J. Morris

OBJECTIVES

After completing this exercise, you will be able to:

- Identify and know the names and structures of the major structures of a fetal pig's external anatomy.
- Identify and know the names and structures of the major structures of a fetal pig's internal anatomy including the following systems:
  - Digestive
  - Circulatory
  - Respiratory
  - Excretory
  - Reproductive
  - Nervous

INTRODUCTION

There are many similarities between the anatomy of humans and pigs. There are several reasons for this. Both humans and pigs are vertebrates, meaning that we have backbones and skulls. We are both mammals, meaning that we are warmblooded, have a four chambered heart, closed circulatory system, hair, and mammary glands. We are also both omnivores, meaning that our teeth and digestive systems are capable of regularly consuming both plant and animal material. Our reproductive systems are similar in that the female gonads (the ovaries) are an internal organ while the male gonads (the testes) are external structures contained within the scrotum.

The purpose of today's lab is to get an introduction to dissection tools and dissection techniques along with identifying some of the common structures of the pig's internal anatomy. In order to understand the instructions for the dissection, there are certain anatomical terms that must be learned. These terms are used to direct and navigate as you view the internal organs of the fetal pig.

- **Dorsal:** The back surface, toward the spine
- **Ventral:** The front surface, toward the abdomen and stomach
- **Anterior:** Above, toward the head
- **Posterior:** Below, toward the tail
- **Lateral:** Away from the center
- **Medial:** Toward the center
- **Superficial:** Toward or along the surface
- **Deep:** Substantially below the surface
- **Longitudinal:** A vertical line that runs anterior and posterior.
- **Horizontal:** A line that runs right and left
- **Dextral:** To the right of
- **Sinistral:** To the left of

PROCEDURE

OBSERVING THE EXTERNAL ANATOMY

1. Obtain a dissecting tray, fetal pig, and a complete set of the dissecting tools that you will need.

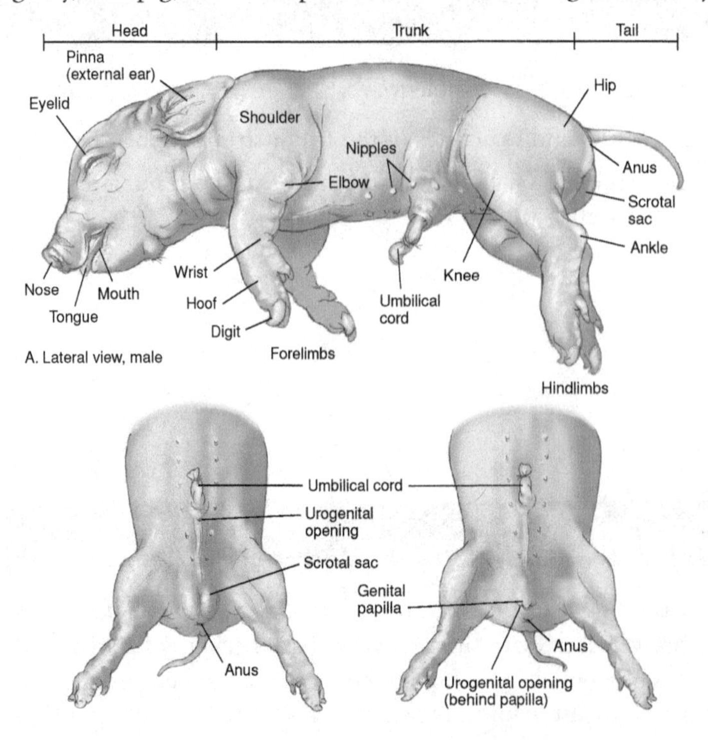

2. The preserving solution is very strong so open the bag and rinse off the pig in the sink along with pouring out of the residual preserving solution (it is biodegradible and nontoxic when diluted, so rinse with an abundance of water, then return to your groups working area with your washed pig on the tray).
3. Make a thorough examination of your pig, including its facial features, limbs, and gender (male pigs have scrotum anterior to the anus and a urogenital opening just posterior of the umbilical cord, female pigs have urogenital opening anterior of the anus that is enclosed by small folds of skin called labia that form the genital papilla, a small projection. If your pig has a visible protrusion near the anus, it is likely a female.)
4. Place the pig with its ventral side up to prepare to open the abdominal and thoracic cavities.

Opening the Abdominal and Thoracic Cavities

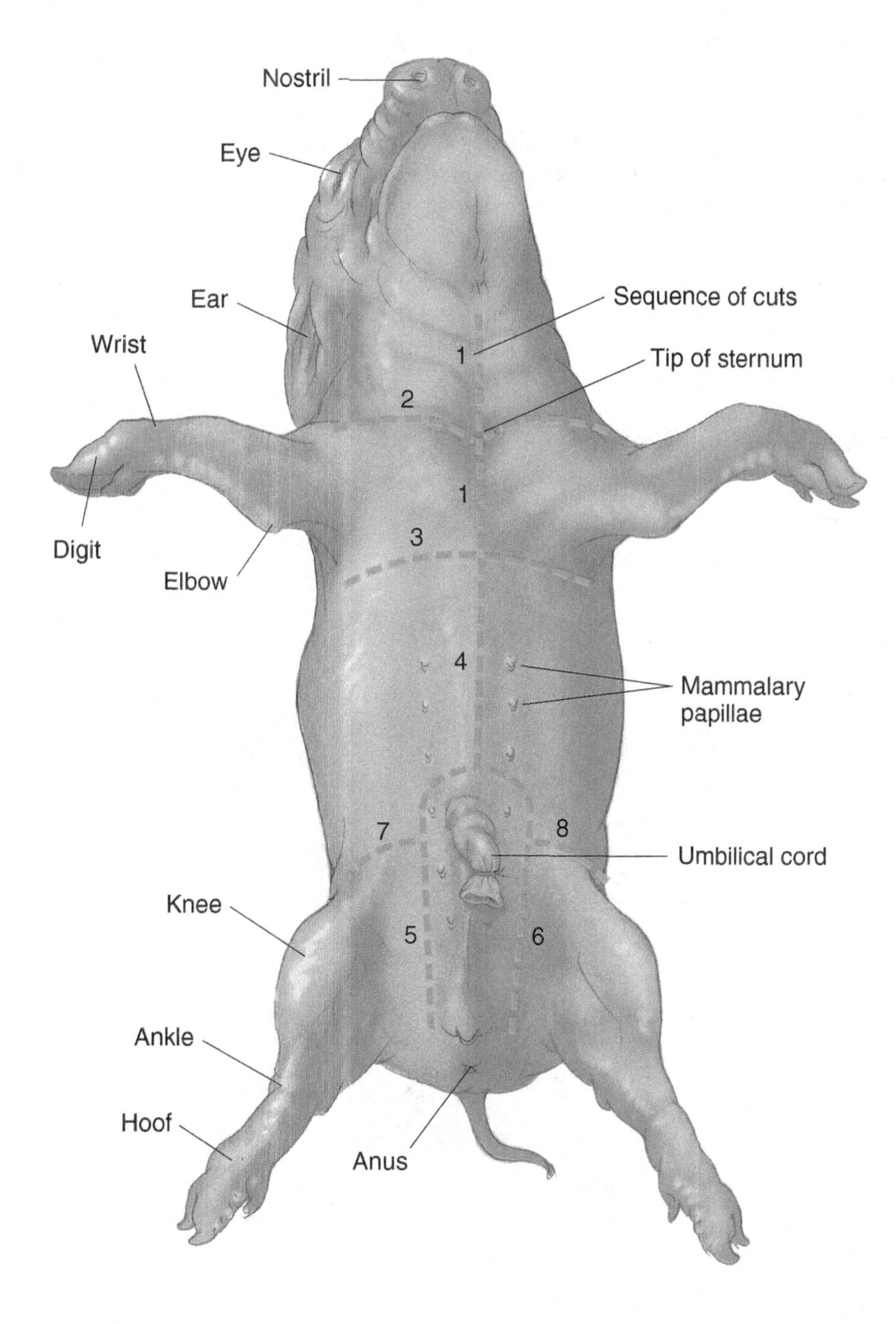

1. Using the scalpel, make a superficial longitudinal incision starting below the chin and cutting posteriorly towards and then across the surface of the rib cage. This first incision should be cutting through the skin and muscle, but not yet the bones of the rib cage.
2. Pull the skin and muscle laterally to reveal the sternum of the rib cage. After the surface of the rib cage is exposed, use the scissors to cut the bones of the rib cage, taking care to not smash and destroy the structures of the deeper tissues. Make sure that this cut is lateral to the sternum, as the breastbone is harder to cut through than the neighboring ribs. Continue cutting through the skin, ribs, and muscles until you reach the posterior edge of the rib cage and the diaphragm (an internal, horizontally-aligned muscular sheet used during breathing).
3. Cut a horizontal line anteriorly and posteriorly to the forelimbs, along the edges of the ribcage. Move the forelimbs laterally to obtain a better view. Two longitudinal incisions on the lateral edges of the ribcage can be performed to remove the surface of the rib cage from obstructing the view. Separate the edge of the diaphragm from the abdominal wall if necessary.
4. Continue the center longitudinal incision from the diaphragm to a location just anterior of the umbilical cord.
5. From the umbilical cord, continue the incision dextrally and sinistrally in a posterior manner to the genital region. There is an umbilical vein that is anterior to the umbilical cord that may still be connecting the umbilical cord to the anterior portion of the abdomen. After this vein is severed, the entire skin flap containing the umbilical cord can be folded posteriorly to increase the view of the abdominal organs.
6. Make a horizontal incision on the body wall just anterior of the hind limbs on both the right and the left hand side. The skin and muscle along the sides of the abdominal cavity can now be moved out of the way or removed using a longitudinal incision on the lateral edges of the abdominal wall.

Organs of the Digestive System

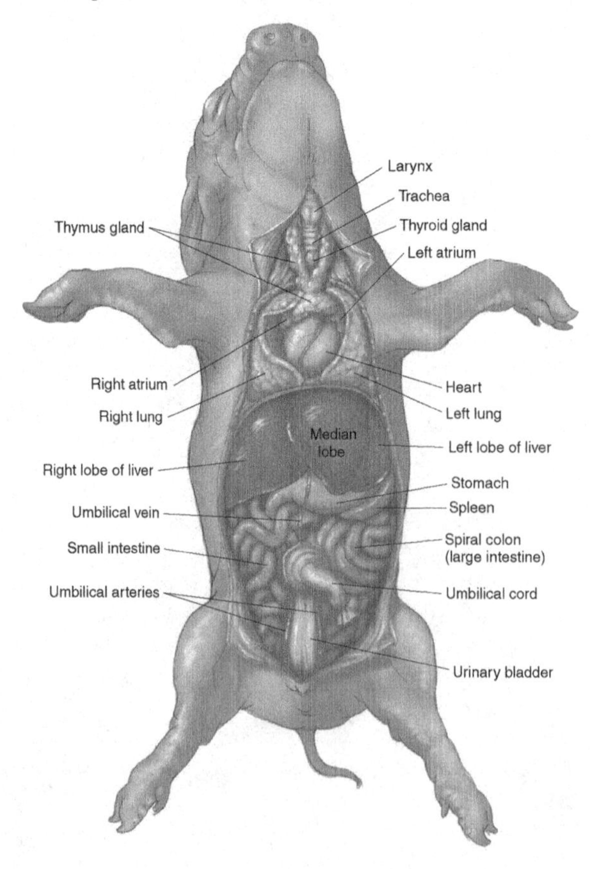

1. The organs of the digestive system in pigs are very similar in shape and orientation to the digestive system in humans. Begin your observation by identifying the stomach, small intestines, and large intestines. These structures will all be found below the diaphragm.
2. From the stomach, search anteriorly to find the esophagus and posteriorly to find the duodenum (first part of the small intestine). Near where the stomach and duodenum meet, deeper into the abdominal cavity, there will be a grainy organ connected to (and possibly surrounded by) the mesenteries known as the pancreas. This will be found between the duodenum and the spleen, behind the stomach.
3. The liver will be a dark prominent multi-lobed organ below the diaphragm and near the stomach; count the number of lobes of the liver present in your pig and see if you can identify the gallbladder, usually located between a few of the lobes of the liver and transparent or greenish in color.
4. The small intestines will transition to the large intestines. Continue to follow this path to the large intestines near the anus. Cutting the lower portion of the colon and the most anterior portion of the esophagus posterior of the diaphragm will allow this entire portion of the alimentary canal to be removed (with the removal of the connective tissue holding it in place), providing a view of the back wall of the pig's abdomen, including the kidneys. The liver may also be removed by cutting the anterior connective tissue holding it in place.
5. The kidneys will be revealed, attached to the dorsal wall of the abdomen. Some connective tissue may need to be removed in order to clearly view the kidneys.
6. Unwind the small intestines and large intestines by cutting them loose from the mesenteries that are holding them, and compare their relative lengths and diameters.

Organs of the Circulatory and Respiratory Systems

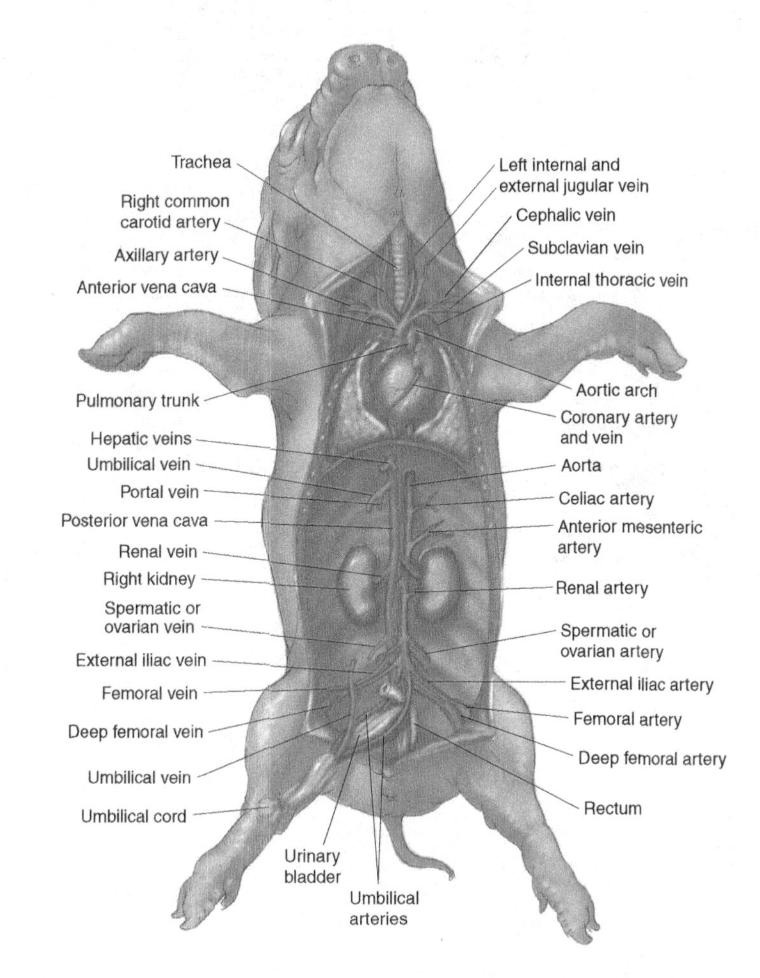

1. The internal organs of the chest cavity include the major organs of the circulatory and respiratory systems. Upon opening the chest cavity, identify the heart and the lungs. The heart may likely be located under a layer of connective tissue known as the pericardium. This pericardium may be removed to allow for a clearer view of the heart.
2. Like humans, the heart of the pig has four chambers; two upper chambers known as the atria and two lower chambers known as the ventricles. See if you can identify where these structures are found in the heart and draw an image of the external view of the fetal pig's heart.
3. Identify the lungs and the cartilaginous trachea (windpipe). This may require the moving of or removal of the thymus gland.

Organs of the Reproductive System

1. Your pig will be male or female, which you had established with the observation of the external anatomy. The internal reproductive anatomy will be different for the two sexes.

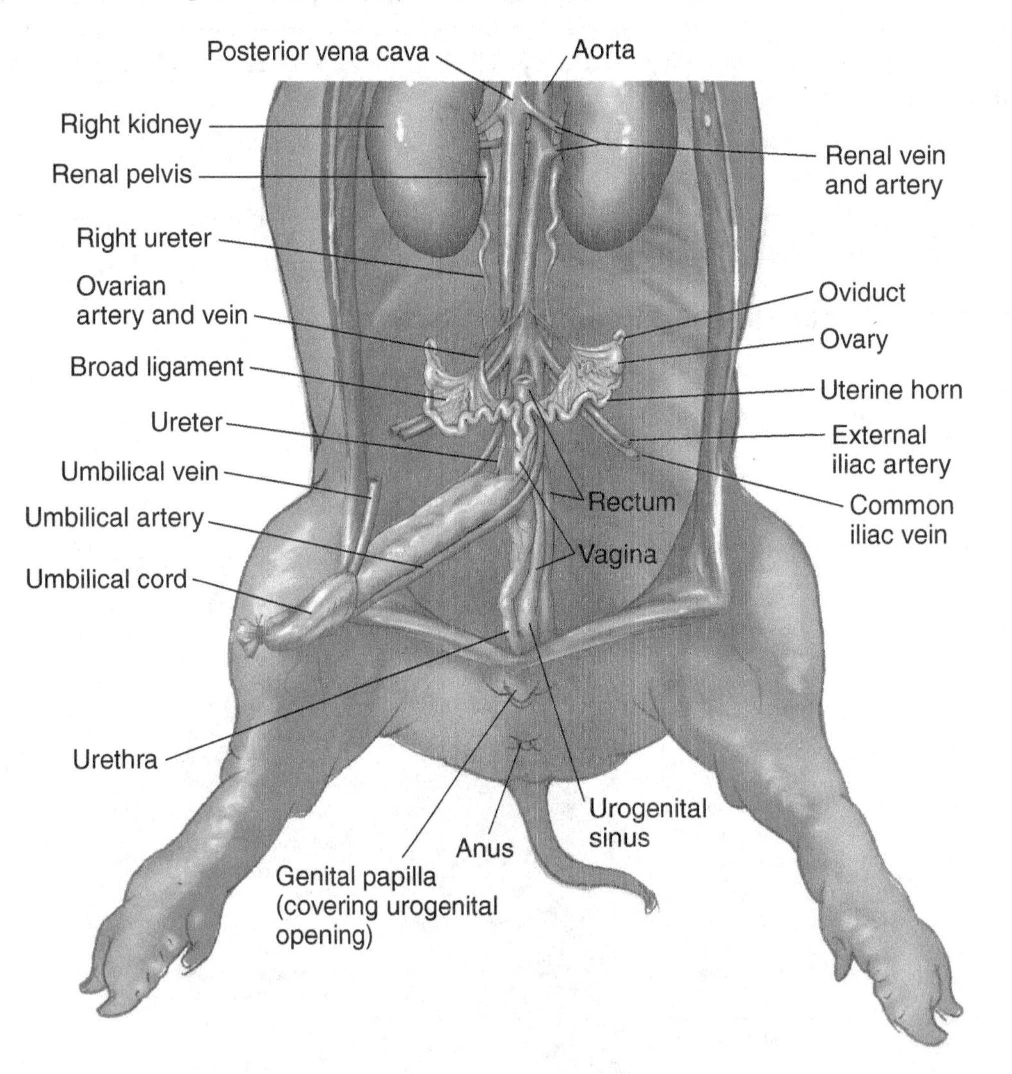

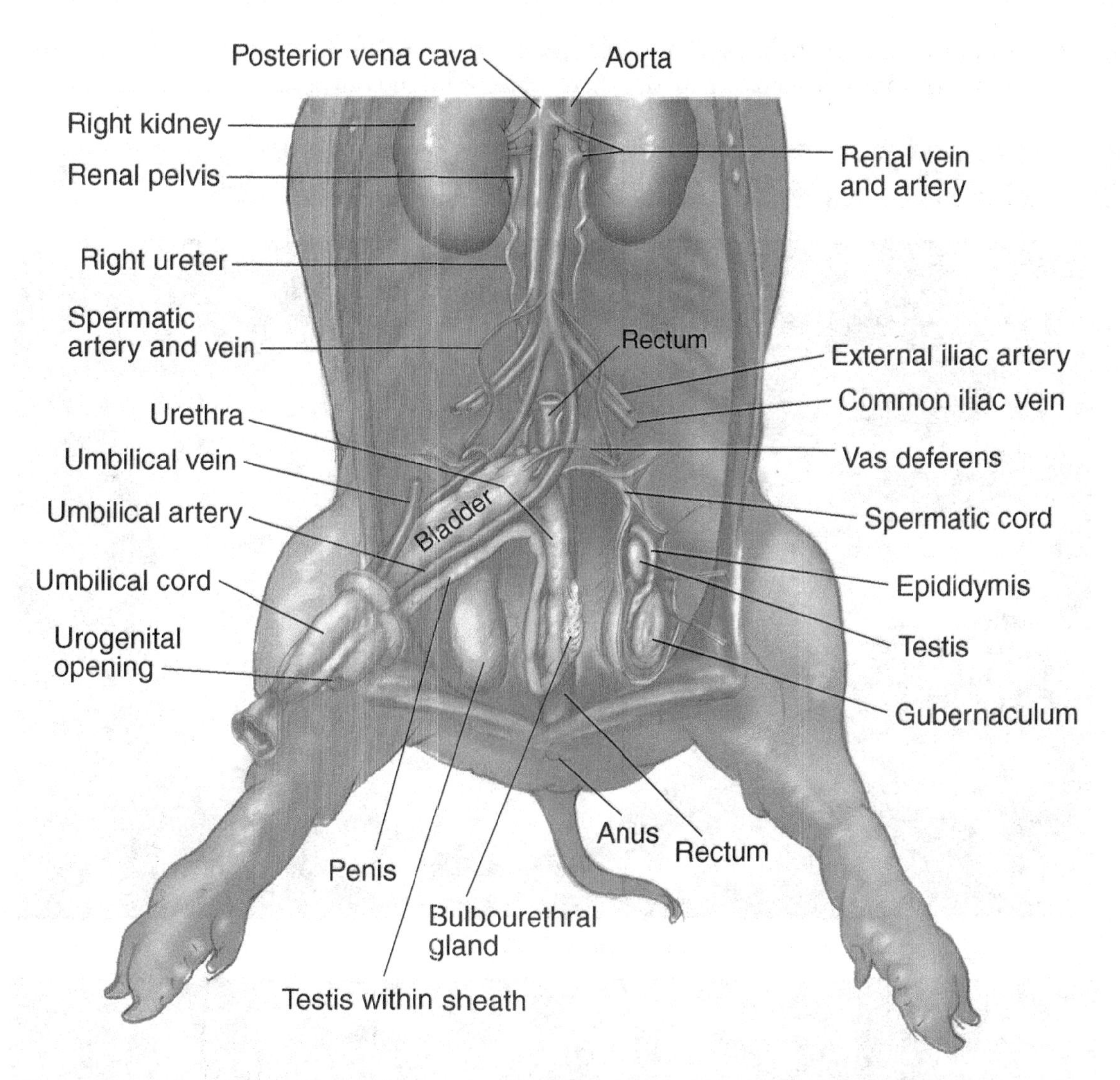

2. If you have a female pig, identify the urinary bladder, vagina, uterine horns, and ovaries.
3. If you have a male pig, using a scalpel or pair of scissors, open one of the sheaths surrounding a testis and identify the testis, epididymis, penis, and urinary bladder.
4. After completing your observation of the organs of the reproductive system, find a lab group with a pig that is of a different gender than your own and observe the above listed structures.[2]

THE CENTRAL NERVOUS SYSTEM

The central nervous system consists of the brain and the spinal cord. Before we dissect a sheep brain, we will conclude our fetal dissection by observing the fetal pig's spinal cord.

The spinal cord serves as a relay center between the brain and the body. However, remember that the spinal cord also processes neural information locally. To examine the spinal cord:

1. Reposition your fetal pig dorsal side up in the dissecting tray.
2. Cut the skin along the midline from an area behind the base of the skull for a distance of 5 cm. Then remove the skin from the left and right sides of your midline cut to expose the vertebral column.
3. Using a forceps, open the vertebral column at the base of the skull and expose the juncture between the medulla oblongata (the continuation of the spinal cord within the skull which forms the lowest part of the brainstem) and the spinal cord.

4. Open additional vertebrae to observe the spinal cord and the spinal nerves. Note that the spinal nerves are continuous with the spinal cord on both sides of the spinal cord and run between two successive vertebrae.

SHEEP BRAIN DISSECTION

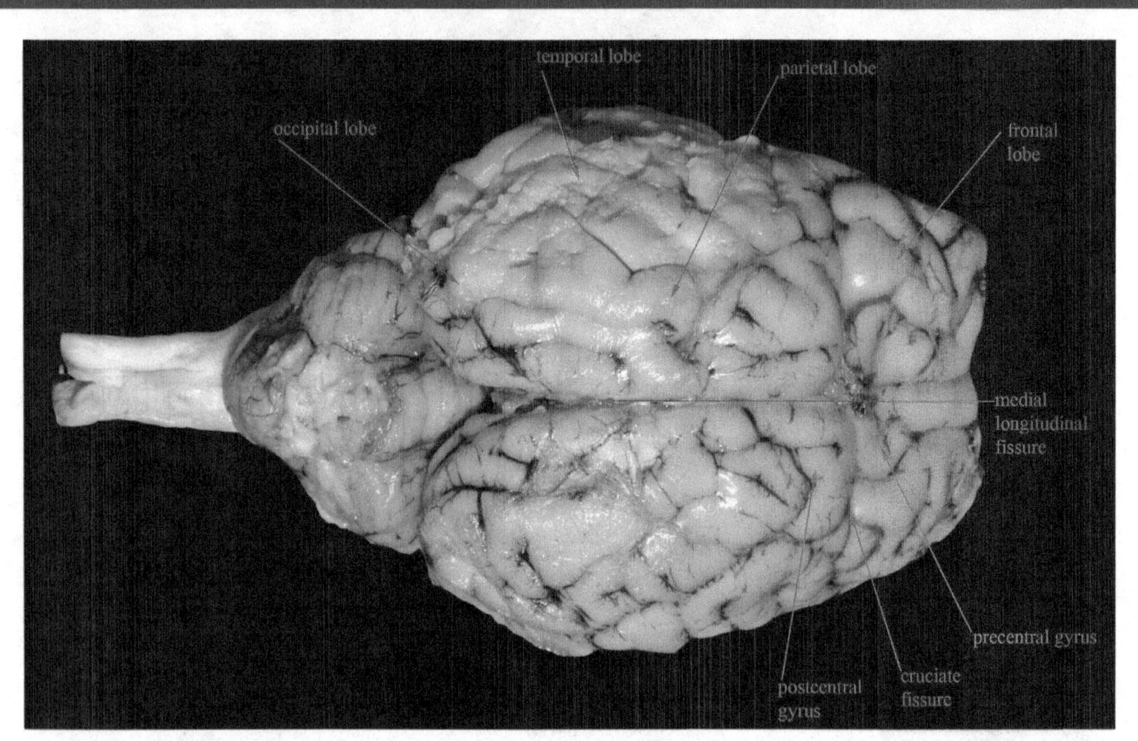

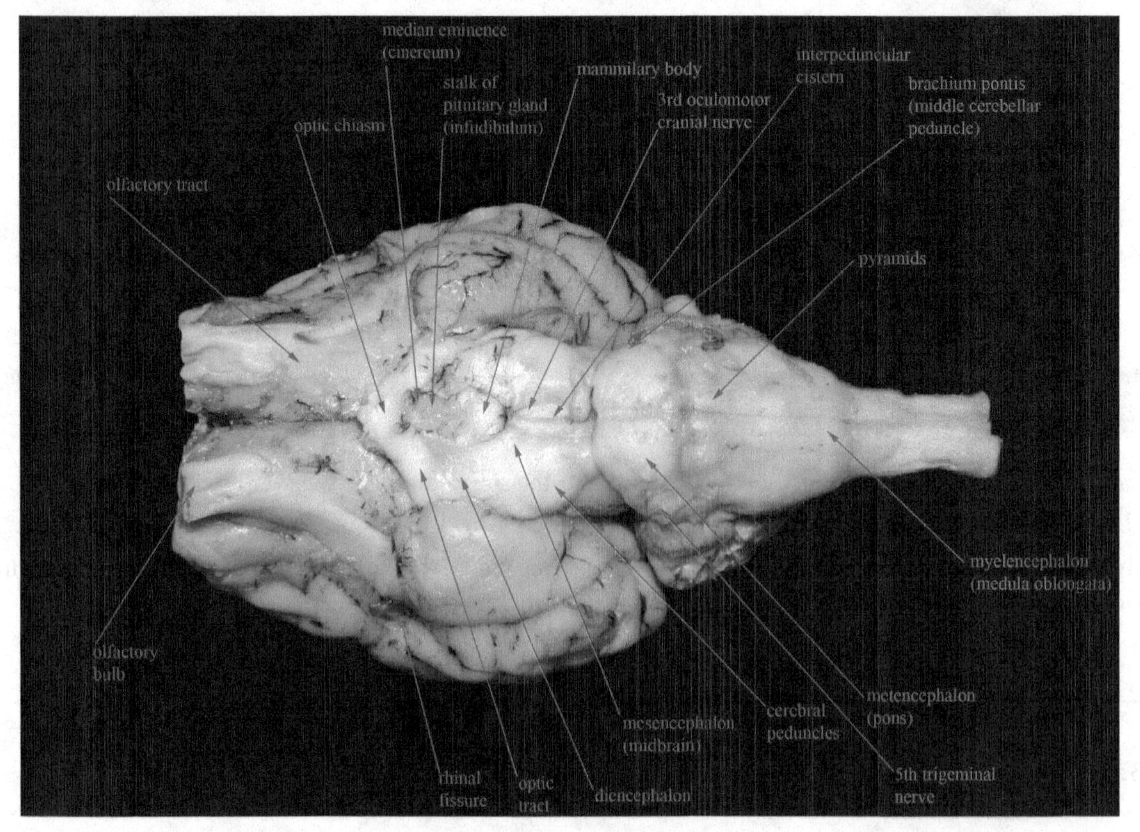

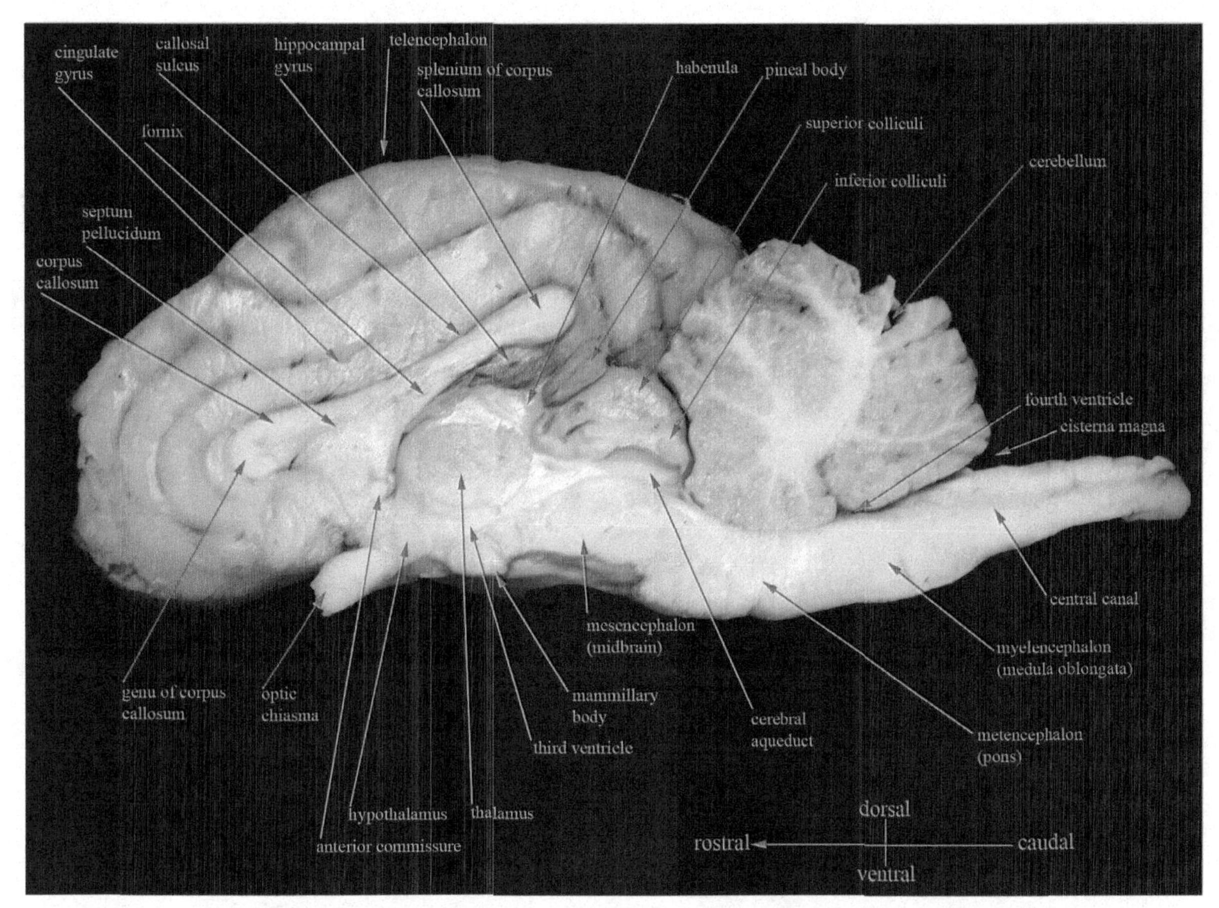

Sheep Brain (Dorsal View, Ventral View, Sagittal View) (© Mark Yarchoan/Shutterstock.com)

Place the sheep brain in a dissecting tray. Note that brain tissue is very delicate and care needs to be taken to ensure that this tissue is not damaged in the dissection

1. Examine the exterior of the entire brain. The meninges are the brain's protective covering. The meninges is composed of 3 layers: dura mater - the tough outer layer; arachnoid layer – the middle layer; and pia mater is the inner layer. The pia mater is a thin transparent inner layer that follows with the gyri (ridges) and sulci (depressions) on the cerebral cortex. Before you begin your examination of the brain, remove the dura meter and the arachnoid layer as well as any associated blood vessels.
2. Identify the main regions of the brain:
   - Cerebrum (Located at the front end of the brain and consists of 2 hemispheres separated by the longitudinal fissure.)
   - Cerebellum (Located behind the cerebrum and separated from the cerebrum by the transverse fissure.)
   - Brain stem (Composed of the pons, medulla oblongata, and the cerebellum.)
3. With the dorsal side up, identify the features of the cerebrum including the hemispheres, gyri, sulci, the longitudinal fissure which separates the 2 hemispheres, and the transverse fissure which separates the cerebrum and the cerebellum. Also, identify the frontal, parietal, occipital, and temporal lobes of the cerebrum.
4. With the ventral side up, identify the olfactory bulb, optic chiasm, midbrain, pons, and medulla oblongata.
5. With the dorsal side up, use a scalpel to cut the entire brain from the front of the brain along the longitudinal fissure down through the cerebellum, midbrain, pons, medulla oblongata, and spinal cord. Separate the two halves of the brain and lay them with the inside facing up.
6. Identify the corpus callosum, cerebellum, pons, medulla oblongata, and spinal cord.

7. Within the cerebellum, observe the lighter tissue surrounded by darker tissue. The lighter tissue is the white matter and is composed of nerve axons while the darker tissue is gray matter which is composed of nerve cell bodies. You can also see white and gray matter within the cerebrum.

CLEAN UP

1. After completing the directed dissections and any additional investigatory dissections, the lab instructor will explain the disposal procedures for the fetal pig and the sheep brain.
2. Rinse the dissecting trays and tools with water, then dry and return them to their storage area.
3. The countertop should be sprayed with disinfecting spray (Wavicide) and wiped down with a paper towel.[2]

CHAPTER 32

Pond Ecology

Photo courtesy of Holly J. Morris

OBJECTIVES

After completing this exercise, you will be able to:

- Distinguish between abiotic and biotic factors in the environment.
- Identify each trophic level of a food chain as a producer, consumer, or decomposer and provide an example of each.
- Understand the importance of adequate levels of oxygen and carbon dioxide.
- Measure and understand the effect on a pond of:
  - pH
  - Acidity
  - Alkalinity
  - Carbon dioxide
  - Dissolved oxygen
  - Hardness
- Understand the difference between a food chain and a food web.
- Prepare a food web for the pond.

INTRODUCTION

The study of the relationships between living things and their environment is known as ecology. Limnology is the branch of ecology that studies all of the interrelationships of inland water bodies such as lakes, reservoirs, rivers, streams, wetlands, and groundwater. An ecosystem can be defined as a biological community of interacting organisms and their physical environment. Since ponds are a stable environment of interacting organisms and their physical environment, a pond is considered to be an ecosystem.

Freshwater habitats are divided into two groups, standing water and flowing water. Flowing water habitats are divided into slow streams and rapid streams. A standing water habitat includes lakes, ponds and swamps. In this exercise, we will study the features of a pond ecosystem.

A pond is a small, shallow standing body of water, usually less than 15 feet deep, which allows sunlight to reach the bottom, thereby permitting the growth of rooted plants throughout the bottom of the pond.

Like all ecosystems, the pond ecosystem consists of living (biotic) and nonliving (abiotic) components. The biotic component includes all the living organisms in the pond while the abiotic component includes both organic and inorganic substances such as water, salts, nitrogen, oxygen, carbon dioxide, and amino acids, as well as temperature, light, and pH.

Ponds contain a variety of plants and animals that are generally in ecological balance, though the nature of that balance changes as the pond matures. As the pond ages, the variety of species increases, and larger plants and algae, along with the accumulation of waste results in turning the pond into a marsh, or causes the pond to dry up. This is ecological succession.

A pond has several areas, or habitats, in which organisms live. The surface film habitat is located on top of the water and is inhabited by organisms that float or can stride on the surface of the water.

The area of water near the shore is the *littoral zone*. This area will have plants emerging from the water, as well as submerged and floating plants. Farther into the water is the *limnetic zone*. This is the open water.

The bottom of the pond is called the *benthic zone*. The organisms that live in the benthic zone are dependent upon the type of bottom that the pond may have. For example, a shallow pond with a sandy bottom could be inhabited by sponges, earthworms, snails and insects. The bottom with quiet, standing water is usually muddy or silty, and life represented in these types of ponds are crayfish, and the nymphs of mayflies, dragonflies, and microorganisms.

Some common pond organisms include:

Common Pond Organisms

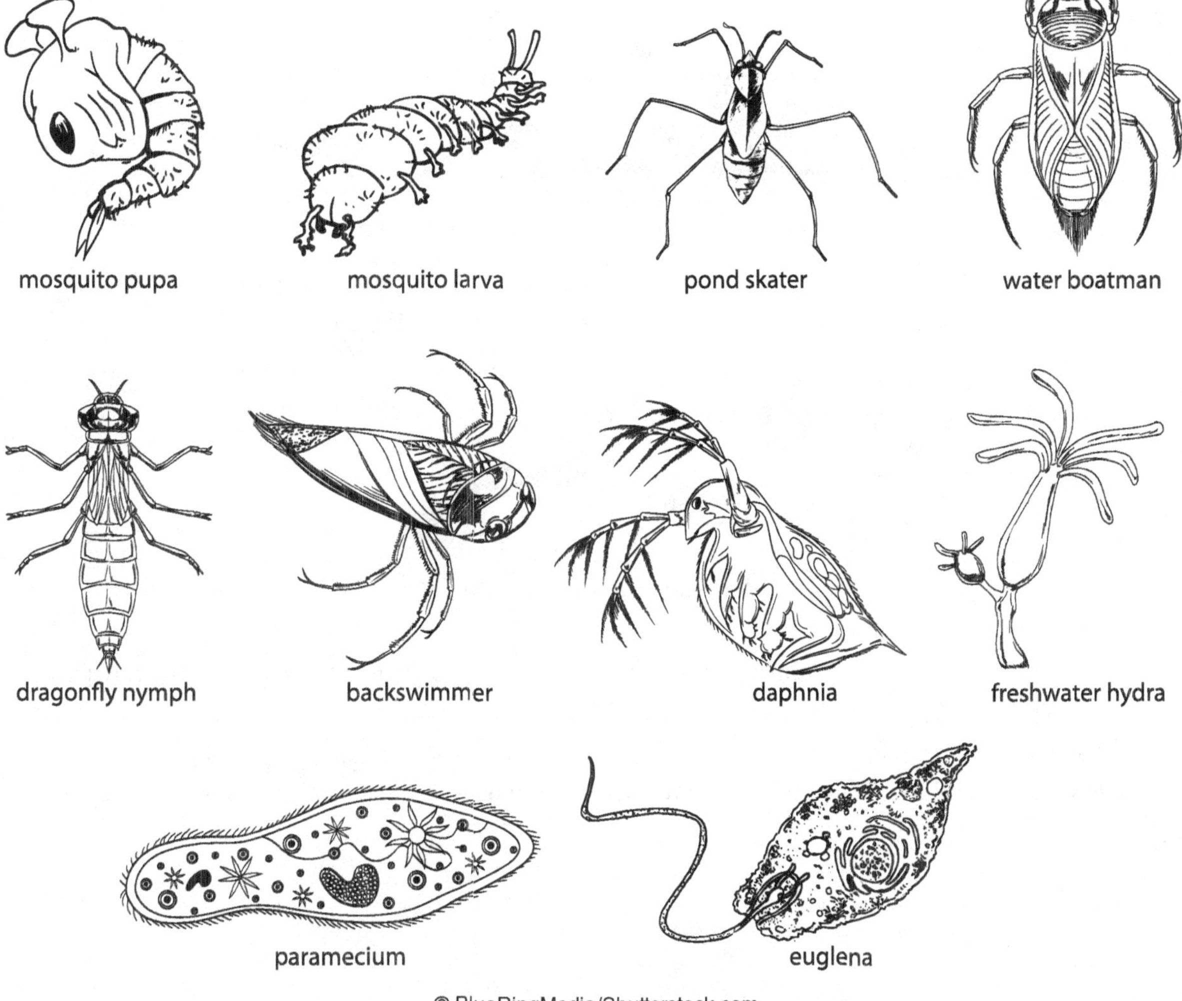

The components of a pond ecosystem can be categorized as follows:

- Abiotic substances
- Producers
- Consumers
- Detritivores (decomposers)

Energy and nutrients flow from organism to organism in what is called a food chain. Food chains divide types of organisms into trophic levels. The base of the food chain, the first trophic level, consists of the *producers*, organisms such as plants, algae, and cyanobacteria that can capture energy from the sun through photosynthesis.

There are two different kinds of producers typically found in a pond:

- Phytoplankton (small and/or microscopic floating plants, which can include algae and cyanobacteria).
- Rooted/large floating plants found in the littoral zone.

The second trophic level consists of *primary consumers*, herbivores that consume producers. The third tropic level consists of *secondary consumers*, animals that eat primary consumers.

An example of a food chain:

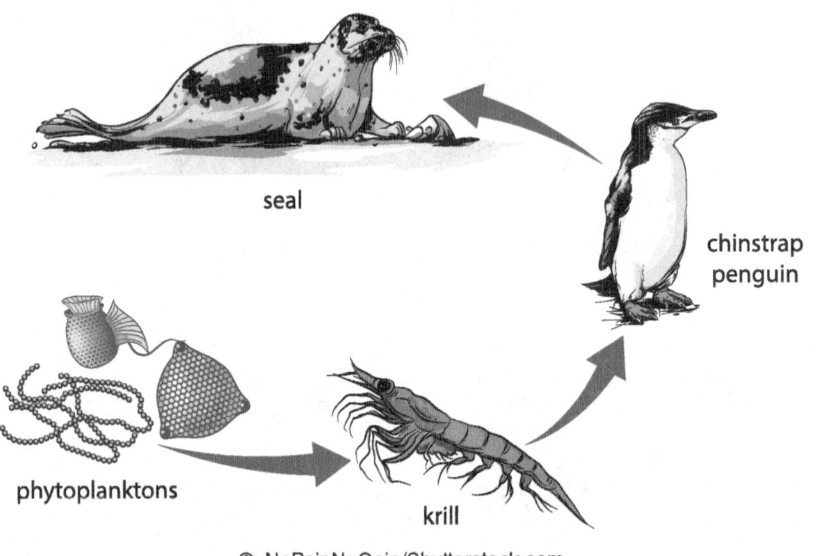

© NoPainNoGain/Shutterstock.com

People sometimes talk about an animal that is at the top of the food chain. Another name for this animal is "apex predator." Apex predators have no natural predators.

Food chains are simplified interactions. In reality, interactions in ecosystems are complex and interconnected, creating food webs, which are essentially a representation of what-eats-what in that ecological community.

An example of a food web:

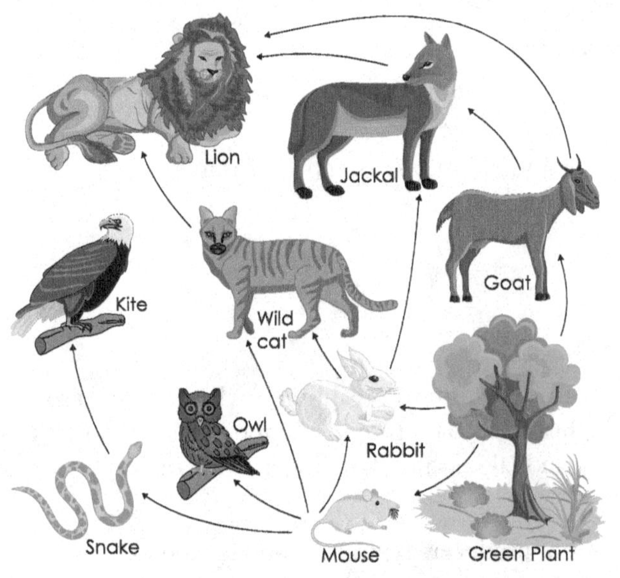

© snapgalleria/Shutterstock.com

A separate trophic level, the *decomposers*, consists of organisms that break down dead or dying organisms or waste material. The terms detritivore and decomposer are sometimes used for two groups of organisms, however, we will use them interchangeably. Included in this group are organisms such as earthworms, mil- lipedes, sea stars, and dung flies, as well as bacteria and fungi. Decomposers consume dead or dying organ- isms, as well as waste from all trophic levels. In the process of breaking down detritus, nutrients not used by the decomposer are recycled back into the ecosystem.

EXERCISE 1 Observing and Analyzing the Characteristics of a Pond Ecosystem

In this exercise you will make observations of a pond ecosystem and the adjacent terrestrial ecosystem. You will look at both macroscopic and microscopic components of both the pond and the surrounding land.

- Make a list in your lab notebook as to what you observe about 5 feet into the water, at 1 foot into the water, and about 1 foot away from the water's edge.
- Is the land immediately surrounding the pond consistent in its composition around the circumference of the pond? Describe what you see.

Purpose

To determine the characteristics of a pond ecosystem.

Materials taken to the pond

- Petri plates (1 or 2)
- Hygrometer
- Anemometer
- Light meter
- Thermometer
- Plankton net
- Grab
- *Pond life Guides*
- 1 Gallon plastic bags (3-4)
- Empty jars with screw top lids (2-3)
- Bucket for waste products

Material needed for the lab

- Light microscope
- Stereoscope
- Petri plates
- Disposable pipettes
- Slides
- Cover slips
- Water testing kit

Methods

At the site:

1. Record the following data for the pond in the **Climatic Conditions** table:
    - Sampling date
    - Starting time
    - Ending time
    - Cloud cover

- Wind speed
- Precipitation
- Relative humidity
- Light level
- Air temperature
- Water temperature
- Sampling locations at the pond. Draw a picture of the pond (an oval is sufficient) and place an X on the approximate location where you collected each sample.

2. Collect a sample of the surface and open water habitat with a plankton net.
3. Collect a sample of the benthos habitat with a grab.
4. Collect a sample of the littoral habitat by grabbing several handfuls of rooted plants.
5. Collect water samples at the pond inlet (near side), middle, and outlet (far side). In the lab:

1. Make wet mounts of the samples obtained with the plankton net and using a compound light microscope, examine the sample looking for phytoplankton and zooplankton. Identify the types of phytoplankton found and make a sketch of the organisms in the **Phytoplankton** table. Likewise, identify the types of zooplankton found and make a sketch of the organisms in the **Zooplankton** table.
2. Put a sample from the plankton net samples in a Petri plate and using a stereoscope, examine the sample looking for larger zooplankton. Identify the types of zooplankton found and make a sketch of the organisms in the **Zooplankton** table.
3. Examine the samples obtained from the pond bottom for zooplankton. Identify any zooplankton found and make a sketch of the organisms found in the **Benthic Organisms** table.
4. Examine the samples obtained from the littoral habitat for zooplankton. Identify any zooplankton found and make a sketch of the organisms in the **Animals Found Among the Macrophytes** table.
5. Using the *Pond Life Guide*, try to identify the organisms you sketched in all of the tables.
6. Perform the following water quality tests on the water samples from the inlet, middle, and outlet:
   - pH
   - Acidity
   - Alkalinity
   - Carbon dioxide
   - Dissolve oxygen
   - Hardness
7. Record your results in the **Water Quality Test Results** table.

You will be sharing data collected from all regions of the pond with other groups in the class.

Name:

Exercise 1 Results

1. Compete the tables with your observations:

Climatic Conditions	
Sampling date	
Starting time	
Ending time	
Cloud cover	
Wind	
Precipitation	
Relative humidity	
Light level	
Air temperature	
Water temperature	
Sampling locations (draw a picture)	

Phytoplankton			
Organism Name	**Sketch**	**Organism Name**	**Sketch**

Zooplankton			
Organism Name	**Sketch**	**Organism Name**	**Sketch**

Benthic Organisms			
Organism Name	**Sketch**	**Organism Name**	**Sketch**

Animals Found Among the Macrophytes			
Organism Name	**Sketch**	**Organism Name**	**Sketch**

Water Quality Test Results			
Test Performed	**Inlet**	**Middle**	**Outlet**
pH			
Acidity Multiply drops by 2.28			
Alkalinity Multiply drops by 6.84			
Carbon dioxide Multiply drops by 5.0			
Dissolved Oxygen Multiply drops by 0.5			
Hardness Multiply drops by 17.1			

2. **Pond Food Web:** Using the data you gathered, construct a food web for the pond which includes all of the organisms listed in the above tables (**Phytoplankton, Zooplankton, Benthic Organisms,** and **Animals Found Among the Macrophytes)**. Include the flow of energy with arrows from source to consumer.

Exercise 2

1. Complete the **Trophic Levels** table.

Trophic Levels			
	Trophic Level	**Food Source**	**Example**
Producers			
Herbivores			
Carnivores			
Omnivores			
Apex predators			
Decomposers			

Writing a Formal Lab Report

Photo courtesy of Holly J. Morris

After you finish a scientific experiment, the next step is to document the research in a formal lab report. A formal lab report discusses the experimental design, explains the results, reaches conclusions about the experiment, and discusses the significance of the experiment.

Writing a formal lab report is similar to writing a manuscript for publication in a scientific journal. A formal lab report should be written in the third person and should follow the Council of Science Editors (CSE) guidelines, Name-Year style, for formatting and referencing.

Contractions should not be used.

Do not use direct quotes from published articles. Put the information into your own words BUT you must cite the source within the paragraph.

A properly written lab report contains the following sections:

TITLE

The title of the lab report describes the topic of the experiment. It should clearly reflect the question that is being researched. The title can be written as a statement, or in the form of a specific question.

For example, if you are investigating the effect of various pHs on the enzymatic activity of amylase, a title for the formal lab report could be "The Effect of pH on Amylase." Alternatively, the title could be "How Does pH Affect Amylase?"

INTRODUCTION

This section clearly states the objective of the experiment and why it is being done. The Introduction should provide the reader with the relevant background information about the topic being investigated, explain why the topic is important, and provide any additional information needed to understand the experiment.

Using the amylase example, the objective might be stated as "The objective of this experiment was to study the effect of pH on the enzymatic activity of amylase." The formal lab report then continues by providing the relevant background information including topics such as what is pH, how does pH affect reactions, what does amylase do, how does amylase work, and why is amylase important.

The Introduction also includes the hypothesis that is being investigated. A hypothesis is your educated guess, based on common sense or the known laws/theories of science, about what the results of the experiment will be. The hypothesis can be included in the same section as the objective of the experiment or it can be an isolated section. Using the amylase example, the hypothesis could be "amylase will optimally break down starch at pH 7".

The Introduction also includes the major results of the experiment.

MATERIALS AND METHODS

This section summarizes the materials and methods (procedures) that were used in the experiment. The materials and methods needs to be sufficiently detailed so that another experimenter can redo your experiment. Bullets/numbers should be used to help organize this section.

RESULTS

This section begins by describing what the data shows and summarizes the significant data. The results of the experiment are presented in the form of graphs, tables, diagrams, or pictures.

Tables are a useful tool for organizing data. When constructing a table for a formal lab report, follow these guidelines:

- Number each table in sequence as Table 1, Table 2, etc.
- Has a descriptive title.
- Contains the same kind of data in a column.
- Has a heading for each column describing the data in the column as well as the measurement units.
- Contains the control data.

Graphs are used to give a visual depiction of the results as well as show the relationship between the independent variable and the dependent variable. In the amylase example, a graph would show the relationship between pH (independent variable) and the amount of enzymatic activity (dependent variable).

A graph can be in the form of a line graph, a bar graph, or a pie chart. When constructing a graph for a formal lab report, follow these guidelines:

- Keep all graphs, as well as other figures in the report, separate from tables.
- Number each graph in sequence as Figure 1, Figure 2, etc.
- Contains a title describing what is shown.
- Places the independent variable on the *x*-axis (the horizontal axis).
- Places the dependent variable on the *y*-axis (the vertical axis).

- Contains axes labels that include the units of measurement.
- Contains a legend.

The type of graph to be used depends on the type of data. If the data is continuous, a line graph should be selected. If the data contain discontinuous measurements or nonnumeric categories, a bar graph can be used. Alternatively, a pie chart might be more appropriate.

DISCUSSION

This section is an interpretation of the data in the "Results" section. It contains several items that need to be addressed including:

- Restating the objectives of the experiment as a topic sentence to frame the discussion.
- Summarizing the major findings of your experiment using references to the data in the "Results" section.
- Explaining whether the data leads you to accept or reject your hypothesis.
- Offering plausible explanations discussing why the results occurred. Include data from the "Results" section to support your analysis.
- Acknowledging any deviations from what you expected the results to be.
- Relating your results to previous research.

A technique to conclude your "Discussion" would be to discuss the significance of your experiment with regard to the investigated topic in general.

The "Discussion" section does not discuss educational value, skills learned, or other commentary on the experiment.

CONCLUSION

This section explains the results as they relate to the experiment's objective as stated in the Introduction.

This section should also discuss the implications of the experiment's results. Other items that could be included in this section include discussing any weaknesses in the experimental design, improving the experiment to better test the hypothesis, and extending the experiment for further research.

WORKS CITED

If outside references have been used to prepare any part of a formal lab report, the report must include them in a bibliography according to CSE formatting guidelines.

Credits

FN 1: From Biological Investigations II: Lab Exercises for General Biology II By Edward E. Devine and Gretchen S. Bernard. Copyright © 2014 by Kendall Hunt Publishing Company. Reprinted by permission.

FN 2: From Fundamentals of Life Science: Lab Book For Biology 189 at Nevada State College by Nevada State College. Copyright © 2013 by Board of Regents of the Nevada System of Higher Ed. Reprinted by permission.

FN 3: From BI102 Laboratory Manual: General Cellular Biology by Washburn University. Copyright © 2009 by Kendall Hunt Publishing Company. Reprinted by permission.

FN 4: From Introductory Biology: A Laboratory Exploration of Life by Stacy Pfluger and Taylor Hall. Copyright © 2013 by Kendall Hunt Publishing Company. Reprinted by permission.

FN 5: From Organismal Biology 1108: Laboratory Manual by Elizabeth J. Walsh, Paul E. Hotchkin, Shawn T. Dash, and Susan H. Watts. Copyright © 2013 by Elizabeth J. Walsh, Paul E. Hotchkin, Shawn T. Dash, and Susan H. Watts. Reprinted by permission.

FN 6: From Microbes in Health and Disease Lab Manual by Cynthia W. Littlejohn. Copyright © 2015 by Kendall Hunt Publishing Company. Reprinted by permission.

FN 7: From Microbiology Laboratory Manual by Caren D. Shapiro. Copyright © 2009 by Bent Tree Press. Reprinted by Permission of Kendall Hunt Publishing Company.

FN 8: From Fundamentals of Microbiology for Allied Health by Kathleen Dannelly, Angela K. Chamberlain and William M. Chamberlain. Copyright © 2009 by Kendall Hunt Publishing Company. Reprinted by permission.

FN 9: From Microbiology of Human Diseases: Laboratory Techniques and Applications by Kevin B. Kiser. Copyright © 2014 by Kendall Hunt Publishing Company. Reprinted by permission.

FN 10: From General Microbiology Laboratory Manual by Davis Pritchett, Dale Amos, Debra Jackson, and Allison Wiedemeier. Copyright © 2011 by Kendall Hunt Publishing Company. Reprinted by permission.

FN 11: From Laboratory Exercises in Microbiology by Keith E. Belcher. Copyright © 2012 by Kendall Hunt Publishing Company. Reprinted by permission.

FN 12: From Laboratory Manual for Biology I by Lalitha Jayant, Owen Meyers, Matthew Geddis, and Christine Priano. Copyright © 2014 by Lalitha Jayant, Owen Meyers, Matthew Geddis, and Christine Priano. Reprinted by Permission of Kendall Hunt Publishing

FN 13: From Principles of Biology: Biology 100 Laboratory Manual by Sylvester Allred and Emma P. Benenati. Copyright © 2011 by Kendall Hunt Publishing Company. Reprinted by permission.

FN 14: From Investigating Biology: The Diversity of Life Lab by Paul Florence and Annisa Florence. Copyright © 2013 by Paul Florence and Annisa Florence. Reprinted by permission of Kendall Hunt Publishing Company.

FN 15: From Explorations in General Biology by Betty A. Rosenblatt and Sarah Warrington. Copyright © 2008 by Kendall Hunt Publishing Company. Reprinted by permission.

FN 16: From Explorations in General Biology by Betty A. Rosenblatt and Sarah Warrington. Copyright © 2008 by Kendall Hunt Publishing Company. Reprinted by permission.

FN 17: From Regional Human Anatomy Workbook by Stephen E. Bassett. Copyright © 2013 by Stephen E. Bassett. Reprinted by Permission of Kendall Hunt Publishing Company.

FN 18: From Explorations in General Biolgy by Betty A. Rosenblatt and Sarah Warrington. Copyright © 2008 by Kendall Hunt Publishing Company. Reprinted by permission.

FN 19: From Fundamentals of Microbiology Laboratory Manual by Marlene Demers. Copyright © 2009 by Kendall Hunt Publishing Company. Reprinted by permission.